Physics and Politics

BEITRÄGE ZUR GESCHICHTE
DER DEUTSCHEN
FORSCHUNGSGEMEINSCHAFT

----

herausgegeben von
Rüdiger vom Bruch und Ulrich Herbert

Band 5

Physics and Politics

Research and Research Support
in Twentieth Century Germany
in International Perspective

Edited by
Helmuth Trischler and Mark Walker

Franz Steiner Verlag Stuttgart 2010

Gedruckt mit Mitteln der Deutschen Forschungs-
gemeinschaft

Cover picture: Chinese party-state leaders Zhou Enlai,
Mao Zedong, Lin Boqu, Zhu De, Chen Yun, Nie Rongzhen, and Deng Xiaoping met with scientists making
the Twelve-Year Plan for Science and Technology in
Beijing in 1956.
Source: Wang Yusheng (ed.), Fendou yu huihuang:
Zhonghua keji bainian tuzhi (1901–2000) (struggles
and glories: a pictorial history of Chinese science and
technology in the last century, 1901–2000). Kunming
2002, p. 112.

Bibliografische Information der Deutschen Nationalbibliothek
Die Deutsche Nationalbibliothek verzeichnet diese
Publikation in der Deutschen Nationalbibliografie;
detaillierte bibliografische Daten sind im Internet
über <http://dnb.d-nb.de> abrufbar.

ISBN 978-3-515-09601-0

INHALT

PREFACE

The present volume is part of the research program on the history of the Deutsche Forschungsgemeinschaft (German Research Foundation, henceforth referred to as DFG) from 1920 to 1970. The research program was inaugurated in 2000. In the meantime, about 20 projects have been launched in order to get a deeper understanding of one of the most important institutions within German science and its research and innovation system. The program focuses on National Socialism, but rather than isolating this period, it integrates the "Third Reich" into long-term developments of research and science policy. Furthermore, the program seeks to place the history of the DFG into international contexts.

The same holds true for this volume. It documents the results of a conference that was held at the Harnack-House, a conference venue and guest house of the Max Planck Society in Berlin-Dahlem. The conference was entitled "Physics in Germany from 1920 to 1970: Concepts, Instruments, and Resources for Research and Research Support in International Comparison".

We wish to express our gratitude to the following people who helped us prepare the conference in Berlin: Michael Eckert, Paul Erker, Ulf Hashagen, Karin Orth, Arne Schirrmacher, and Stefan Wolff. Our thanks go to Andrea Lucas for her full support in copyediting this volume and to Angela Schmiegel from Steiner Verlag for her crucial role in the production of the book. Finally, we wish to thank the Deutsche Forschungsgemeinschaft for all its efforts in launching and financing the overall project on the history of the DFG in general as well as for the financial support of this volume in particular.

Munich and Schenectady, August 2009   *Helmuth Trischler and Mark Walker*

PHYSICS AND POLITICS. RESEARCH AND RESEARCH SUPPORT IN TWENTIETH CENTURY GERMANY IN INTERNATIONAL PERSPECTIVE – AN INTRODUCTION

Helmuth Trischler

Physics and Politics – what's new under the sun? When the British businessman, essayist, and journalist Walter Bagehot published his seminal book under that very title in 1872, he aimed at showing the effects of the "sudden acquisition of much physical knowledge" and its transformation in technical inventions like railways and telegraphs on "two old sciences": politics and political economy.[1] Bagehot, who as an economist was associated with the English institutionalist-historicist tradition, particularly struggled with the question, why all the verifiable progress in science and technology of the last decades had not led to a similarly visible progress by the nation and its political bodies. Bagehot and many successors were interested in the effects of physics on politics, and this research agenda has occupied cohorts of historians and sociologists of science alike for decades.

In recent years, this traditional, nevertheless still relevant, angle of research has been replaced by an understanding of the relation between science – with physics center-stage – and politics as two discrete but strongly intertwined subsystems of society. Sociologists of science have developed a number of different conceptual models to frame the interrelations between science on one hand and politics or the state on the other, including the influential model of a triple helix between science, economy and the state as well as the widely discussed mode 1/mode 2-model of knowledge production. Both models argue that science, and here in particular academic research, has experienced a rapidly growing interconnectedness both with economy and politics, leading to a new regime of knowledge production.[2] On the other hand, historians and sociologists of science have questioned the apparent novelty of these models. They have traced its roots back deep in history, and come up with alternative models to explain what happened in the 20th century: namely the growth in density of interrelations between science and politics since the late 19th century, if not the late early modern period, attributing to the First World War a prominent role as a catalyst for the process of accelerating historical change in the relation between science and politics.[3]

---

[1] Bagehot, Physics, p. 5.
[2] Etzkowitz/Leydesdorff, Universities; Etzkowitz/Leydesdorff, Endless Transition; Etzkowitz/Webster/Healey, Capitalizing Knowledge; Gibbons et al., New Production; Nowotny/Scott/Gibbons, Re-thinking Science.
[3] Shinn, Triple Helix; Szöllösi-Janze, Science; Weingart, From Finalization; Audétat, Re-thinking Science.

The asymmetry of the old model has been replaced by deliberate symmetrical concepts of interrelations between science and politics.

One of the most promising of these symmetrical concepts, both in analytical and heuristic perspectives, frames science as a societal force which provides resources for other subsystems of society; in return it consumes resources of other systems.[4] Not by chance, the concept has been developed with the aim of getting to a better understanding of the complex relationship between the sharp political caesuras in German history of the 20th century on the one hand and the manifold continuities in scientific developments of the same period on the other hand. Like no other nation, 20th century Germany experienced a half a dozen different socio-political regimes: the imperial period of the Wilhelmine Empire until 1918, the first democratic experiment of the Weimar Republic from 1918/19 to 1933, the fascist autocracy of the "Third Reich" from 1933 to 1945, the interplay of the Allied occupation between 1945 and 1949, the parallelism of the federal democracy in West Germany and the socialist republic in East Germany from 1949 to 1989/90, and for the last two decades the re-unified Federal Republic of Germany.

For the subject under study in this volume, German physics research and its support during the 20th century in international perspective, the resource model offers particularly favorable perspectives in two dimensions.

Firstly, it enables the historian to frame the organizational systems and performative structures of funding and supporting research as part of the political subsystem of society. Research support systems are not understood as separated from science policy but rather as an integral part of the political governance of research by providing the latter with the resources it needs to perform appropriately.

Secondly, the concept of science and politics as mutual resource systems allows the historian to study the changing interdependencies of science, politics, economy, military, and the public over time. While most other models have interpreted National Socialism as a period of extensive exploitation of scientific knowledge for criminal political ends, the resource model enables us to gain a more complex view of science and politics. In addition, the model facilitates a question that has been crucial for the overall research project on the history of the DFG and furthermore for the history of science and politics in 20th century Germany: how are the continuities on the epistemic-cognitive level and the discontinuities on the political level related?[5]

---

[4] Ash, Wissenschaft und Politik; Nikolow/Schirrmacher, Wissenschaft und Öffentlichkeit. – For the application of the model on the role of the DFG in the natural sciences and engineering sciences see Trischler, Rückstandssyndrom.

[5] See e. g. Bruch/Kaderas, Wissenschaften; Bruch/Gerhardt/Pawliczek, Kontinuitäten; Orth/Oberkrome, Deutsche Forschungsgemeinschaft.

The volume includes five interrelated sets of questions:

A first set of questions focuses on physicists as actors in the political and social realm. To what degree do physicists understand themselves as political actors, as experts with intellectual resources valuable for society, or as scientists who can cross the barriers between science, the economy, politics, and society?

A second set of questions discusses the emergence of main research fields in physics at the crossroads between political agendas and the dynamic cognitive developments in science. Which support systems can be identified through international comparisons and what effects did they have? How were the constellations of resources formed at the local and regional levels? Which formal and informal networks in physics can be identified? How was it possible for a comparatively small group of theoretical physicists to acquire so much influence in setting the agenda for physical research and science funding? What is the relationship between informal connections and the formal processes for setting the goals of scientific policy?

A third set of questions tackles the widely discussed issue of the relationship between physics and the Cold War. Which patterns of interdependency between science (here physics) and political and cultural developments dominated the period of the Cold War? How did physical knowledge function as a cognitive resource for the Cold War, and turning this on its head, how did the Cold War act as a political and social resource for physical research? Did the Cold War influence the research agenda in physics, in particular by separating the world and in particular Germany into two opposing camps?

A fourth set of questions examines the role of physics as the leading science of the 20th century: the central importance of physical methods, discourses and research agendas and their transformation into other disciplines – not only for the natural and engineering sciences but also for the life sciences, social sciences and humanities. What role did physics play in the development of technology? How did the 20th century become the century of physics? This assessment does not only come from historians; physicists themselves consciously anchored their discipline as a leading science in the scientific, political, and public discourse. Why did the "negative" result of the atomic bomb end up "enriching" physics as a discipline?

A fifth set of questions asks for the role of physics as a public science. Physicists like Albert Einstein, Werner Heisenberg, and Niels Bohr became icons of the 20th century. They embodied on one hand the genius, but on the other hand the incomprehensiveness of science in the modern world. However, Carl-Friedrich von Weizsäcker, Werner Heisenberg and Walther Gerlach also stand for the secular assurance of salvation thanks to modern science and the public role allocated to them as experts of knowing answers to the questions of the time. They understood themselves as mediators between science and society, and the public often sought them out in this role. Which instruments and media, and which languages, images, and metaphors were central to this role? How did this change over the course of an increasingly important media and the formation of the modern mass media of the 20th century? How did science and the public function as resources for each other and in which phases in particular were these resources needed?

These overarching questions are tackled by two different types of contributions. A first set of chronologically ordered articles analyzes the structures and practices of physics and physics funding in relation to the various political regimes in Germany. The case studies cover the full period of the research programme on the history of the DFG, i. e. 1920 to 1970, with a deliberate focus on the period after World War II. As Mark Walker in his overarching survey on the scholarly literature about physics and physics support in Germany shows, a considerable number of publications cover the so called golden age of German physics in the 1920s and, thanks to a number of concerted efforts to foster research on science and technology in National Socialism, in the "Third Reich". In contrast, the decades after the war have only recently begun to attract substantial interest from the scholarly community.

The second set of articles places physics in Germany into an international perspective. Like the first sample of case studies this part of the book is not comprehensive but instead very selective. In order to present novel evidence from recent research projects, the comparative articles tackle regions and countries which are usually not in the focus of historians of science. Instead of the U.S., Russia, Great Britain or other (West) European Countries, the comparative contributions aim at shedding light on the scientific periphery: the Ukraine, Japan and China.

Having surveyed the rich literature on science in Germany in the twentieth century, Mark Walker finds much evidence for the assumption that both physics as a body of knowledge production and the politics of supporting physics research were very innovative. A generous support system enabled scientific breakthroughs. The political authorities in return profited a great deal from the stock of innovative knowledge gained by a rising number of state financed academic and non-academic research institutions. Despite some conflicts and tensions between researchers and state authorities, in physics the exchange of resources between science and politics worked quite well. In twentieth century Germany, the historical constellation of mutual benefits dominated over the unparalleled multitude of regime changes. Walker comes to the conclusion that the methods and instruments politics developed to fund science were very important.

While the classical fields of modern physics like relativity theory and quantum mechanics have gained much attention within history of science, applied physics has been overshadowed. Alexander von Schwerin breaks with this pattern by addressing the interrelations between physics and life sciences. He embarks on the practical turn in history of science which has stimulated a rapidly growing body of literature on "experimental systems" (Hans-Jörg Rheinberger) and the material culture of science. Instruments as research technologies and laboratories as spaces of knowledge production were the contact zones in which clinical sciences, biology and physics amalgamated into the emerging field of biophysics. In

Germany, the technical backbone of this slowly emerging field was the X-ray tube. Schwerin develops a narrative that puts techniques and objects of research center-stage and, hence, stresses the significance of resources in the double sense of the term. Firstly, X-ray generators developed into a key resource for medical scientists and biologists, and secondly scientists were forced to search for additional financial resources to cover the high costs of equipping radiation laboratories. Within the diversifying landscape of research funding bodies in Weimar Germany, it was the "Emergency Foundation for German Science", the institutional forerunner of the DFG, which enabled biophysics to establish itself as a new sub-discipline of science between physics, clinical medicine and biology. Morreover, in the case of this field, which transgressed the boundaries between physics and life sciences, the DFG served as an important institutional actor to foster not only knowledge production but also institutional stabilization.

Since Lutz Niethammer published his path-breaking book on the "Mitläuferfabrik" (Fellow Traveller Factory) in 1982, many scholars have investigated the allied attempts to rid German society of the remnants of National Socialism and the responses of German elites to this Allied goal.[6] Meanwhile an overwhelmingly rich body of literature on the overall theme of the "Vergangenheitspolitik" (Politics of the Past), another very influential historiographical concept, has integrated denazification into the wider context of the attempts of the emerging West German society to cope with its Nazi past.[7] It is surprising to see that denazification in the field of science has only gained little scholarly interest so far. All the more welcome is Gerhard Rammer's article, which is based on a rich case study on the nazification and denazification of physics at the University of Göttingen. Politics, manifested in the Allied goal to control German science, played a crucial role for the transformation of physics from a heavily militarized science during the war into a field of research which could help rebuild West Germany as a civil society. Denazification in physics on the one hand was crucially linked with the gradual change in the Allied occupation policy from a politics of controlling science to a politics of fostering research. On the other hand, it expressed itself as a process of self-denazification based on collegial networks and inner-disciplinary cultural values. As Rammer shows with his case studies on the physicists Karl-Heinz Hellwege, Hans König and Kurt Hohenemser, the opportunities for scientists to either avoid the threat of dismissal from academia or, in the case of physicists who had suffered under National Socialism, to be rehabilitated, depended very much on their collegial ties and their individual ability to accommodate themselves to the cultural self-definition of the physics community.

The core of the physicists' self-understanding was the ideology of non-ideology. With this phrase Richard Beyler very elegantly frames what other historians of science have identified as the ideology of science as an a-priori non-political

---

[6] Niethammer, Mitläuferfabrik.
[7] Frei, Vergangenheitspolitik.

realm.[8] This ideology was deeply rooted in the German scientific culture. It was established during the Wilhelmine Empire, when scientists collectively defined themselves as servants of the nation beyond political partition and ideological differences. This cultural self-understanding was reinforced by the experience of the First World War and the Weimar Republic. It developed into a powerful ideological tool after the Second World War, when physicists were confronted with the self-induced necessity to cope with the experience of their self-cooptation in and massive involvement with the politics of the Nazi regime. The discursive re-interpretation of their activities in the recent past as an innocent and inherently apolitical endeavor was all the more difficult for the spokesmen of science because the division of Germany and the Cold War confronted West German scientists with a fundamentally opposing ideology: the socialist world view of science as an intrinsically ideological entity. In consequence, scientists and in particular physicists had to take sides in order to distance themselves from the ideology of science as ideological that prevailed on the other side of the Iron Curtain. As Beyler demonstrates, the DFG and its leading representatives were very prominent in the discursive construction of the ideology of science as a non-ideological field of society.

The interwar period and, at an accelerated pace, the Second World War witnessed the rise of big science. Physics in particular was affected by the rapid expansion of fields of large scale research like rocketry, nuclear science, radar and aeronautics. The most visible and most resource-intensive incarnation of big science in the post-war period was nuclear particle physics. During the early 1950s, several European states had already joined forces to build CERN in Meyrin near Geneva, a central research establishment for high energy physics. The Federal Republic of Germany participated in CERN early on, with Werner Heisenberg as its most prominent scientific figure head. In her contribution Cathryn Carson focuses not on the founding period, but rather on the second phase of CERN's program to design and construct advanced accelerators – machines even bigger than the "big machine" which the prominent science writer and futurologist Robert Jungk had portrayed in 1966.[9] The long and intensive quarrels over the German participation on CERN's second generation accelerator program display, as Carson very convincingly argues, the return of the Federal Republic's science policy to the normalcy of international science funding patterns. After the immediate postwar period of seemingly endless growth under the powerful rhetoric of the necessity to catch up, this normalcy was characterized by financial constraints and limited growth if not budget cuts, in particular after the final end of the postwar prosperity with the economic crises of 1966/67. The need to make choices and to prioritize certain projects with the consequence of killing others, led to fierce fights within the physics community and with colleagues from other fields of science. Physics lost its position as the most privileged discipline, and Heisen-

[8] See e.g. Walker, Legenden; and Gordin et al., Ideologically Correct Science; Trischler, Self-Mobilization.

[9] Jungk, Die große Maschine.

berg his role as undisputed spokesman of science with the ability to steer and balance the flow of financial resources. In reaction to the new challenges resulting from CERN's accelerator program, German science policy entered a new phase. The expanding Federal Research Ministry developed new tools of science policy, including planning schemes, which for a long time had horrified scientists and politicians alike as being antidemocratic and socialist. The ministry's staff also built up its own expertise in order to become more independent from scientific self-administration. As in other fields of politics and society, for science policy the late 1960s marks a watershed which resulted in a lasting regime change.

In the course of the 50s and 60s, physicists like Werner Heisenberg or his Munich colleague Walter Gerlach, long-standing vice-president of the DFG, became spokesmen of science. They developed into public figures, representing the voice – and wisdom – of science in a society dominated by mass media. The mediazation of society stimulated the emergence of a new type of scientist: the public scientist, the scientist as intellectual, as Martin Strickmann argues, who engaged himself as an intellectual in publically led political debates. The most prominent example of such debates was the protest of the Göttingen Eighteen against Konrad Adenauer's plans for the nuclear armament of West Germany. Strickmann uses the striking parallel between the Göttingen Eighteen and Frederic and Irène Joliot-Curie, who struggled against nuclear armament in postwar France, to undertake a transnational comparison between French and German physicists as public intellectuals. In doing so, Strickmann transfers to science the concept of public intellectuals from other fields of societal elites, most visible in philosophy, in order to gain a better understanding of the sociopolitical role of scientists in modern societies. Whether the transfer will work for disciplines other than nuclear science remains open, but physics is obviously the most suitable test case for such a conceptual experiment.

The nuclear arms race, against which the Göttingen Eighteen and other physicists directed their political protest, was stimulated by the Cold War. As the extensive research activities on the history of Cold War-science since the end of the Cold War have demonstrated, the gigantic amount of financial resources which in particular the Department of Defense in the US channeled into academic science led to a number of scientific breakthroughs, among them the laser. Helmuth Albrecht points out that the laser did not come out of the blue. The invention of the laser was based on a stockpile of academic knowledge on high-frequency physics. In Germany, the DFG played a crucial role in fostering laser-related research in the foundational period of this field, which was quickly perceived by science, industry and politics alike as seedbed for new key technologies. Not surprisingly, the laser ranked among those fields that deserved prioritized financial support when the Federal Research Ministry in 1969 launched its novel research support program with the aim of stimulating the creation of "new technologies". High expectations existed long before technical applications. Only in the late 70s did laser development lead to commercially successful products. Nevertheless, Albrecht underlines the importance of the 60s as a transitory phase, in which academic research strongly supported by DFG-programs laid the cognitive basis in several areas of laser research for later practical applications – not to mention the role of universities for educating new cohorts of scientific experts in the emerging field.

Depending on age, physicists in twentieth century Germany most likely experienced more than one or two different regimes. More or less the same holds true for Ukrainian physicists. Thanks to a number of recent scholarly studies we have gained a rich picture of Stalinist science policy and its devastating effects on the opportunities for scientists to perform well. Like other intellectuals, scientists suffered heavily from the interference of Bolshevist party authorities and Stalinist purges.[10] To be situated at the periphery of the gigantic Soviet empire did not save physicists from political invention, as Paul Josephson and his co-authors Yury Ranyuk, Ivan Tsekhmistro and Karl Hall show. Their well chosen case study on the Ukrainian Physical Technical Institute (UFTI) in Kharkiv, an international leading centre for nuclear and low temperature physics, illustrates how ideology hampered the scientists' ability to do excellent research. Its international reputation did not prevent the institute from undergoing severe purges, which included the execution of leading physicists during the "Great Terror" of the years 1936 to 1938. To the contrary: perhaps more than any other research institute in the USSR, the UFTI suffered from ideological and bureaucratic strictures. In addition, when the German Army attacked the Soviet Union and conquered Kharkiv, the UFTI was almost fully destroyed and totally exploited by German troops. After the Second World War, the institute profited in part from the vast resources of the Soviet nuclear weapons program, but in the eyes of the Soviet authorities it remained peripheral, discriminated against in comparison with the centres of Soviet physics in and around Moscow and Leningrad as well as with the newly established science cities like Akademgorodok near Novosibirsk. All in all, the UFTI illustrates the dark side of the relations between physics and politics in the "age of extremes" (Eric Hobsbawm).

Like in Germany, physics in postwar Japan was deeply influenced by wartime experience. Research activities that could lead to military applications were almost taboo. The retreat to pure science which seemed to be farthest removed from any further threat of scientific knowledge being used for political purposes can be explained as a collective psychological response to the deep shock of the military defeat. In addition – and unlike their German counterparts – Japanese physicists were deeply affected by the experience of their home country being massively destroyed by American nuclear bombs which were obviously the result of research in physics. Against this background of discontinuities with long lasting effects, Morris Low emphasizes a number of important continuities which rooted deeply in the Japanese national culture of science and innovation. Influential spokesmen of science like the Nobel laureate Tomonaga succeeded in rebuilding a state sponsored funding network that even enabled Japanese scientists to engage in resource-intensive fields of physical sciences such as space research and nuclear physics. In contrast to the emergence of large scale national research laboratories in the US and many European countries, including Germany, the national in-

---

[10] Cf. Beyrau, Dschungel; Josephson, Totalitarian Science; Josephson, Red Atom.

novation system of Japan was characterized by joint-use research institutes, which were perceived as an acceptable way to receive state funding without surrendering the idea of the autonomy of science from political interference.

In the postwar era shaped by the hegemony of the two superpowers US and USSR, Japan sided with the West as both state and society and followed the Western democratic model. Although Chinese physicists suffered from devastating political and ideological purges similar to their Russian colleagues in the Stalinist era, China under the leadership of Mao tried to remain as independent as possible from Soviet influence. Political independence included becoming autonomous from Soviet knowledge in nuclear science and technology. Zuoyue Wang argues that the struggle for political independence led to a culture of techno-nationalism. The institutional nucleus of this scientific-technical culture was the Chinese Academy of Sciences, which – like the national academies of science in the USSR and in the GDR – provided the communist regime both with scientific and ideological resources. The successfully nuclear weapons program and other big science projects provide evidence that the ideological concept of techno-nationalism served as a mediator between scientists and state authorities. Wang draws the conclusion that science and technology could prosper under an authoritarian regime. But the social costs were high, and the scientists had to pay as high a price as other parts of Chinese society.

The massive abuse of scientists and other intellectuals both in Maoist China and Stalinist Russia point to the analytical limits of the conceptual model to frame the relations between science and politics as a mutual exchange of resources. While the model has proved its explanatory power to get a better understanding of the dynamic relation between physics and the state in Germany – and this holds true also for the autocratic period of the National Socialism – it has to be complemented with other conceptual and methodological tools for other states. Transnational comparative studies which have made the second part of this volume rank among the most valuable tools in this enlarged tool chest of historians of science.

BIBLIOGRAPHY

Ash, Mitchell: Wissenschaft und Politik als Ressource füreinander, in: Bruch, Rüdiger vom and Brigitte Kaderas (eds.): Wissenschaften und Wissenschaftspolitik. Bestandsaufnahmen zu Formationen, Brüchen und Kontinuitäten im Deutschland des 20. Jahrhunderts, Stuttgart 2002, pp. 32–51.

Audétat, Marc: Re-Thinking Science, Re-Thinking Society, in: Social Studies of Science 31, 2001, pp. 950–956.

Bagehot, Walter: Physics and Politics or Thoughts on the Application of the Principles of 'Natural Selection' and 'Inheritance' to Political Society, London 1872 (Reprint: Kitchener/Ontario 2001).

Beyrau, Dieter (ed.): Im Dschungel der Macht. Intellektuelle Professionen unter Stalin und Hitler, Göttingen, Zürich 2000.

Bruch, Rüdiger vom and Brigitte Kaderas (eds.): Wissenschaften und Wissenschaftspolitik. Bestandsaufnahmen zu Formationen, Brüchen und Kontinuitäten im Deutschland des 20. Jahrhunderts, Stuttgart 2002.

Bruch, Rüdiger vom, Ute Gerhardt, and Alexandra Pawliczek (eds.): Kontinuitäten und Diskontinuitäten in der Wissenschaftsgeschichte des 20. Jahrhunderts, Stuttgart 2006.

Etzkowitz, Henry and Loet Leydesdorff (eds.): Universities and the Global Knowledge Economy. A Triple Helix of University-Industry-Government Relations, London 1997.

Etzkowitz, Henry and Loet Leydesdorff: The Endless Transition. A „Triple Helix" of University-Industry-Government Relations: Introduction, in: Minerva 36, 1998, pp. 203–208.

Etzkowitz, Henry, Andrew Webster, and Peter Healey (eds.): Capitalizing Knowledge. New Intersections of Industry and Academia, Albany 1998.

Frei, Norbert: Vergangenheitspolitik. Die Anfänge der Bundesrepublik und die NS-Vergangenheit, München 1996.

Gibbons, Michael et al.: The New Production of Knowledge. The Dynamics of Science and Research in Contemporary Societies, London 1994.

Gordin, Michael, Walter Grunden, Mark Walker, and Zuoyue Wang: 'Ideologically Correct' Science, in: Walker, Mark (ed.): Science and Ideology: A Comparative History, London 2003, pp. 35–65.

Josephson, Paul: Totalitarian Science and Technology, Atlantic Highlands 1996.

Josephson, Paul: Red Atom, New York 1999.

Jungk, Robert: Die große Maschine. Auf dem Weg in eine andere Welt, München 1966.

Niethammer, Lutz: Die Mitläuferfabrik. Die Entnazifizierung am Beispiel Bayerns, 2nd edition Berlin et al. 1982 (first ed. Frankfurt am Main 1972 under the title: Entnazifizierung in Bayern. Säuberung und Rehabilitierung unter amerikanischer Besatzung).

Nowotny, Helga, Peter Scott, and Michael Gibbons: Re-thinking Science. Knowledge and the Public in the Age of Uncertainty, London 2001.

Nikolow, Sybilla and Arne Schirrmacher (eds.): Wissenschaft und Öffentlichkeit als Ressourcen füreinander. Studien zur Wissenschaftsgeschichte im 20. Jahrhundert, Frankfurt am Main, New York 2007.

Orth, Karin and Willy Oberkrome (eds.): Die Deutsche Forschungsgemeinschaft 1920–1970. Forschungsförderung im Spannungsfeld von Wissenschaft und Politik, Stuttgart 2010.

Shinn, Terry: The Triple Helix and New Production of Knowledge: Prepackaged Thinking on Science and Technology, in: Social Studies of Science 32, 2002, pp. 599–614.

Szöllösi-Janze, Margit: Science and Social Space: Transformations in the Institutions of Wissenschaft from the Wilhelmine Empire to the Weimar Republic, in: Minerva 43, 2005, pp. 339–360.

Trischler, Helmuth: Self-mobilization or Resistance? Aeronautical Research and National Socialism, in: Renneberg, Monika and Mark Walker (eds.): Science, Technology and National Socialism, Cambridge 1994, pp. 72–87.

Trischler, Helmuth: Das Rückstandssyndrom. Ressourcenkonstellationen und epistemische Orientierungen in Natur- und Technikwissenschaften, in: Orth, Karin and Willy Oberkrome (eds.): Die Deutsche Forschungsgemeinschaft 1920–1970. Forschungsförderung im Spannungsfeld von Wissenschaft und Politik, Stuttgart 2010.

Walker, Mark: Legenden um die deutsche Atombombe, in: Vierteljahrshefte für Zeitgeschichte 38, 1990, pp. 45–74.

Weingart, Peter: From "Finalization" to "Mode 2". Old Wine in New Bottles?, in: Social Science Information 26, 1997, pp. 513–519.

GERMAN PHYSICS AND ITS SUPPORT

Mark Walker

Germany and German physicists were of course very important for the development of modern physics in the twentieth century, including relativity theory and quantum mechanics. However, German scientists and science administrators also made important innovations with regard to science policy in general and funding in particular. This book seeks to place this innovation in perspective by means of international comparisons with physics policy and funding, both under different political regimes in Germany and in other countries. The juxtaposition of these international case studies also brings out the main theme of this book, the interaction between physics (and physicists) on one hand and politics on the other.

Just as the intellectual successes of modern physics have their roots in the German Empire, so do the science policy and funding innovations. Before German unification in 1871, the various German states had incorporated the "research imperative" into their universities: scholars, including physicists, were expected to research and publish new knowledge, not just teach.[1] This contributed to a flowering of German physics in the second half of the nineteenth century, including Hermann von Helmholtz and Heinrich Hertz.[2]

However, it was also already clear in the nineteenth century that experimental physics was an expensive proposition, and that private industry and the German state would have to intervene in order to facilitate the research Germany needed as a modern industrial power. The industrialist and entrepreneur Werner von Siemens worked with Helmholtz and the new German Empire to create the Imperial Physical-Technical Institute (Physikalisch-Technische Reichsanstalt, PTR), an institution devoted to performing physical research for the benefit of the German state and private German industry.[3] Although the main mission was thus practical and applied research, the excellent facilities and staff also contributed to fundamental research. PTR physicists' experiments with black-body radiation provided the data the theoretical physicist Max Planck used to create his quantum hypothesis, that light was not continuous, rather made up of discrete pieces or quanta.[4]

Similar efforts for German chemistry[5] led to the other major science policy and funding innovation during the Empire, the Kaiser Wilhelm Society (Kaiser-

---

[1] Pfetsch, Entwicklung; Turner, Growth; Ben-David, Role; Lenoir, Revolution.
[2] Harman, Energy; Jungnickel/McCormmach, Intellectual Mastery.
[3] Cahan, Institute.
[4] Kuhn, Black-Body Theory; most recently, see Hoffmann, Entstehung.
[5] Johnson, Chemists.

Wilhelm-Gesellschaft, KWG).[6] This society was created in order to liberate a small number of first-rate scientists from the obligation of teaching and to provide them with excellent equipment and staff in order to facilitate their research. The institutes that made up this society were also financed (at least initially) jointly by the German state and private industry. As a result, the first Kaiser Wilhelm Institutes (KWI) were devoted to scientific fields of direct interest to particular industries willing to fund research. Some of these were certainly close to physics, such as Fritz Haber's KWI for Physical Chemistry.

There was also a KWI for Physics, at least on paper. Planck managed to lure the young and already influential Albert Einstein to Berlin with a combination of appointments that allowed Einstein to focus on his research. Eventually this combination included a KWI for Physics that consisted of money Einstein could pass on to other researchers. When Max von Laue became the assistant director, Einstein did not even have to concern himself with the administration of this paper institute. However, Planck, who in 1930 became the second president of the KWG, had a long-term goal of creating a real, world-class KWI for Physics that included experimental research. The physicists and physical chemists connected to the KWG certainly enjoyed international prestige: Planck, Haber, Einstein, and Laue were all Nobel laureates.

Although many German physicists served in World War I, German physics was not really mobilized for the conflict as a science. Indeed at first the German military leadership was contemptuous of the contribution that science could make towards winning the war. It soon became clear, however, that chemistry in particular helped stave off an early defeat by allowing the synthetic production of the nitrates Germany had previously imported, but were now cut off by the British naval blockade. Scientists also helped develop innovative weapons, including poison gases by the physical chemist Fritz Haber[7] and airplanes by the aerodynamic expert Ludwig Prandtl.

A few physicists served their country as scientists, for example James Franck worked on chemical weapons, but many more were common soldiers. On the home front, older physicists like Philipp Lenard, Max Planck, and Wilhelm Wien participated in nationalistic manifestos that defended the German war effort.[8] Albert Einstein was one of the very few scientists inside or outside of Germany who advocated pacifism and internationalism, drawing heavy criticism from many of his colleagues.[9]

In the end, the mobilization of German science, and in particular physics for the "Great War" was spotty and inconsistent. Perhaps one of the most important legacies of this inconsistency was that by the armistice influential figures in the German military had become convinced that science-based technology would be an important part of any future conflict. It was also significant that scientists

---

[6] Vierhaus/vom Brocke (eds.), Forschung.
[7] Haber, Poisonous Cloud; Szöllösi-Janze, Haber.
[8] Heilbron, Dilemmas; Wolff, Physicists.
[9] See Rowe/Schulman (eds.): Einstein.

in the few fields that had experienced profound growth due to generous funding during the war, for example, aeronautics, were subsequently disappointed with what the postwar German republic, hampered by the restrictions of the Versailles treaty, could offer them.

With very few exceptions German scientists, like most imperial elite groups, were shocked, dismayed, and bitter about how the war ended. Self-criticism or critical reflections on who was responsible for the defeat, let alone why the war was started or how it had been managed, were rare. Instead critics focused on the new German democracy as well as old foes or scapegoats like socialists and Jews when assigning blame. Although these attitudes proved to be quite compatible with the new National Socialist movement, during the Weimar Republic very few scientists openly supported Hitler's political movement. Here the two conservative physics Nobel laureates, Philipp Lenard and Johannes Stark, were the exceptions that proved the rule.[10]

Many academic scientists instead indulged in what Brigitte Schröder-Gudehus and Paul Forman have described as: "science as a replacement for political power" (Wissen als Machtersatz).[11] The great power status Germany had lost after the war would be compensated by the continued excellence of German scholarship. Speaking before the Prussian Academy of Sciences (Preußische Akademie der Wissenschaften, PAW) on November 14, 1918, Max Planck made this argument clear:

> If the enemy has taken from our fatherland all defense and power, if severe domestic crises have broken in upon us and perhaps still more severe crises stand before us, there is one thing which no foreign or domestic enemy has yet taken from us: that is the position which German science occupies in the world. Moreover, it is the mission of our academy above all, as the most distinguished scientific agency of the state, to maintain this position and, if the need should arise, to defend it with every available means.[12]

Ironically this rhetoric was accompanied by assertions that precisely these scientists were being "apolitical," as all scientists should be.[13] However, rhetoric alone could not realize this quite self-indulgent characterization of the role scientists wanted to play in Weimar Germany, rather financial and material support were also required. Fortunately for German scientists, the democratic Reichstag (parliament) proved to be generous in providing support despite the "more or less veiled hostility of the academic scientists to the Weimar Republic."[14]

[10] Kleinert, Lenard; Kleinert, Spruchkammerverfahren; Kleinert (ed.), Stark; Kleinert, Weyland; Kleinert, Briefwechsel; Walker, Nazi, pp. 14–15.

[11] Schröder-Gudehus, Argument, p. 551; Schröder-Gudehus, Challenge; Forman, Internationalism, pp. 161–165.

[12] Forman, Internationalism, p. 163.

[13] Ibid., pp. 169–171.

[14] Schröder-Gudehus, Argument, p. 555.

The interwar years have often been characterized as the "golden age" of modern physics in Germany (or German-speaking Europe).[15] Albert Einstein had published his general theory of relativity during the war. Shortly after the end of the conflict, a British expedition delivered experimental verification of one of its key predictions. This result was popularized and taken up by the press around the world. Einstein became a celebrity, which also led to an anti-Semitic backlash against him, his politics, and his science.[16] The quantum physics begun during the first decades of the twentieth century culminated in the creation of quantum mechanics during the 1920s. But these brilliant theoretical advances overshadowed the fact that German experimental physics, much more dependent on significant investments of capital, was having a much harder time.

Shortly after the end of the war, the already weakened German economy was hit by hyperinflation as well. Institutional endowments and individual savings evaporated, putting pressure both on universities in particular and younger scientists without secure positions. Influential industrialists, politicians, and scientists began calling for new funding institutions for science. The scientists and their allies sought to obtain financial support from either both industry and the state, while at the same time trying to remain free of the influence of either.[17] The German state founded the Emergency Foundation for German Science (Notgemeinschaft der deutschen Wissenschaft, NGW), while industrialists created the Helmholtz Foundation (Helmholtz Gesellschaft, HG).[18]

Friedrich Schmidt-Ott, an experienced bureaucrat and first NGW president, tried to forestall the HG and persuade the private sector to donate money to the NGW, but the industrialists preferred to donate money and provide support through the HG or similar organizations where they could control how the money was spent.[19] The two industrialists who initiated the HG, Carl Duisberg and Albert Vögler, apparently feared that the NGW would neglect applied research and technology and wanted to support university scientists, not the Kaiser Wilhelm Society.[20] Industrialists also set up the Donors' Union of the NGW (Stifterverband der Notgemeinschaft, SN) as a mechanism to funnel money to the NGW while retaining some say over how it was spent.[21]

The Weimar central government not only provided the bulk of the funds, up until the late 1920s it allowed the NGW "almost complete autonomy."[22] In the end, the NGW was "the dominant source of funds for the Weimar Period ..."

---

[15] See for example Beyerchen, Scientists, pp. 6–8; the most comprehensive study of physics in Weimar remains Paul Forman's Ph. D. dissertation, Forman, Practice; in a subsequent influential article, Forman, Culture, he uses "German speaking Europe" as opposed to Germany.

[16] See the Historical Introduction to Rowe/Schulman (eds.), Einstein, pp. 1–59.

[17] Schröder-Gudehus, Argument, pp. 538, 550.

[18] Richter, Forschungsförderung; Schröder-Gudehus, Argument; Forman, Financial Support; Marsch, Notgemeinschaft; Hammerstein, Forschungsgemeinschaft.

[19] Forman, Financial Support, p. 63.

[20] Marsch, Notgemeinschaft, pp. 86–87.

[21] Ibid., p. 88.

[22] Forman, Financial Support, p. 61.

Together the NGW and HG "led to at least a doubling of the real expenditures for the direct support of research in physics compared with the pre-war period."[23] In comparison with the state contribution, research funds from private sources or industry remained modest.[24]

The need for financial support of science was so great in Germany, but the available resources so meager, that the NGW was forced to innovate. First of all, NGW Fritz Haber explained in 1921 that:

> The Emergency Foundation can in general neither erect buildings, nor bear the costs of heating them, nor pay salaries to individual investigators. It must, except in very special cases, presuppose that the expense of keeping the machinery of scientific research idle is defrayed by some other agency, primarily by the governments of the states.

Instead the NGW could only "raise productivity" by making grants to particular research projects.[25] It also worked together with the German Physical Society (Deutsche Physikalische Gesellschaft, DPG) to acquire and make available foreign publications.[26]

Instead of the long-standing practice of providing money to the small number of full professors or institute directors, who would then pass it on as they saw fit, the NGW introduced the concepts of individual scientists at all levels applying directly for project-grants, and peer review of the various applications in order to determine who received the funding.[27] These research grants were supposed to:

> ... provide a minimum level of continuing support for one or several years for scholars, especially younger ones, who are suitable for scientific research, but for economic reasons threaten to be lost to science.[28]

The level of support was supposed to allow a scientist with "modest expectations" to devote himself to science without having to earn money on the side.[29]

Schmidt-Ott refused to fund researchers with jobs at university – that was up to the German states. Instead he wanted to support scientists who did not yet have a secure position.[30] For the first time, young German scientists could apply directly for their own funding. Experimental apparatus funded by the NGW was tied to the individual researcher, not the institute. Thus if a scientist moved to another institution, he took the NGW equipment with him.[31] These innovations also allowed the NGW to steer support towards important and timely topics.

---

[23] Ibid., pp. 39, 42.

[24] Richter, Forschungsförderung, p. 24.

[25] Haber quoted in Forman, Financial Support, p. 53.

[26] Richter, Forschungsförderung, pp. 25, 30; Marsch, Notgemeinschaft, pp. 100–101.

[27] Forman, Financial Support, pp. 50–54, 61–63.

[28] Quoted in Richter, Forschungsförderung, p. 30; here Wissenschaft has been translated as science.

[29] Richter, Forschungsförderung, pp. 31–32.

[30] Marsch, Notgemeinschaft, pp. 102–103.

[31] Zierold, Forschungsförderung, pp. 86–87.

Indeed, in this way whole fields, such as atomic physics, could be supported.[32] This had a direct and positive effect on physics, for many young scientists were supported at an early stage of their careers and thus given the opportunity to help create modern physics.[33]

Certain physicists, like Max Planck, were especially influential within the network of committees that judged applications for support from the NGW.[34] Not surprisingly, research fields he favored, like modern physics, benefited. Experimental research did benefit, but theoretical physics benefited even more, for it was much less expensive to pay the salary of a post-doc scientist like Werner Heisenberg than to purchase expensive experimental equipment like that found in the United States, which was now becoming the dominant country in physics.

In 1926, the yearly NGW report made clear how important its work had been, both for physics and for Germany.

> In countless cases work has certainly already led to results with a great, sometimes fundamental importance for the development of physics. The experimental work includes … the atomic beam experiment from [Otto] Stern and [Walther] Gerlach and the absorption spectra from [Walter] Grotrian … As far as the theoretical work is concerned, the work from Heisenberg and from Heisenberg together with [Max] Born is especially significant … It is well known that the quantum mechanics is of central interest to physicists in all countries and without the support [from the NGW] the work from Heisenberg and Born in particular would most probably not have taken place in Germany, rather somewhere else.[35]

Although these reforms in the NGW were effectively democratic as far as the scientific community was concerned, this is ironic, for Schmidt-Ott was profoundly autocratic in how he ran the NGW.

> … the Prussian Kultusministerium attacked the very methods by which Schmidt-Ott had secured his considerable independence: namely, the withholding of complete information from both the public and the inner circle of the NGW, and the restriction of participation in the making of decisions. If Fritz Haber had not forced the issue, Schmidt-Ott would not even have agreed to elections, rather than appointments, to the selection panels. Schmidt-Ott avoided holding such elections between 1922 and 1929, and when he was obliged to do so in 1929, the presidential office did not ask for new nominations.[36]

Haber clearly spelled out to Schmidt-Ott where they differed:

> The difference between a self-governing body and a well-meaning autocracy is, in my eyes, that such figures (i. e., the budget) are discussed and decided upon by the competent body (the principal committee) on the basis of a detailed proposal, and are not determined by the chairman.[37]

---

[32] Richter, Forschungsförderung, p. 12.
[33] Ibid., p. 34.
[34] Ibid., pp. 35–37.
[35] Ibid., p. 37, also see pp. 44–46.
[36] Schröder-Gudehus, Argument, p. 563.
[37] Quoted in Schröder-Gudehus, Argument, p. 569.

At the time, these innovations were perceived and justified as necessary given the shortage of funds, and were not seen as the future model for the support of science. In the United States, for example, the most important sources of research support were private institutions like the Rockefeller Foundation, which had its own programs and criteria for supporting research.

During the last years of the Weimar Republic, German physics was certainly world-class. Einstein was in Berlin,[38] the young physicists who had created quantum mechanics and continued its development like Heisenberg, Pascual Jordan, Wolfgang Pauli,[39] and Erwin Schrödinger[40] had become professors throughout German-speaking Europe,[41] and the most talented and ambitious foreign physicists like Robert Oppenheimer[42] had traveled to German to study and sometimes do their Ph. D. Even if experimental research was not so robust, German physics certainly appeared to be flourishing. This changed dramatically in early 1933, when Adolf Hitler was appointed German chancellor and the National Socialist movement began to transform German society.

Physics became a high-profile target for some National Socialists because of Einstein, who was out of the country when Hitler came to power and criticized the new regime's racial policies.[43] This culminated when a Prussian Academy of Sciences spokesman essentially announced that the Academy was glad that Einstein – who had already resigned – was gone.[44] Physics was purged when the National Socialists forced almost all Jews out of the civil service, leading to a large emigration of physicists out of Germany.[45] The "Aryan Physics" movement now attacked modern theoretical physics as the product of Jewish spirit.

German physics certainly thereby lost a great deal of talented human capital – including many scientists whose careers had been supported by the NGW[46] – and a few teaching positions were "lost" to the Aryan Physicists.[47] However, there was no corresponding global cut in support for physics. Instead funds were redistributed, not reduced. There were also sufficient competent scientists who were racially-acceptable to the National Socialist government and willing to work for the Third Reich to fill most of the vacant positions.

Applied physics and scientific fields close to physics and immediately relevant for the National Socialists' long-term goals of rearmament were generously funded from the start of the Third Reich. Once again Ludwig Prandtl and his

---

[38] Hoffmann, Berlin.

[39] Hermann/von Meyenn/Weisskopf (eds.), Pauli.

[40] Moore, Schrödinger.

[41] See Cassidy, Beyond Uncertainty.

[42] Bird/Sherwin, Prometheus; Cassidy, Oppenheimer.

[43] For physics under National Socialism, see Hentschel (ed.), Physics.

[44] Wolff, Ausgrenzung, as well as Hentschel, Physics, pp. 18–21; Heilbron, Dilemmas, pp. 155–159; Renn/Castagnetti/Damerow, Einstein, pp. 349–351; Hoffmann, Akte.

[45] Hentschel, Physics, pp. liii–lxiv, and Fischer, Emigration; Wolff, Vertreibung; Wolff, Lindemanns; Wolff, Ausgrenzung.

[46] Richter, Forschungsförderung, p. 57.

[47] See Litten, Mechanik, pp. 377–398.

aerodynamic research program stands out. He was pleased to realize that, once Hitler's government came to power, he could not ask for too much.[48] Other scientists in fields related to physics, such as metals research, also found the new government accommodating.[49]

A few years later, when rearmament began in earnest and culminated in the four-year-plan, fundamental physics also benefited. Planck was able to overcome opposition from DFG (see below) President Johannnes Stark,[50] hold on to previously promised funding from the American Rockefeller Foundation, and gain new support from National Socialist officials to build and open a new KWI for Physics in place of Einstein's former paper institute.[51] In 1936 the Dutch physicist and recent Nobel laureate Peter Debye became the first director of an institute with state-of-the-art equipment, a large staff, and generous funding.[52] Three years later Army Ordnance commandeered this KWI for nuclear weapons research,[53] and Debye left for the United States.

Schmidt-Ott voluntarily implemented many National Socialist policies, including using the "Heil Hitler" greeting in correspondence and stopping any payments to "non-Aryans," but was nevertheless fired.[54] In 1933 Johannes Stark was rewarded for his early and consistent support of Hitler with the presidencies of the PTR and the NGW, now renamed the German Research Foundation (Deutsche Forschungsgemeinschaft, DFG).[55] As a result, Stark now had a great deal of influence over the support of physics outside of the budgets of university institutes.

Stark was autocratic about dispensing NGW funds, had to personally approve any application, and transferred large amounts to the PTR.[56] Outside reviewers were occasionally used for proposals. It is clear that some applications were rejected because the applicant was Jewish, although this was not the official reason given.[57] Applicants had to satisfy political and racial criteria before it was passed on to the experts, including filling out a "Aryan questionnaire."[58] Although it is reasonable to assume that the DFG now did not fund, or provided less funds to theoretical research in modern physics, in fact up until now historians have as-

---

[48] For Prandtl see Trischler, Self-Mobilisation, and Trischler, Luft; Epple, Rechnen; Epple/Remmert/Karachalios, Aerodynamics.

[49] Maier, Waffe.

[50] Mertens, DFG, p. 257.

[51] Kant, Einstein; Kant, Debye; Macrakis, Wissenschaftsförderung.

[52] For Debye and the recent controversy surrounding his actions during the Third Reich, see Kant, Debye; Altschuler, Convictions; Hoffmann, Debye; Eickhoff, Name.

[53] Walker, Quest, pp. 17–21.

[54] Mertens, DFG, p. 61.

[55] Walker, Nazi, pp. 16–21.

[56] Zierold, Forschungsförderung, p. 186; Mertens, DFG, p. 34.

[57] Richter, Forschungsförderung, p. 21.

[58] Mertens, DFG, pp. 24, 127.

sumed, not demonstrated this.[59] The example of the DFG makes clear that the total amount of money flowing to physics was not reduced significantly, rather redirected, especially to applied and military-relevant research.

In 1936, coincidentally in the same year the four-year-plan was announced, Stark was forced to resign his presidency of the DFG because of infighting within the National Socialist hierarchy.[60] At around the same time Stark also lost control over the budget of the PTR. His replacement at the DFG was the chemist Rudolf Mentzel, and three years later, when Stark retired from the PTR, the technical physicist Abraham Esau succeeded him. Both Esau and Mentzel continued Stark's policies of favoring more applied research, but not cutting off money to fundamental research. It is probably true that less money, or at least when compared to the Weimar Republic, less money was flowing to theoretical research in modern physics, but the National Socialist purge of the German civil service had also decimated the ranks of theoretical physicists, so that there may well have been fewer around to receive such funding. Once again, this remains speculative because we do not know the actual breakdown by field of research funding for physics and how it changed over the course of the Third Reich.

The war had a profound effect on all aspects of German society and also changed how and why physics was supported. The Reich Research Council (Reichsforschungsrat, RFR) was created in 1937 to coordinate academic research for the war effort, and included thirteen different disciplinary sections, each headed by an influential scientist.[61] Eventually the RFR essentially took over the task of deciding where the DFG funds should go, with a single individual making the decisions for each discipline. Esau was responsible for "physics and mechanical engineering."[62] When the war began and some scientists began working on classified research with DFG funds, this sometimes led to the DFG being asked to approve funding for projects without knowing what was being done or, if it was a renewal, what progress had been made.[63] The RFR was subsequently reorganized in 1942 under the authority of Reichmarshall Hermann Göring, after the German "lightning war" (Blitzkrieg) strategy had failed and victory no longer appeared so easy. When the RFR took over the German nuclear weapons project over Army Ordnance in 1942, Esau was given additional authority as the "Reich Plenipotentiary for Nuclear Physics". At the start of 1944, the experimental physicist Walther Gerlach replaced Esau.

On paper, Esau and Gerlach both had available and dispensed large sums of money for physics research, but that is only part of the story. As the war wore on,

---

[59] See Richter, Forschungsförderung, who on p. 52 assumes but does not specify a drop in funding for nuclear physics, and on p. 58, states that theoretical work was no longer supported by the DFG and that fundamental research was increasingly turned down; however, these judgements appear to be based on memoirs, not archival research; Mertens, DFG, notes on p. 92 that theoretical physics was rejected in favor of experimental physics, but also provides no details.

[60] Flachowsky, Reichsforschungsrat, pp. 163–200.

[61] See Flachowsky, Reichsforschungsrat.

[62] See Zierold, Forschungsförderung, p. 220, and Richter, Forschungsförderung, p. 21.

[63] Richter, Forschungsförderung, p. 21.

the amount of money became less important than the availability of scientific manpower, other labor, materials, and equipment. For example, the experimental physicist Walther Bothe at the KWI for Medical Research in Heidelberg received funding from the HG for a large magnet for the cyclotron he was building. This is an example of equipment usually considered relevant for fundamental research now being justified by it potential military uses, here for the nuclear weapons project. However, the relatively low priority of the project led to repeated delays.[64] Even after the magnet was delivered, Bothe struggled to get the cyclotron running, in part because of the ever-deteriorating state of the home front.[65]

The German nuclear weapons project also provides many examples of research being slowed down for a myriad of reasons.[66] Nuclear reactor research was slowed down by an insufficient priority classification, until Armaments Minister Albert Speer personally intervened and granted the highest status. The Auer Society's production of large metal uranium plates was delayed until another firm was able to provide hardened metal tools; both companies had more orders with the highest possible priority classification than they could possibly satisfy. The state-of-the-art underground bunker laboratory being constructed at the KWI for Physics was delayed because of lack of materials and of labor – both unskilled and skilled labor like an electrician – until 1944. Finally, the research on nuclear reactors and isotope separation was hampered and in some cases ended by the evacuation of institutes in large cities to the German countryside. None of these problems were caused by lack of funds, and more money alone could not have solved them.

Beginning in 1943, after the catastrophic German defeat at Stalingrad, the Third Reich enlisted its scientists and engineers in the search for "wonder weapons," military hardware that could either stave off defeat or bring victory. There were physicists scattered throughout the German war economy working on rockets, aircraft, submarines, radar, and other potential wonder weapons. Nuclear weapons were certainly such weapons, even though by the last years of the war most of the physicists involved with this research did not think that atomic bombs could be created before the end of the war. A small subset of the German nuclear weapons project apparently tested a nuclear weapon design – not an atomic bomb – during the very last years of the war with the strong support of the Plenipotentiary for Nuclear Physics Walther Gerlach.[67] In fact, this was typical of all branches of science and engineering during the last desperate months of the war: some scientists and engineers took whatever expertise and materials were available and tried to make something that could help the German war effort.

When Germany was defeated in 1945, the four victorious allied powers (Britain, France, the Soviet Union, and United States) divided the country and the

---

[64] Eckert/Osietzki, Wissenschaft, pp. 46–54.

[65] Osietzki, Ideology; Walker, Quest, p. 134.

[66] See Walker, Quest; Walker, Nazi, and Walker, Reactor; for alternative interpretations of the German nuclear power project, see Powers, War, and Rose, Heisenberg.

[67] See Karlsch, Bombe, and Karlsch/Walker, Light.

capital Berlin into four zones of occupation. Support of physics now depended on the occupying power, because German funding institutions like the DFG were dissolved.[68] The different powers had their own objectives and methods. Scientific and technological reparations were taken from Germany – although the word "reparation" was avoided. The Soviets called individuals with valuable skills "specialists."[69] Whereas this was most blatant in the Soviet occupation zone, where thousands of scientists, engineers, and technicians were forced to go to the Soviet Union, each of the other powers also exploited their zone. Wernher von Braun and the other rocket researchers were the most famous example of "specialists" brought to the United States.[70] France also carried scientists and engineers to research labs across its border with Germany.[71]

In contrast, the British pursued a policy of exploiting German society in Germany. One important consequence of this was Britain's willingness to rebuild German science in its own zone of occupation.[72] This included the universities, which were being rebuilt in all zones, as well as the newly founded Max Planck Society (Max-Planck-Gesellschaft, MPG), the successor to the KWG. Both the MPG and the Max Planck Institute (MPI) for Physics were established in Göttingen in the British zone. Walther Bothe's physics section of the MPI for Medical Research remained in Heidelberg in the American zone.

Although the British were quite generous about funding the MPG and MPI for Physics, all physicists in Germany were hampered by restrictions on their research. Anything that appeared related to what had been military research during the war was now controlled – including many types of nuclear physics. German scientists responded by systematically reinterpreting their wartime work as fundamental research. This can be clearly seen in the FIAT (Field Information Allied Technical) reports on wartime German science.[73] The Allies commissioned these reports, and experts in various fields wrote surveys of the various branches of scientific research. Here, for example, the various branches of the German nuclear weapons project were misleadingly reduced to fundamental experiments on nuclear physics.

During the four years of occupation the physicists, like all Germans, were preoccupied with very basic concerns: housing, food, ability to work. Housing was in short supply because of the destruction of German cities during the last few years of the war and the very many German refugees who had been expelled from the newly-defined or -expanded territories of Poland, Czechoslovakia, and the Soviet Union. Even in a city like Göttingen, which had been largely spared

---

[68] Beyerchen, German Scientists; Cassidy, Controlling I; Cassidy, Controlling II.

[69] Albrecht, Heinemann-Grüder, and Wellmann, Spezialisten.

[70] See Gimbel, Reparations; Neufeld, Rocket; Neufeld, Braun; Ciesla/Trischler, Legitimation.

[71] Walker, Quest, pp. 179–191.

[72] See Vierhaus/vom Brocke, Forschung; Walker, Quest, pp. 188–191; Hentschel, The Mental Aftermath; Hentschel/Rammer, Kein Neuanfang; Hentschel/Rammer, Nachkriegsphysik; Hentschel/Rammer, Physicists; Rammer, Göttinger Physiker; Rammer, Sauberkeit.

[73] See for example the volume on nuclear physics and cosmic rays, Bothe/Flügge, Kernphysik.

from destruction, there was a shortage of housing. At least one of the physicists who asked for permission to come to Göttingen was told that there was no room. Scientists also had to pass through the denazification process before being sure that they could retain their positions or get new ones. Although in the end this process proved flawed and with few exceptions most people passed through as "fellow travelers," this was not clear at the time, and the whole ordeal took up a great deal of time and energy.[74]

In 1949 the two German postwar states were founded, one capitalistic and democratic, the other communist. At first some physicists moved back and forth between the two states in the search of better positions and support, but the barriers between east and west gradually became more and more difficult, until the Berlin Wall closed off the communist German Democratic Republic (Deutsche Demokratische Republik, DDR) from the Federal German Republic (Bundesrepublik Deutschlands, BRD). Even after the two states were separated, there was still an element of competition between the two regimes that scientists and physicists could exploit, although the resources of the DDR were always significantly inferior.[75]

In West Germany, one set of scientists, including Walther Gerlach, worked to revive the DFG, while another group, led by Werner Heisenberg, proposed a new German Research Council (Deutscher Forschungsrat, DFR) that would influence, if not control the distribution of funding. After a few years of infighting, the more elite DFR essentially lost and was folded into the DFG. The DFR was hampered both by the elitist attitudes of its advocates and comparisons with the wartime RFR. Although Heisenberg lost this battle, he emerged in the early years of the BRD and an influential science policymaker because of his role as an informal advisor to BRD chancellor Konrad Adenauer.[76]

Although the DFG resumed operation, this was not where the big money was in physics. The legacy of Hiroshima and Nagasaki and the developing Cold War created an atmosphere of "atomic euphoria" that was very favorable for all aspects of nuclear research. The BRD, eager to catch up with the United States, Britain, and France, poured immense sums of money into nuclear "Big Science" (Großforschung) and thereby opened up great opportunities for physics.[77] The BRD secretly funded preparations for this work even before West Germany had regained full sovereignty in 1953 and thereby was allowed to pursue such research. Shortly thereafter the Ministry of Nuclear Questions and the German Atomic Commission were established.

Generous support from this ministry (as well as some from the Ministry of Defense and West German industry) flowed into the new research centers for nu-

---

[74] Walker, Nazification; Ash, Denazifying Scientists; Hentschel, The Mental Aftermath; Hentschel/Rammer, Kein Neuanfang.

[75] See Macrakis/Hoffmann, Socialism.

[76] See Eckert, Kernenergie; Carson, Framework; Carson, Development; Carson/Gubser, Science Advising.

[77] Szöllösi-Janze/Trischler, Grossforschung.

clear reactors in Karlsruhe, for nuclear ship propulsion in Geestacht, and for various aspects of nuclear research in Jülich. Heisenberg's institute moved to Munich and was expanded, eventually leading to a new MPI for Plasma Physics. Even physicists at universities benefited, as the example of Heinz Maier-Leibnitz's research with a nuclear reactor at the Technical University in Munich demonstrates.[78] Finally, the Federal German Republic also built a large particle accelerator near Hamburg and joined the new collaborative high-energy physics center CERN (Centre Européenne pour la Recherche Nucléaire, European Center for Nuclear Research) in Geneva. This last project also served to promote the reintegration of West German physics into the international community. Although this generous funding stream eventually did dry up, for many years German physicists were able to ride the wave of atomic euphoria.

The German Democratic Republic tried to keep up by focusing on a few prestige projects, but in retrospect had significantly few resources.[79] Scientific institutions were created according to the Soviet model. In particular, a plethora of research institutes were connected to the new DDR Academy of Sciences. These institutes were the most prestigious research centers in the DDR. Research was sometimes overshadowed by the competition between the superpowers. The Soviet Union helped build nuclear reactors and research centers, but this dependence on the Soviet Union was a double-edged sword. Soviet scientists were only interested in East German colleagues playing a subordinate, supporting role. This was compensated a little by collaboration within the Soviet satellites. However, East German physicists like Gustav Hertz, Robert Rompe, Peter Adolf Thiessen, and Max Volmer were able to play a dominant role in DDR science policy analogous to their colleagues in the West.

As the example of CERN above demonstrates, the West German physicists worked to reintegrate themselves into the international community and the Federal Republic of Germany spent money in order to make this possible. The understandable reluctance of foreign colleagues to put the National Socialist persecution of Jews, the war, and finally the Holocaust behind them had a distinct generational character. Whereas scientists like Gerlach and Heisenberg were treated with respect, but also with reservations, the next generation of their students found it much easier to study and even work abroad. Many members of this younger generation of physicists eventually returned to West Germany and helped reconnect German physics to the rest of the world. All of this cost money, and much if not most of it was provided by West German sources, including the German Academic Exchange Service (Deutsche Akademische Austauschdienst, DAAD), DFG, and the Humboldt Foundation – with Heisenberg serving as its first president.

Finally the end of the Cold War in 1989 and the "turning point" (Wende) reintegrated East German physics into the Federal German Republic – on the West's terms. The communist structures like the Academy institutes were dis-

---

[78] Eckert, Neutrons.
[79] Macrakis/Hoffmann, Socialism.

solved. The best East Geman scientists were hired by the MPG or other leading West German institutions, while the East German universities were transformed according to the West German model and often ended up hiring West German scientists to work in the East.[80]

Physics has flourished in Germany thanks to its physicists, but they would not have been sufficient. The financial support of German physics demonstrates how this science has been well integrated into the political evolution from the Empire, through unstable democracy, fascism, and communism, to the present. Breakthroughs in physics have gone hand in hand with generous, and perhaps most important, innovative support. The lesson of this story is that science funding matters, and that it is provided for by different types of regimes for various reasons.

The preceding survey of German physics under different political regimes begs the question of how this compares with other countries? This is the question the rest of this book aims to answer, with contributions both on German and non-German topics.

BIBLIOGRAPHY

Albrecht, Ulrich, Andreas Heinemann-Grüder, and Arend Wellmann: Die Spezialisten: Deutsche Naturwissenschaftler und Techniker in der Sowjetunion nach 1945, Berlin 1992.

Altschuler, Glenn: The Convictions of Peter Debye, in: Daedalus 135 (4), 2006, pp. 96–103.

Ash, Mitchell G.: Denazifying Scientists and Science, in: Judt, Matthias and Burghard Ciesla (eds.): Technology Transfer out of Germany, Amsterdam 1996, pp. 61–80.

Ash, Mitchell G.: Wissenschaftswandlungen in politischen Umbruchszeiten – 1933, 1945 und 1990 im Vergleich, in: Acta Historica Leopoldina 39, 2004, pp. 75–95.

Ben-David, Joseph: The Scientist's Role in Society: A Comparative Study, Chicago 1984.

Beyerchen, Alan: German Scientists and Research Institutions in Allied Occupation Policy, in: History of Education Quarterly 22 (3), 1982, pp. 289–299.

Beyerchen, Alan: Scientists under Hitler: Politics and the Physics Community in the Third Reich, New Haven 1977.

Bird, Kai and Martin J. Sherwin: American Prometheus: The Triumph and Tragedy of J. Robert Oppenheimer, New York 2005.

Bothe, Walther and Siegried Flügge (eds.): Kernphysik und kosmische Strahlen. Naturforschung und Medizin in Deutschland 1939–1946, Weinheim 1948.

Cahan, David: An Institute for an Empire: The Physikalisch-Technische Reichsanstalt 1871–1918, Cambridge 1989.

Carson, Cathryn: Heisenberg and the Framework of Science Policy, in: Fortschritte der Physik 50 (5), 2002, pp. 432–436.

Carson, Cathryn: Nuclear Energy Development in Postwar West Germany: Struggles over Co-operation in the Federal Republic's First Reactor Station, in: History and Technology 18 (3), 2002, pp. 233–270.

Carson, Cathryn and Michael Gubser: Science Advising and Science Policy in Post-War West Germany: The Example of the Deutscher Forschungsrat, in: Minerva 40 (2), 2002, pp. 147–179.

Cassidy, David: Beyond Uncertainty: Heisenberg, Quantum Physics, and The Bomb, New York 2009.

---

[80] See Ash, Wissenschaftswandlungen.

Cassidy, David: Controlling German science I: U.S. and Allied forces in Germany, 1945–1947, in: Historical Studies in the Physical and Biological Sciences 24, 1994, pp. 197–235.

Cassidy, David: Controlling German science II: Bizonal occupation and the struggle over West German science policy, 1946–1949, in: Historical Studies in the Physical and Biological Sciences 26, 1996, pp. 197–239.

Cassidy, David: J. Robert Oppenheimer and the American Century, New York 2005.

Ciesla, Burghard and Helmuth Trischler: Legitimation through Use: Rocket and Aeronautic Research in the Third Reich and the U.S.A, in: Walker, Mark (ed.): Science and Ideology: A Comparative History, London 2003, pp. 156–185.

Eckert, Michael: Die Anfänge der Atompolitik in der Bundesrepublik Deutschland, in: Vierteljahrshefte für Zeitgeschichte 37, 1989, pp. 115–143.

Eckert, Michael: Kernenergie und Westintegration. Die Zähmung des westdeutschen Nuklearnationalismus, in: Herbst, Ludolf, Werner Bührer, and Hanno Sowade (eds.): Vom Marshallplan zur EWG. Die Eingliederung der Bundesrepublik Deutschland in die westliche Welt, Munich 1990, pp. 313–334.

Eckert, Michael: Neutrons and Politics: Maier-Leibnitz and the Emergence of Pile Neutron Research in the FRG, in: Historical Studies in the Physical and Biological Sciences 19, 1988, pp. 81–113.

Eckert, Michael and Maria Osietzki: Wissenschaft für Macht und Markt. Kernforschung und Mikroelektronik in der Bundesrepublik Deutschland, Munich 1989.

Eickhoff, Martijn: In the Name of Science? P.J.W. Debye and His Career in Nazi Germany, Amsterdam 2008.

Epple, Moritz: Rechnen, Messen, Führen. Kriegsforschung am Kaiser-Wilhelm-Institut für Strömungsforschung 1937–1945, in: Maier, Helmut (ed.), Rüstungsforschung im Nationalsozialismus. Organisation, Mobilisierung und Entgrenzung der Technikwissenschaften, Göttingen 2002, pp. 305–356.

Epple, Moritz, Volker Remmert, and Andreas Karachalios: Aerodynamics and Mathematics in National Socialist Germany and Fascist Italy: A Comparison of Research Institutes, in: Sachse, Carola and Mark Walker (eds.): Politics and Science in Wartime: Comparative International Perspectives on the Kaiser Wilhelm Institutes, Chicago 2005, pp. 131–158.

Fischer, Klaus: Die Emigration von Wissenschaftlern nach 1933. Möglichkeiten und Grenzen einer Bilanzierung, in: Vierteljahrshefte für Zeitgeschichte 39, 1991, pp. 535–549.

Flachowsky, Sören: Von der Notgemeinschaft zum Reichsforschungsrat. Wissenschaftspolitik im Kontext von Autarkie, Aufrüstung und Krieg, Stuttgart 2008, pp. 163–200.

Forman, Paul: The Environment and Practice of Atomic Physics in Weimar Germany: A Study in the History of Science, Ph. D. dissertation, University of California, Berkeley, 1967.

Forman, Paul: The Financial Support and Political Alignment of Physicists in Weimar Germany, in: Minerva 12, 1974, pp. 39–66.

Forman, Paul: Scientific Internationalism and the Weimar Physicists: The Ideology and Its Manipulation in Germany after World War I, in: Isis 64 (2), 1973, pp. 150–180.

Forman, Paul: Weimar Culture, Causality, and Quantum Theory, 1918–1927: Adaptation by German Physicists and Mathematicians to a Hostile Intellectual Environment, in: Historical Studies in the Physical Sciences 3, 1971, pp. 1–115.

Gimbel, John: Science, Technology, and Reparations. Exploitation and Plunder in Post-war Germany, Palo Alto 1990.

Haber, L. F.: The Poisonous Cloud: Chemical Warfare in the First World War, New York, Oxford 1986.

Hammerstein, Notker: Die Deutsche Forschungsgemeinschaft in der Weimarer Republik und im Dritten Reich, Munich 1999.

Harman, Peter: Energy, Force and Matter: The Conceptual Development of Nineteenth-Century Physics, Cambridge 1982.

Heilbron, John L.: The Dilemmas of an Upright Man: Max Planck as Spokesman for German Science, Cambridge, MA 2000.

Hentschel, Klaus (ed.): Physics and National Socialism: An Anthology of Primary Sources, Basel 1996.

Hentschel, Klaus: The Mental Aftermath: The Mentality of German Physicists 1945–1949, Oxford 2007.

Hentschel, Klaus and Gerhard Rammer: Kein Neuanfang: Physiker an der Universität Göttingen 1945–1955, in: Zeitschrift für Geschichtswissenschaft 48, 2000, pp. 718–741.

Hentschel, Klaus and Gerhard Rammer: Nachkriegsphysik an der Leine. Eine Göttinger Vogelperspektive, in: Hoffmann, Dieter (ed.), Physik in Nachkriegsdeutschland, Frankfurt am Main 2003, pp. 27–56.

Hentschel, Klaus and Gerhard Rammer: Physicists at the University of Göttingen, 1945–1955, in: Physics in Perspective 3 (1), 2001, pp. 189–209.

Hermann, Armin, Karl von Meyenn, and Victor Weisskopf (eds.): Wolfgang Pauli: Scientific Correspondence with Bohr, Einstein, Heisenberg, vol. 1, 1919–1929. Berlin 1979.

Hoffmann, Dieter: Einsteins Berlin. Auf den Spuren eines Genies, Weinheim 2006.

Hoffmann, Dieter: Einsteins politische Akte, in: Physik in unserer Zeit 35 (2), 2004, pp. 64–69.

Hoffmann, Dieter: Max Planck: Die Entstehung der modernen Physik, Munich 2008.

Hoffmann, Dieter: Peter Debye (1884–1966). Ein Dossier, in: Max-Planck-Institut für Wissenschaftsgeschichte Preprints 314, 2006.

Johnson, Jeffrey: The Kaiser's Chemists: Science and Modernization in Imperial Germany, Chapel Hill 1990.

Jungnickel, Christa and Russell McCormmach: Intellectual Mastery of Nature: Theoretical Physics from Ohm to Einstein: I. The Torch of Mathematics 1800–1870, II. The Now Mighty Theoretical Physics, 1870–1925, Chicago 1986.

Kant, Horst: Albert Einstein, Max von Laue, Peter Debye und das Kaiser-Wilhelm-Institut für Physik in Berlin (1917–1939), in: vom Brocke, Bernhard and Hubert Laitko (eds.): Die Kaiser-Wilhelm-/Max-Planck-Gesellschaft und ihre Institute. Studien zu ihrer Geschichte: Das Harnack-Prinzip, Berlin 1996, pp. 227–243.

Kant, Horst: Peter Debye und das Kaiser-Wilhelm-Institut für Physik in Berlin, in: Albrecht, Helmuth (ed.): Naturwissenschaft und Technik in der Geschichte. 25 Jahre Lehrstuhl für Geschichte der Naturwissenschaft und Technik am Historischen Institut der Universität Stuttgart, Stuttgart 1993, pp. 161–177.

Karlsch, Rainer: Hitlers Bombe, Munich 2005.

Karlsch, Rainer and Mark Walker: New Light on Hitler's Bomb, in: Physics World (6), 2005, pp. 15–18.

Kleinert, Andreas: Der Briefwechsel zwischen Philipp Lenard (1862–1947) und Johannes Stark (1874–1957), in: Leopoldina-Jahrbuch 46, 2000, pp. 243–261.

Kleinert, Andreas (ed.): Johannes Stark: Erinnerungen eines deutschen Naturforschers, Mannheim 1987.

Kleinert, Andreas: Lenard, Stark und die Kaiser-Wilhelm-Gesellschaft. Auszüge aus der Korrespondenz der beiden Physiker zwischen 1933 und 1936, in: Physikalische Blätter 36, 1980, pp. 35–43.

Kleinert, Andreas: Paul Weyland, der Berliner Einstein-Töter, in: Albrecht, Helmuth (ed.): Naturwissenschaft und Technik in der Geschichte. 25 Jahre Lehrstuhl für Geschichte der Naturwissenschaft und Technik am Historischen Institut der Universität Stuttgart, Stuttgart 1993, pp. 198–232.

Kleinert, Andreas: Das Spruchkammerverfahren gegen Johannes Stark, in: Sudhoffs Archiv 67, 1983, pp. 13–24.

Kuhn, Thomas: Black-Body Theory and the Quantum Discontinuity, 1894–1912, Oxford 1978.

Lenoir, Timothy: Revolution from above: The Role of the State in Creating the German Research System, 1810–1910, in: The American Economic Review 88 (2), 1998, pp. 22–27.

Litten, Freddy: Mechanik und Antisemitismus. Wilhelm Müller (1880–1968), Munich 2000.

Macrakis, Kristie: Wissenschaftsförderung durch die Rockefeller Stiftung im 'Dritten Reich'. Die Entscheidung, das Kaiser-Wilhelm-Institut für Physik finanziell zu unterstützen, 1934–1939, in: Geschichte und Gesellschaft 12, 1986, pp. 348–379.

Macrakis, Kristie and Dieter Hoffmann (eds.): Science under Socialism: East Germany in Comparative Perspective, Cambridge, MA 1999.

Marsch, Ulrich: Notgemeinschaft der Deutschen Wissenschaft. Gründung und frühe Geschichte 1920–1925, Frankfurt am Main 1994.

Maier, Helmut: Forschung als Waffe. Rüstungsforschung in der Kaiser-Wilhelm-Gesellschaft und das Kaiser-Wilhelm-Institut für Metallforschung 1900–1945/48, 2 vols., Göttingen 2008.

Mertens, Lothar: DFG-Forschungsförderung 1933–1937, Berlin 2004.

Moore, Walter J.: Schrödinger: Life and Thought, Cambridge 1992.

Neufeld, Michael: The Rocket and the Reich: Peenemünde and the Coming of the Ballistic Missile Era, New York 1995.

Neufeld, Michael: Von Braun: Dreamer of Space, Engineer of War, New York 2007.

Osietzki, Maria: The Ideology of Early Particle Accelerators: An Association between Knowledge and Power, in: Renneberg, Monika and Mark Walker (eds.): Science, Technology, and National Socialism, Cambridge 1993, pp. 255–270, 396–400.

Powers, Thomas: Heisenberg's War, New York 1993.

Pfetsch, Frank R.: Zur Entwicklung der Wissenschaftspolitik in Deutschland, 1750–1914, Berlin 1974.

Rammer, Gerhard: Göttinger Physiker nach 1945. Über die Wirkung kollegialer Netze, in: Göttinger Jahrbuch, 2003, pp. 83–104.

Rammer, Gerhard: "Sauberkeit im Kreise der Kollegen": Die Vergangenheitspolitik der DPG, in: Hoffmann, Dieter and Mark Walker (eds.): Physiker zwischen Autonomie und Anpassung. Die Deutsche Physikalische Gesellschaft im Dritten Reich, Weinheim 2007, pp. 359–420.

Renn, Jürgen, Giuseppe Castagnetti, and Peter Damerow: Albert Einstein. Alte und neue Kontexte in Berlin, in: Kocka, Jürgen (ed.): Die Königlich Preußische Akademie der Wissenschaften zu Berlin im Kaiserreich, Berlin 1999, pp. 333–354.

Richter, Steffen: Forschungsförderung in Deutschland 1920–1936. Dargestellt am Beispiel der Notgemeinschaft der Deutschen Wissenschaft und ihrem Wirken für das Fach Physik, Düsseldorf 1972.

Rose, Paul Lawrence: Heisenberg and the Nazi Atomic Bomb Project, 1939–1945: A Study in German Culture, Berkeley 1998.

Rowe, David and Robert Schulman (eds.): Einstein on Politics: His Private Thoughts and Public Stands on Nationalism, Zionism, War, Peace, and the Bomb, Princeton 2007.

Schröder-Gudehus, Brigitte: The Argument for the Self-Government and Public Support of Science in Weimar Germany, in: Minerva 10 (4), 1972, pp. 537–570.

Schröder-Gudehus, Brigitte: Challenge to Transnational Loyalties: International Scientific Organizations after World War I, in: Science Studies 3, 1973, pp. 93–118.

Szöllösi-Janze, Margit: Fritz Haber, 1868 bis 1934. Eine Biographie, Munich 1998.

Szöllösi-Janze, Margit and Helmuth Trischler (eds.): Großforschung in Deutschland, Frankfurt am Main 1990

Trischler, Helmuth: Self-Mobilisation or Resistance? Aeronautical Research and National Socialism, in: Renneberg, Monika and Mark Walker (eds.): Science, Technology, and National Socialism, Cambridge 1993, pp. 72–87.

Trischler, Helmuth: Luft- und Raumfahrtforschung in Deutschland 1900–1970. Politische Geschichte einer Wissenschaft, Frankfurt am Main 1992.

Turner, R. Steven: The Growth of Professorial Research in Prussia, 1818–1848 – Causes and Context, in: Historical Studies in the Physical Sciences 3 (2), 1971, pp. 137–182.

Vierhaus, Rudolf and Bernhard vom Brocke (eds.): Forschung im Spannungsfeld von Politik und Gesellschaft – Geschichte und Struktur der Kaiser-Wilhelm/Max-Planck-Gesellschaft, Stuttgart 1990.

Walker, Mark: German National Socialism and the Quest for Nuclear Power, 1939–1949, Cambridge 1989.

Walker, Mark: Nazi Science, New York 1995.

Walker, Mark: The Nazification and Denazification of Physics, in: Judt, Matthias and Burghard Ciesla (eds.): Technology Transfer out of Germany, Amsterdam 1996, pp. 49–59.

Walker, Mark: Nuclear Weapons and Reactor Research at the Kaiser Wilhelm Institute for Physics, in: Heim, Susanne, Carola Sachse, and Mark Walker (eds.): The Kaiser Wilhelm Society during National Socialism, Cambridge 2009, pp. 339–369.

Wolff, Stefan: Die Ausgrenzung und Vertreibung der Physikergemeinschaft im Dritten Reich, in: Hoffmann, Dieter and Mark Walker (eds.): Physiker zwischen Autonomie und Anpassung. Die Deutsche Physikalische Gesellschaft im Dritten Reich, Weinheim 2007, pp. 91–138.

Wolff, Stefan: Frederick Lindemanns Rolle bei der Emigration der aus Deutschland vertriebenen Physiker, in: Yearbook of the Research Center for German and Austrian Exile Studies 2, 2000, pp. 25–58.

Wolff, Stefan: Physicists in the "Krieg der Geister": Wilhelm Wien's "Proclamation", in: Historical Studies in the Physical and Biological Sciences 33 (2), 2003, pp. 337–368.

Wolff, Stefan: Vertreibung und Emigration in der Physik, in: Physik in unserer Zeit, 24, 1993, pp. 267–273.

Zierold, Kurt: Forschungsförderung in drei Epochen. Deutsche Forschungsgemeinschaft: Geschichte – Arbeitsweise – Kommentar, Wiesbaden 1968.

THE ORIGINS OF GERMAN BIOPHYSICS IN MEDICAL PHYSICS. MATERIAL CONFIGURATIONS BETWEEN CLINIC, PHYSICS AND BIOLOGY (1900–1930)

Alexander von Schwerin

INTRODUCTION

Today biophysics is an established biological discipline. It may be described as a spectrum of physics-based techniques applied to biological problems. This becomes more evident when one goes back in the history of biophysics. Within Germany, the X-ray tube stood at the center of the disciplinary formation of biophysics.

This article traces back the origins of biophysics to the very beginnings of radiology in the first two decades of the 20th century. Rather than biology itself, it was this early context of the practical use of the X-ray generator that initially introduced physicists to medical and biological problems. X-rays defined an innovative field of research and medical practice, particularly radiotherapy, binding physicians, including gynecologists, surgeons, radiologists, and physicists together. Physicists working on medical problems were sometimes called "medical physicists". While radiology developed later into a medical discipline, institutions of medical physics existed from the early 1920s on. This development can be shown in the work of Friedrich Dessauer and Walter Friedrich who became outstanding biophysicists. Less known today is the physicist Richard Glocker who later became famous for his work on the structure of metals. But all three of them shared not only interests in the application of X-rays to medicine but also an interest in more fundamental related problems. Their experimental and theoretical work became the starting point for a more intense collaboration between physicists and biologists from the late 1920s.

This article will emphasize the constitutive role played by epistemic things in the formation of research fields and disciplines. The technical problems posed by the application of X-rays to radiological therapy continually pointed towards a fundamental biological problem. What are the effects of radiation on living organisms? To that end, physicists even developed a new theoretical approach known as "quantum biology".[1] With respect to the history of biophysics, these epistemic things – the researched biological processes – formed the constitutive problem determining the field of radiotherapy, the institutional formation of

---

[1] Dessauer: Kontrapunkte, stands pars pro toto for several approaches towards a theory of biological radiation effects made by physicists. See also Fischer, Licht.

medical physics in the twenties, and its conversion towards biology from the late twenties on.

This chapter will also highlight the mutual constitution of techniques and objects of research, and, hence, the significance of resources. After sketching the cooperative relation of radiology and radio-electric industry, (1) it will depict how the application of X-ray techniques in radiotherapy and the experiments on biological effects of radiation intersected in the early years of medical physics. A prominent example (2–3) is the collaboration of the physicist Friedrich and the physician Bernhard Krönig at Freiburg (im Breisgau), which reveals the technical configurations of radiotherapy as the driving force of medical physics up to the early 1920s. It shows that both the techniques and the problems of understanding the biological effects of radiation were the main challenge of radiotherapy. In general, the technical requirements of radiotherapy ensured that radiobiological research defined the objectives of early biophysics in Germany.

The sections following this example (4–6) will cover the 1920s and illustrate how medical physics managed its material resources: money, technical devices and model animals. Given that X-rays laboratories were expensive, medical physicists needed financial support when they wanted to work independently of physicians. The "Emergency Foundation for German Science" (Notgemeinschaft der deutschen Wissenschaft, NGW), was the main German research organization at that time. The relevance of its research policy for medical physics will be sketched.[2] Actually, another resource, biological models, was more relevant for the development of medical physics towards biophysics. These models will be discussed at the end of this paper.

1. HOW PHYSICISTS JOINED THE FIELD OF RADIOLOGY (1900–1920)

Both physicists and physicians were engaged in the application of X-rays to medicine early on from the first trials in the late 1890s. X-rays, or "Röntgen radiation" in the German context, were systematically explored in order to extend their ability to penetrate the body and generate pictures of the inner structures.[3] Radiography was the main purpose of the medical application of X-rays in the early years of radiology. However, it was not the field where questions on the biological effects of radiation arose. This was different with respect to the application of X-rays to therapy, which faced with some additional difficulties.

Initially, only pathological conditions of the skin had been scrutinized by radiotherapists.[4] This work was pursued mainly through trial and error. The next

---

[2] A detailed evaluation of NGW's research policy will be the topic of an upcoming study by the author on radiation biology and biophysics.

[3] For a comprehensive study of radiography and the interaction of physicians and physicists see Dommann, Durchsicht.

[4] Schinz, Jahre, pp. 149–155.

step was more invasive, focusing the radiation on tumors lying beneath the skin. Beginning in 1905 the so-called "deep therapy" (Tiefentherapie), was developed and became the biggest challenge radiotherapy ever faced. It was generally not accepted prior to the 1920s.[5]

Gynecologists were amongst the first who experimented systematically with deep therapy, beginning around 1909.[6] In some ways the organs of the female body mediated the transition from dermal therapy to deep therapy when gynecologists placed capsules filled with radium through the vagina into the interior of the body. Thus, gynecologists broadened therapy to an increasing spectrum of diseases such as uterine myomes, hemorrhages or genital tuberculosis.[7] When they started to combat malignant tumors of the uterus and ovaries from outside the body using high-energy X-rays, optimism grew and many postulated that X-rays would be a universal therapeutic. "Danger of cancer is banished!" and other such comments of physicians answered the promising results of the Freiburg gynecologist Bernhard Krönig (1863–1917) and his school presented at the congress of gynaecologists in 1913.[8] Among them was a surgeon claiming: „Yesterday was the last time I will have had to touch a knife!"[9] However, since radiotherapy competed with surgery, the most appreciated method in the therapy of cancer at that time, most surgeons remained skeptical about the achievements.

Other obstacles of invasive radiotherapy lay within the technique. In general during the first years of radiology, existing X-ray generators were not designed for a common use in clinical practice. Physicians who wanted to apply X-rays often relied on the assistance of physicists.[10] The "promised healing" (Heilsversprechen) of radiotherapy depended on the modulation of techniques to increase and direct the destructive power of radiation.[11] These techniques required technical skills, physical knowledge and the industrial production of machines. Thus, physicists, engineers and industry were part of the developmental efforts of "deep therapy" from the beginning.

Physicists provided physicians with technical support. But their work also embraced experimentation on technical improvement and on the biological effects of radiation.[12] An outstanding example is found with the physicist Walter Friedrich (1883–1968), a well-trained researcher who had been the assistant of Max v. Laue, but had subsequently been thrilled by the problems of radiotherapy.[13] In 1914, he was hired by Krönig in Freiburg in the first physical assistantship established at a German gynecological hospital. Subsequently, he led a newly established

[5] Schinz, Jahre.
[6] Willers/Heilmann/Beck-Bornholdt, Jahrhundert, pp. 54–55.
[7] Gauß, Geschichte, p. 637; Frobenius, Röntgenstrahlen, pp. 72 ff.
[8] Friedrich, Geschichte, p. 1; Gauß, Geschichte, p. 642.
[9] Cited in Frobenius, Röntgenstrahlen, p. 84 (translated by AS).
[10] Dommann, Durchsicht, p. 223.
[11] Frobenius, Röntgenstrahlen, pp. 84 and 94 ff.
[12] Ibid., p. 156.
[13] Schierhorn, Friedrich, pp. 18–22 and 25. Friedrich had described the X-ray interference together with Paul Knipping.

radiological laboratory.[14] Other laboratories founded afterwards in Würzburg, Erlangen and Munich were modeled after the laboratory in Freiburg.[15]

Another feature of the early intersection of physics and clinic was the collaboration with industry. The Freiburg Institute was supported by the prominent company for X-ray technology, Reiniger, Gebbert & Schall at Erlangen (RGS).[16] This kind of support was neither an exception nor a marginal aspect of laboratory work at the universities.[17] The strategy of RGS was to help ambitious clinicians in their efforts to build up an X-ray laboratory. In addition to the clinic in Freiburg, the company provided X-rays generators to the clinics in Munich, Erlangen and Berlin with the aim of stimulating research into radiotherapy and furthering collaboration. For example, the RGS and the X-ray laboratory at the gynecological hospital in Erlangen were concerned with crucial clinical and technical problems that could result in the innovation of new products.[18] One result was the construction of a device that would produce high voltage (not merchandised before 1916), a notorious problem with deep therapy.[19] The company provided not only material but additionally employed two postdoctoral physicists for the X-ray laboratory of the clinic.[20]

Another leading company in the field of X-ray techniques was the Veifa in Frankfurt (Main).[21] Founded by the technician Friedrich Dessauer in 1907, its business was to improve and construct new generators intended strictly for the medical use. Dessauer was interested in constructing generators for invasive radiotherapy immediately after the first 1905 publication claimed it was possible. He was ambitious to expand X-ray techniques to ordinary physicians who were not pioneers.[22] From 1906, Dessauer's "Elektrotechnisches Laboratorium" announced quite popular courses that introduced more than hundreds of physicians per year to the physical-technical basics of X-rays generators.[23] Dessauer

[14] Gauß, Geschichte, p. 639.

[15] Ibid. The physicists Theodor Neeff, Walther Rump, and Friedrich Voltz joined these clinics.

[16] Frobenius, Röntgenstrahlen, p. 356.

[17] Ratmoko, Hormone, highlights the pharmaceutical industry. Similar to the collaboration of industry and universities the network of collaboration within industry has to be taken into account. In Germany, the networking was favoured by concentration and cartelization of the radio-electric industry beginning with WWI and continuing through the 1920s. See company chronicles in http://www.med-archiv.de/geschichte.php (accessed 12 July 2006).

[18] Frobenius, Röntgenstrahlen, pp. 99–104. The gynaecologist Hermann Wintz established a centre for gynaecological radiotherapy in 1913. The technical collaboration with the Institute of Physics of the University of Erlangen and the RGS was crucial. A cooperative contract guaranteed both sides benefits from patents. Ibid., pp. 117–126.

[19] For the so-called "Symmetrie-Apparat", see Frobenius, Röntgenstrahlen, pp. 165 ff., for other examples pp. 159–171.

[20] Ibid., p. 165.

[21] "Veifa" stands for "Vereinigte Elektroinstitute Frankfurt-Aschaffenburg".

[22] Goes, Dessauer, pp. 212 ff.

[23] Anonymous: Ärztliche Unterrichtskurse im Röntgenverfahren in Aschaffenburg, in: Archiv für physikalische Medizin und medizinische Technik 1, 1906, pp. 196–197.

was able to finish his studies on the effects of X-rays, completing a dissertation in physics during World War I that paved the path for his subsequent career as medical physicist.[24]

Similar to other firms, Veifa benefited from the introduction of X-ray technology to military medicine. Although gynecologists complained that many of them had been killed in action,[25] the war proved to be a driving force for the development of radiology.[26] The construction of mobile X-ray devices ranked high in importance among the industrial efforts undertaken to facilitate the use of X-rays in the field hospitals.[27] Additionally, the physicists benefited from the military X-ray services through acquiring new skills. One example is the physicist Richard Glocker (1890–1976), a pupil of Wilhelm Conrad Röntgen, who worked as a "military X-ray technician" (Feldröntgenmechaniker). Glocker's later interests in the problems of radiology originated from this military engagement.[28] Thus, war accelerated the construction of robust X-ray devices, but also guided the attention of physicists towards the problems of medical physics.

In general, at the end of WWI, X-rays had been quite well established as a medical therapy. There were many clinical protocols that guided the physician in planning radiological therapy for cancer and other diseases. As early as 1912, the RGS had presented an apparatus that was constructed specifically for deep therapy; the Veifa developed their own version one year later.[29] Thus, since the days of WWI, the ordinary therapist could rely on industrially produced and standardized generators that were simple to handle. Subsequently, physicians increasingly no longer needed technical support from physicists.[30] However, the expansion of the radio-therapeutic method into the every day practice of clinics did not mean that physicists became superfluous. Around 1920, the main radiological sites within university clinics still drew on the support of physicists.[31] This was due to the fact that the conquest of the body by radiotherapy still faced crucial difficulties.

---

[24] Goes, Dessauer, p. 217.

[25] Frobenius, Röntgenstrahlen, pp. 148–149.

[26] Dommann, Durchsicht, pp. 211 ff.

[27] See the technical section of the series "Röntgen-Taschenbuch", vol. VII-VIII (ed. by Ernst Sommer). Even before WWI, German radiologists cultivated their relationship to the military. The "Kaiser-Wilhelm-Akademie für das Militärische Bildungswesen" functioned additionally in this area as conduit between science and military. Goercke, Jahre, p. 38; see for other examples Maier, Forschung.

[28] Breitling, Glocker, p. 149.

[29] See the reports of the companies in: Röntgen-Taschenbuch IV, 1912, pp. 329–334 (for RGS) and V, 1913, pp. 326–329 (for Veifa).

[30] Dommann, Durchsicht, p. 223.

[31] This is at least true for southern Germany, Frobenius, Röntgenstrahlen, p. 154.

2. "DEEP THERAPY" AS A PROBLEM FOR MEDICAL PHYSICS
(THE FREIBURG SCHOOL 1904–1913)

Deep radiotherapy was a challenge because no one knew how to effectively irradiate an organ. From the beginning, the experiments on deep therapy were more systematic than the trials on diagnostics and radiography.[32] Radiotherapists in particular took advantage of experiments by physicians and biologists on plants and animals. Biological experimentation showed some influence of radiation on reproductive organs, the individual development and the cell anatomy within the first years of the century.[33] One of the experimenters was the surgeon Georg Perthes (1869–1927). He experimented with eggs of the see-urchin (Ascaris) and chickens before he took the first steps towards deep radiotherapy. In 1905, he tried filters when radiating his objects.[34] This invention tackled the fundamental problem of the transforming vision of the radiotherapists: to invade the depth of the body without harming the surface. Only radiation with high energy was able to invade deeper parts of the body. However, the X-ray generators produced radiation consisting of a wide range of frequencies. Perthe's solution was to use filters in order to absorb low energy radiation and allow only radiation through that passed the skin.[35]

This strategy was immediately copied by Dessauer in Frankfurt and utilized for the technical development of X-rays techniques at the workshops of the Veifa beginning in 1906.[36] However, it was quite complicated to determine which materials or combination of materials was best suited for filtering. The answer depended on finding the correct spectrum of radiation for a certain locus within an organism and, hence, biological parameters. When Dessauer described the technical complexity, he also expressed the mechanical precision that should become the ideal of medical physics: the challenge lay in "transforming the deep areas of human body into a radiation field while not harming the surface and also understanding exactly how much radiation was required for a specific point. Furthermore, this would need to be reproducible under the same conditions".[37] From a biological perspective, it was not at all clear whether the effects of radiation changed not only with the depth in the body, but also with the type of tissue and other biological parameters. In short, the engineering efforts proved to be closely tied to the question of how radiation affected biological substance.[38]

---

[32] Schinz, Jahre, p. 155.

[33] Ibid., pp. 77–86. The significance of human experimentation for the development of radiology can not be evaluated here in detail. However, the Swiss radiologist Hans Schinz was correct to point to the so-called "accidents", that showed the invasive effects of radiation on human beings. Ibid., pp. 78 and 155.

[34] Ibid., pp. 82, 86 and 158.

[35] Willers/Heilmann/Beck-Bornholdt, Jahrhundert, pp. 56–57.

[36] Dessauer, Therapie, p. 2.

[37] Ibid., p. 23 (translated by AS).

[38] Schinz, Jahre, p. 76.

Thus, the practical context of radiotherapy influenced further efforts in radiobiological research.

An example of these dynamics is the aforementioned research on "deep therapy" at the gynecological hospital in Freiburg, one of the pioneering centres for deep therapy. Beginning in 1904, Krönig experimented with X-rays, although he was also an advocate of surgical methods.[39] Initially he attempted radiating carcinomas of the uterus and the intestines after opening the abdominal wall in order to better target the radiation.[40] Following another laboratory's announcement of the effects of radiation on the reproductive function of ovaries of rabbits, Krönig's coworkers were sent to that laboratory.[41] Additionally, they were trained in X-ray technology joining Dessauer's courses. In 1909, they began parallel experiments with beans (Vicia) and frogs (Rana) studying the histological and developmental effects of different doses, filters and distances of the X-ray source.[42] They examined the effects of new filters combined with the calibration of the X-ray tube. Mainly based on these biological experiments, but also by examining clinical trials, the Freiburg group developed a new method that they hoped would cure female chronic hemorrhaging.[43] In 1912, they announced having significantly reduced the harm to the skin.

Another problem faced was that X-ray generators at that time were not capable of producing high-energy radiation. In order to increase the radiation dose on tumors, gynecologists started combining X-ray generators with the application of radium encapsulated and deposited in the vagina.[44] Although Krönig and his co-workers consulted some specialists of radium therapy, it became clear that the handling of radium and the dosimetry became too complicated. That was the moment when Krönig decided to employ the physicist Walter Friedrich.[45]

Over the course of this early collaboration, the direction of medical physics was crystallised. Krönig and Friedrich continued the biological experiments on beans and the eggs and larvae of frogs in the following years. Because most of the experimental work and clinical data produced by physicians was not comparable, they concluded that an exact dosimetry was necessary in order to improve

[39] Gauß, Geschichte, pp. 633–634.

[40] Ibid., p. 634.

[41] The Hamburg gynaecologist Heinrich E. Albers-Schönberg was a pioneer when he sterilised male and female rabbits and guinea-pigs in 1903 and treated women 1908–1909. Schinz, Jahre, pp. 77–78; Frobenius, Röntgenstrahlen, p. 75. The effects of X-rays on ovaries known as "Amenorrhoe" became a prior target of gynaecologists when developing radiotherapy targeting several diseases including haemorrhages. Gauß, Geschichte, pp. 634–635; Schwerin, Experimentalisierung, p. 120.

[42] The experiments were performed by the gynaecologist Carl Joseph Gauß who worked intensively on radiotherapy for years together with the gynaecologist Hermann Lembcke. Gauß, Geschichte.

[43] Gauß/Lembcke, Röntgentiefentherapie. The method combined the use of the X-ray source at close range, the use of an aluminum filter and radiation from multiple points ("Mehrfeldbestrahlung" or "Freiburger Methode").

[44] Schinz, Jahre, p. 164; Gauß, Geschichte, pp. 638 ff.

[45] Gauß, Geschichte, p. 639.

radiotherapy.[46] After showing the failures of several commercial dosimeters, they developed probes in order to measure the quality, the intensity and the absorption of radiation at all loci of a biological object. Engineers of the RGS supported them by providing data from experiments carried out at the company's laboratory.[47] Using the new dosimetry they started a long series of experiments in order to measure the biological effects of different types of radiation. These experiments revealed a certain "biological factor" in the effects of radiation.[48] The situation became even more complicated when the researchers compared the biological experiments with clinical data derived from different kinds of tissue (skin, ovarian, cancer). Several hundred patients were included in the trials and, it should be noted, the researchers did not hesitate to risk the patient's health in some series. But in the end, it appeared possible to deduce a biological formula from the data that would allow researchers to predict the variations of the effects that were dependent on tissue.[49]

The results of the Freiburg group were published in a monograph on "The Physical and Biological Basis of Radiation Therapy" (Physikalische und biologische Grundlagen der Strahlentherapie) in 1919.[50] Krönig and Friedrich enthusiastically announced that they would use physical precision in order to seek for the laws ("Gesetzmäßigkeiten") that determined the biological effects of radiation. They felt only such laws would allow the rational use of the X-ray techniques.[51] Biological rules would dictate to technical engineers and industry what the needs of radiotherapy were in detail and provide medical practitioners with formulas allowing control of the effects of radiation in a certain tissue. In fact, the Freiburg results facilitated the improvement of the practice of deep therapy through the use of filters and dosimeters.[52]

---

[46] Krönig/Friedrich, Grundlagen, pp. 4 and 11. Only then did the German radiologists start a commission to standardize the measurement of radiation. An international unit (1 Röntgen) was established in 1928.

[47] Gauß, Geschichte, pp. 639–640. In turn, the Freiburg group provided the company with clinical and biological knowledge to improve radiotherapeutic devices. Frobenius, Röntgenstrahlen, pp. 81–92. In a similar way, the "Oberingenieur" Leonard Baumeister handled the collaboration with Wintz in Erlangen, ibid., pp. 117 and 165 ff.

[48] Krönig/Friedrich, Grundlagen, p. 166.

[49] Ibid., pp. 262 ff. A variant of the soon to be widely used "Hautdosis" was also developed by the gynaecologists Ludwig Seitz and Wintz in Erlangen. Frobenius, Röntgenstrahlen, p. 366.

[50] Krönig/Friedrich, Grundlagen, pp. 3–4.

[51] Ibid., pp. 3–4 and 135.

[52] Frobenius, Röntgenstrahlen, p. 150.

3. X-RAY GENERATORS AND THE INCORPORATION OF PHYSICS INTO RADIOBIOLOGICAL RESEARCH (1913–1920)

Up until now, it has been shown that the technical development of X-rays technique was dependent upon the data of radiobiological research and, therefore stimulated research into the biological effects of radiation. However, it also raised new radiobiological problems, such as the case when a new type of X-ray generator was invented on the cusp of WWI. In 1913, the American chief engineer of the General Electric Company, Schenectady, William D. Coolidge, made a decisive invention: the hot-cathode tube ("Glühkathode"). The name referred to the new principle used by the tube. It produced X-rays by heating tungsten while running under a high vacuum. An advantage of the new principle was that the current and voltage of the tube could be better controlled independently from each other and, subsequently, the energy of the X-rays generated.[53]

By comparison, early machines used ions of gas to generate X-rays. The constructions were not durable and produced varying results. It was therefore difficult to both increase the energy of radiation needed for the therapy of tumours and to homogenize the results.[54] This reflected technical limitations exemplified by the problem faced by Friedrich and Kröning in their experiments, that at that time there was no common dosimetric measurement. The generators differed so much in their performance that it was difficult to compare X-ray delivery. Instead, radiotherapists expressed the delivery by enumerating a number of parameters that depended upon the device used, such as the type of generator and tube, distance from tube, and the exposure time.[55]

One radiologist expressed the expectations generated by the Coolidge tube when he explained that the simple principle of the tube would promise the "mechanisation of radiotherapy".[56] The Coolidge tube not only worked more consistently, but also withstood high currents for a long time.[57] Now hard radiation, or homogenous X-rays of high energy, became available. Radiotherapists who concentrated more and more on the treatment of carcinoma were particularly interested in this kind of radiation because it easily penetrated different types of tissue.[58] In fact, the so-called Coolidge tube would replace the old X-ray generators by the late 1920s.[59]

---

[53] Ibid., pp. 95–96.

[54] Dessauer, Dosierung, p. 16; Frobenius, Röntgenstrahlen, p. 184.

[55] Del Regato, Unfolding, p. 6.

[56] Blum, Apparat, p. 183.

[57] The production of high energy X-rays was also depended on the transformers used. Dessauer's company was busy developing a transformer that would be small and robust enough to steadily produce 200 kV. In 1922, he reported developing a tube designed for 230 kV. Dessauer, Therapie, pp. 15–22 and 24. In the US, the cascade generator came into use so that radiation on the order of 200 kV became available. Del Regato, Unfolding, pp. 7–8.

[58] Schinz, Jahre, p. 165.

[59] Blum, Apparat, p. 5.

The ideal of mechanisation suited the strategy of Dessauer's company, Vei-fa. He took over the rights for the Coolidge tube from the AEG (Allgemeine Elektrizitäts-Gesellschaft, German General Electric) in 1916 in order to develop an apparatus that was easy to handle, flexible and, above all, stable and efficient.[60] Because they would yield reproducible results, Dessauer was convinced that the new tubes would finally provide X-rays that were a sort of "physical medicament".[61] Dessauer pointed as example to the Freiburg school that had used the Coolidge tube since 1914.[62] In fact, Krönig and Friedrich preferred to use this tube for their biological experiments because of its more homogenous output.[63] However, the historian Martina Blum clearly illustrates that the new technique was not the ideal step of progress as Dessauer had praised it. Rather its establishment went hand in hand with some widespread misunderstandings about its working principles and true performance.[64]

Nevertheless, it was true that the new tube permitted production of high-energy radiation and this was enough to present radiotherapists with new problems. Friedrich and the Freiburg physicians were the first to realize this. When they started their dosimetric experiments they were confronted with some irregularities that were due to secondary emissions of the tube.[65] But Friedrich noticed that these technical problems did not fully explain what occurred and that there must also be a biological cause for the irregularities. In fact, the absorption of hard radiation by tissue followed other rules than the absorption of radiation used and measured until then.[66] The more the performance of the tubes increased (soon to be increased to 200 kV) the more the intensity and distribution of radiation in organic matter did not perform as expected.[67] It seemed that a new principle influenced the biological effects at high energies. In fact, Friedrich assumed that the radiation was multiplied and dispersed when it reacted with the organic material. Scattered radiation ("Streustrahlung") now became a main problem to handle in deep therapy. Dessauer subsequently argued that this problem, which arose during Friedrich's experiments, became the origin of the "complete transformation that deep therapy has witnessed in the recent years".[68]

As the techniques produced new biological effects of radiation, the problems of radiobiological research became increasingly complicated and of critical im-

---

[60] Ibid., p. 139.

[61] Dessauer, Problem, p. 1396.

[62] Blum, Apparat, p. 139. In 1914 the AEG provided the Freiburg group with one tube for testing purposes immediately after it had acquired the production rights from General Electrics. Ibid., p. 180.

[63] Krönig/Friedrich, Grundlagen, p. 137.

[64] Blum, Apparat, pp. 183 ff.

[65] Krönig/Friedrich, Grundlagen, pp. 236 ff. All X-ray generators produced secondary radiation, but the radiation produced by the Coolidge-tube was extraordinary high. Blum, Apparat, p. 189.

[66] Krönig/Friedrich, Grundlagen, pp. 101–102 und 257.

[67] Dessauer, Therapie, pp. 29 ff.; Schierhorn, Walter Friedrich, pp. 25–27.

[68] Dessauer, Therapie, p. 29.

portance for the foundation of radiology.[69] But since the medical physicists were active in research on the effects of radiation, the mechanistic understanding of it was increased. They suggested that the complicated configurations of the apparatuses and the technical equipment would be replaced by one machine and radiotherapy would be only a matter of pressing one button. Physicians remained skeptical, arguing that the skills of the medical practitioner could not be replaced by a machine because, even if the dose used remained constant, the reaction to radiation often differed between two patients.[70] It was true that even in the rigid dosimetric regime of Krönig and Friedrich, there was individual variability that could not be eliminated.[71] Thus, most physicians were convinced that the reaction to radiation was not simply a physical reaction, but instead a physiological one that involved processes of the living cells.

Since they were trained in physics, medical physicists shared a different style of thinking and experimenting. They assumed that the witnessed variability would conceal a type of fundamental law of radiation effects. In order to integrate variability, Friedrich and Dessauer were at the forefront of those to use statistical methods extensively.[72] Around 1920 Dessauer came up with a serious theoretical approach. According to Dessauer, the only standardization of the radio-therapeutic techniques that had been achieved gave evidence that the primary effects of radiation in organic matter were of a probabilistic nature. In order to explain the processes he relied on newly developed quantum theory.[73] Dessauer's "Point Heat Theory" (Punktwärmetheorie) stimulated a new wave of experimentation and was the first of several theoretical approaches to follow in the 1920s.

To briefly summarize: medical physicists relied on the close collaboration with physicians and were almost exclusively concerned with the application of X-rays (and radioactivity) to medicine. Radiotherapy invading the body (deep therapy) was the driving force behind these social and technical developments. It is worth noticing here that the source of radiation played almost no role in the early configuration of physics and medicine.[74] The technology of radiotherapy was improved only step-by-step through the calibration of a bundle of technical problems. Therewith, the technical and practical questions of radiology drove the radiobiological research field that had been dominated by physicians and some biologists up until then. Medical physicists strengthened their influence within that field through the technical relevance of their work. Concurrently, the processes developed in order to study the biological effects of radiation formed the epistemic object that became the crystallizing core for a disciplinary emancipation of medical physicists. This became clear when Dessauer transformed radiation biology into an original topic of (atomic) physics.

---

[69] Dessauer, Kontrapunkte, p. 16; Frobenius, Röntgenstrahlen, p. 82.
[70] Blum, Apparat, p. 184.
[71] Krönig/Friedrich, Grundlagen, pp. 269 ff.
[72] In general for the mathematisation of radiotherapy see Blum, Apparat, pp. 204 ff.
[73] Dessauer, Wirkungen, p. 37.
[74] Frobenius, Röntgenstrahlen, p. 240.

4. A BUNDLE OF (FINANCIAL) RESOURCES FOR MEDICAL PHYSICS (EARLY 1920s)

Radiation physics and atom physics were already well-developed sciences in terms of theory. Thus, the understanding of the interaction between radiation and matter was well underway. However, the message of medical physicists was that the effects of radiation on organic material remained somewhat mysterious. Since Walter Friedrich and Krönig had also tried to model organic material using simple systems, the attention of physicists like Dessauer were drawn to liquid matter.[75] The understanding of the physics and chemistry of the radiation of liquids became one of the main objectives of medical physics in the 1920s. Thus, medical physics did not only refer to quantum physics, but steadily extended its references outside medicine (e. g. towards colloid chemistry and photochemistry).

The conceptual and experimental emancipation from medicine was institutionalized when medical physicists began their own institutions. Within three years three institutes were founded that were led by physicists. In 1920, Friedrich Dessauer became the director of the "Institute for Physical Foundation of Medicine" and at the same time full professor for "Physical Foundation of Medicine" at the University of Frankfurt, the first chair for medical physics in Germany.[76] Dessauer's dismissal from Veifa was the consequence of his long endeavor to promote research and teaching in the field of medical physics.[77] Only a short time before, the Technical University of Stuttgart had founded a "X-ray Laboratory" just in time for its first director, the physicist Richard Glocker. Finally, in 1923, Walter Friedrich became full professor for medical physics at the University of Berlin where he began to build up the "Institute for Radiation Research" at the Charité University Hospital. There were other smaller institutions for medical physics as well, such as the "Laboratory for Medical Physics" in Göttingen. However, the three institutes mentioned became crucial with respect to the developmental direction of medical physics towards biophysics. This section will focus the institutes of Dessauer and Glocker and will describe the interests and financial resources that brought them to life.

It is worth mentioning that the debates about the theoretical foundation of radiation biology were not by any means abandoned by physicians. The theoretical challenges of the physicists provoked vivid debates that persisted through the 1920s between medical physicists and radiologists about the mechanisms of radiation effects.[78] The field of biological radiation research became institutionally more heterogeneous as the number of professional radiologists increased. One example may illustrate this: the laboratory of the Hamburg radiologist Hermann

---

[75] Dessauer, Therapie, p. 31.

[76] Goes, Dessauer, pp. 218–219.

[77] Dessauer's company, the Veifa, merged into the tightly integrated German "Röntgenindustrie" that was shaped by so-called "Interessengemeinschaften", cartels and successive concentration. See also footnote 17.

[78] Debates are documented in the "Strahlentherapie", vol. 1924–1925.

Holthusen (1886–1971) met all challenges of medical physics. Holthusen had been trained in experimental work by the physicist Philipp Lenard at the medical clinic in Heidelberg.[79] In 1925, he was asked by the editors of the interdisciplinary journal "Strahlentherapie" to comment on Dessauer's "Points of Heat" (Punkt-wärme) theory.[80]

In order to draw a clear distinction with regard to radiology, Dessauer claimed that his own institute was a "absolute institute for physics and applied physics".[81] However, since the new institutions of medical physics were settled within medical faculties, physicists had to define their function within medicine. Dessauer was clearest in doing so when he claimed that medical physicists were not only competent with respect to X-rays and natural radioactivity, but also regarding "physical pharmaceuticals" in general.[82] The claim of developing medical techniques was central to the disciplinary legitimization of medical physics. Physicists expected that physicians of all faculties would seek technical knowledge and physical understanding from medical physicists.[83]

These far-reaching claims made by Dessauer reflected the wider public interest in and support of his institute than has previously been recognized. The Frankfurt institute was formally affiliated with the university, but largely financed by the "Oswalt Foundation" established by several financiers in 1920.[84] The foundation was named after its founder, the Frankfurt philanthropist and local politician Dr. Henry Oswalt. Contributors to the foundation were private parties and some electro-technical companies such as the AEG.[85] After the inflation had destroyed the capital stock of the foundation, the Reich Ministry of Internal Affairs, the Prussian Ministry for Culture and the city council of Frankfurt intervened. The latter had already been successful in convincing the institute to settle in Frankfurt instead of Berlin. From 1928 onward the institute managed to finance its 100,000 Reichsmark budget largely by selling scientific devices that were produced in the institute's workshops. Thus, the workshops of the institute became the heart centre. Dessauer's dictum was: "A good mechanic is often of more value than a mediocre scientific assistant."[86] Permanent employees included five mechanics and technical assistants, but only two physicists. As Dessauer pointed out proudly, most of the staff members were financed by grants.[87]

Dessauer maintained good contacts with international physico-medical research centers, namely in the United States. These connections were sparked after American radiologists visited Germany following the war in order to study the

---

[79] Holthusen, Vortrag, pp. 142–143.
[80] Holthusen, Punktwärmehypothese.
[81] Dessauer, Institut, p. 1524.
[82] Dessauer, Problem, p. 1397.
[83] Dessauer, Institut; Dessauer, Jahre, pp. 4–8.
[84] Dessauer, Jahre, pp. 12 and 16.
[85] Ibid., pp. 4–9.
[86] Ibid., p. 9.
[87] Ibid., p. 5.

progress in deep therapy. Dessauer was invited to the US thereafter.[88] However, the international contacts resulted, not in a direct financial support, but in an astonishing brain drain of Dessauer's pupils throughout the 1920s.[89]

The conditions of the X-ray laboratory in Stuttgart were somewhat similar to the institute in Frankfurt with respect to the financial sources that Richard Glocker mobilized. Glocker had his first experiences with medicine when he became the chief of the X-ray department of the "reserve military hospital" (Reservelazarett) Stuttgart during World War I.[90] After he received his venia legendi for physics in 1919 he immediately took over the institute for physics of the Technical University, albeit initially as a substitute.[91] In 1920, the X-ray laboratory began work within this institute. However, the initiative for such a laboratory dated from the war years when Glocker was included in the plans of a Swiss medical physicist and a local homeopathic doctor, both of whom believed in radiotherapy and planned an institute dedicated to the physical basics of the effects of X-rays and its measurements.[92] However, much more interest was needed before the plan became a reality. The original plan to have the laboratory at the new homeopathic clinic founded by the industrialist Robert Bosch (later "Robert-Bosch-Hospital") did not materialize due to delays.

Similar to the case of Dessauer, whose appointment as a physicist to the medical faculty met massive resistance from the faculty members, the new plan to develop a laboratory for radiotherapy at the Technical University was not welcomed by some members of the staff who were afraid of losing material support.[93] For this reason, its supporters employed a strategy of financing the laboratory through a foundation that was inaugurated in late 1919 by the influential patron Bosch, well-known clinical researchers and the mayor of Stuttgart. Financing was undertaken by several private individuals, the city council of Stuttgart and by foundations run by Bosch.[94] A newly designed X-ray laboratory that the State of Württemberg had paid for was christened in 1922.[95] The "Robert Bosch War Foundation" (Robert-Bosch-Kriegsstiftung) provided 50% of the current budget in the first years while the Helmholtz Society, a foundation that financed re-

---

[88] Dessauer, Problem, p. 1395. In 1921, Dessauer was invited to the US Society for Radiology. He visited the laboratory of William D. Coolidge (a pupil of the Leipzig physicist Paul Karl Ludwig Drude), the State Cancer Research Laboratory at Buffalo and institutes in New York, Detroit and Chicago. Dessauer, Jahre, pp. 48 ff.

[89] Friedrich Vierfelder became Professor for physics in Buenos Aires, the physicist Albert Bachem Professor for Biophysics at the University of Illinois, College of Medicine, Chicago, Otto Glaser head of an institute of radiation research in Cleveland, the physician Ernst Pohle Professor for Radiology at University of Wisconsin, Madison, the physicist Egon Lorenz (the only Rockefeller grant holder) assistant at Harvard University, Cambridge. Dessauer, Jahre, pp. 313–317.

[90] Breitling, Glocker, p. 149.

[91] Glocker, Gründungsgeschichte.

[92] Ibid.

[93] Maier, Forschung, p. 238; Goes, Dessauer, pp. 218–219.

[94] Glocker, Gründungsgeschichte, p. 573; for more details on the context of the X-ray laboratory and the role of Bosch see Maier, Forschung.

[95] Glocker, Gründungsgeschichte, p. 573.

search of interest to industry, paid grants.[96] Problems of inflation soon arose and were bridged with the help of the Reich Ministry for Interior Affairs and the income received from sales from the institute's workshops. However, even after the institute had its own university budget from 1923 on, the foundation continued its activity.[97]

Helmut Maier has recently described the accelerated secular trend in the budgeting of research institutes during and after WWI.[98] Institutes relied more and more on multiple financial sources. Since at that time institutes fostered a combination of technical and natural science, industrial foundations did not hesitate to pay for basic research. Medical physics was destined to fit best into this trend because of its hybrid origins based in techniques and experiments. The Union of German Engineers (Verein Deutscher Ingenieure) cited Glocker when its regular news report summarized that those new institutions of X-ray physics had originated from the incorporation of physics into X-ray therapy: "That meant a shift from pure medical problems towards the field of technical-physical enterprises."[99] Both Glocker and Dessauer were reluctant to announce that their institutes were working at the edge of basic and applied science.[100] To include "the whole X-ray industry" (Glocker), Dessauer was elected to the council of the foundation for Glocker's laboratory.[101] The Reiniger-Veifa company was the first to donate new machines and other firms then followed this example.[102]

The institutionalized medical physics still participated in the established structures of industrial-academic collaboration between clinics and the X-ray industry, but expanded it. On the one hand medical physicists used health policy successfully in order to mobilize financial resources. On the other hand, new fields of innovation and application came into focus. This especially was true for Glocker, who could rely on his own engineering facilities. He worked on the development of X-ray crystallography in order to use it for the expanding science of evaluating materials ("Werkstoffprüfung").[103] On the basis of a diversified working spectrum, Glocker's X-ray laboratory managed to make a profit thanks to the growing expectations for the application of X-ray technology. It mobilized a variety of different supporters including: the municipal policy-makers, the state of Württemberg (which paid the salaries of the staff, including instrument makers), generous patronage from industry, donations from different foundations and research organizations such as the "Robert Bosch War Foundation", the Helmholtz Society and, finally, the grant-giving NGW. The policy of the NGW will be covered shortly in the next section in order to evaluate whether it had special expectations with respect to medical physics.

---

[96] Maier, Forschung, pp. 236 and 241.
[97] Glocker, Gründungsgeschichte, p. 573.
[98] Maier, Forschung, chapter 3.
[99] Cited by ibid., p. 237 (translated by AS).
[100] Ibid., pp. 237–239; Dessauer, Institut.
[101] Dessauer held no official post in the industry after founding his institute.
[102] Maier, Forschung, pp. 236 and 238.
[103] Glocker, Gründungsgeschichte, p. 573.

5. MEDICAL PHYSICS AND THE RESOURCES OF THE
EMERGENCY FOUNDATION FOR GERMAN SCIENCE (1920–1930)

In the first years of the German republic, new structures were established for financing research. The Helmholtz Society and "Emergency Foundation for German Science", potent research organisations, were founded in 1920.[104] The NGW did not hesitate to help support the foundations of both Dessauer's and Glocker's institutes.[105] However, since the NGW financed several hundred research projects each year, it is important to ask whether it devoted any special attention to medical physics or biological radiation research. The published lists of grants given between 1920 and 1925 clearly show that the NGW did not.[106] However, there are some interesting aspects.

The reluctance towards financing medical physics was not, as suggested in literature, because the NGW only favored human sciences.[107] On the contrary, even in the initial years its research policy suited the demands of medical physics quite well. In its very first report the NGW stressed that physics and natural sciences in general were becoming increasingly technical.[108] For this reason, right from the beginning the NGW tried to strengthen the technical impact of experimental sciences. This can be illustrated by their global expenditures. In 1922, one of the main instruments of research policy was to finance the printing of books and to fund libraries. While grants for research projects made up the smallest portion of the budget, the purchase of foreign-language literature consumed nearly 13.4 million Reichsmark, equivalent to about 32 % of the entire budget.[109] These expenditures corresponded to the need to "catch-up" with the international science community again.[110] However, almost the same amount of money (roughly 13.7 million Reichsmark) was invested in the material and technical support of the experimental sciences.[111]

To coordinate the material and technical support, special commissions ("Sonderausschüsse") were appointed with one focusing on tools ("Werkzeugausschuss"). Since the shortage of money and inflation were restricting resources for research at that time, the commissions' task was to enable laboratories and universities to produce instruments and tools by themselves.[112] The success in facilitating the acquisition of machine tools was overwhelming and was duplicated when the NGW began purchasing various industrial products needed for research.[113] Hence, another special commission ("Materialausschuss") dealt with

---

[104] Hammerstein, Forschungsgemeinschaft, pp. 49–55.
[105] Dessauer, Jahre, pp. 12 and 16; Maier, Forschung, p. 241.
[106] Notgemeinschaft, Berichte 1922–1926.
[107] Hammerstein, Notgemeinschaft, p. 64.
[108] Notgemeinschaft, Bericht 1922, p. 16.
[109] Ibid., pp. 12–13.
[110] Hammerstein, Notgemeinschaft, pp. 69 ff.
[111] Notgemeinschaft, Bericht 1922, p. 12.
[112] Ibid., p. 11.
[113] Notgemeinschaft, Fünfter Bericht, p. 131.

the acquisition of materials such as radioactive substances, chemicals or expensive metals such as platinum and mercury.

The last of these commissions was busy with purchasing and distributing devices and instruments for experimental work. The decisions of the so-called "Apparateausschuss" ranked high in the grant policy of the NGW throughout the 1920s because a large proportion of the budget was used for technical equipment. Five years after beginning its work, the commission counted more than 4,000 devices that belonged to the NGW and were provided to researchers.[114] Of these, many of the devices and equipment were used in various types of radiation research. In fact, the NGW administrators considered X-rays a particularly important area and special attention had been given to this area since 1922.[115] However, neither medical physics nor radiation biology were mentioned in the review.[116] The X-ray generators and other equipment that was distributed suited the demands of physical radiation research and atom physics in the first place.[117] This did not mean a general preference of basic research or theoretical physics. An increasing amount was spent for research in X-ray crystallography and spectrography. Within the NGW, research projects using X-rays or light in order to study the "structure of matter" combined innovative physics, the incorporation of technology into science and the anticipation of useful results.[118] The NGW strongly emphasized the practical use of X-ray spectroscopy and crystallography for the analysis of natural fiber, viscose silk and, last but not least, metals.[119]

In 1926, the NGW came to the conclusion that the incorporation of physical and chemical methods into medicine should be supported through research policy. This took place as part of a general restructuring of NGW's research policy. Instead of only paying for research, the NGW decided to encourage research that would serve "national economy, public health and the general welfare".[120] In order to strengthen its governing power, the NGW installed co-operative programs ("Gemeinschaftsarbeiten") that were problem-focused and pooled projects at different research locations. Biological radiation research was one of the main problems selected by the NGW for a cooperative endeavor. This program marked a significant change of the NGW's research policy and will be studied in detail elsewhere. However, with respect to the history of medical physics and biophysics, the agenda of the research program was as important as it was clear in focus. The outline of the program culminated in the statement that the "large field of X-ray and radium biology" would finally converge in the research on "the physical processes following the adsorption of radiation" in living matter.[121] In other

---

[114] Ibid., p. 79.
[115] Ibid., p. 200.
[116] See the overview for the years 1922–1926, ibid., pp. 129–252.
[117] Notgemeinschaft, Vierter Bericht, p. 53, and see the chart on pp. 69–75.
[118] Overview on the support of physics during 1922–1926, in: Notgemeinschaft, Fünfter Bericht, pp. 194–202, here: p. 201.
[119] Ibid., pp. 212–213.
[120] Hammerstein, Notgemeinschaft, pp. 73 ff.
[121] Notgemeinschaft, Fünfter Bericht, p. 240 (emphasis AS).

words: the policy of the NGW reflected the standpoint of medical physicists, namely, that physical research was the only appropriate approach for the complex problem of radiation biology. Consequently, the laboratories of Glocker, Dessauer and Friedrich in Berlin were on the top list of institutions to be financed in the coming years.[122]

6. BIOLOGICAL MODELS AS RESOURCE – MEDICAL PHYSICISTS TURN TOWARDS BIOPHYSICS

Around the time when the NGW presented the co-operative program, "Radiation Research", some medical physicists began cooperating with biologists. This turn of events now draws attention to another resource that had been part of the experimental configuration of medical physics since its beginnings. In addition to expensive technical devices, the radiation experiments of medical physics relied upon biological models. While the whole discourse of radiotherapy had been concentrated on the techniques of X-ray generation, most of the objects of biological experiments – animals and plants – were only mentioned within an article's methodological description. In reality, this did not mean that these biological models were an uncomplicated aspect of the experiments. On the contrary, the historiography of biology has depicted the influence of the choice of animals and plants on the contents of biological problems and concepts.[123] However, when physicists entered the field of radiotherapy, they also entered an existing world of biological experimentation. It was easy for them to simply rely on the experimental systems that had been used by physicians until then. This was the case when Walter Friedrich joined the group of Krönig in Freiburg.

The first years of physico-medical radiation research were characterized by the use of specific organisms: plants, worms and amphibians or, more precisely, special developmental stages of these organisms, such as seeds and germ buds of beans or eggs and the larvae of frogs.[124] The choice of these objects was influenced by biologists, who themselves had begun research on the effects of radiation in the first decade of the 20th century.[125] Thus, the radiologists at Freiburg based their research on specimens that had been introduced by biologists.[126] At the beginning they also experimented with mammals such as mice.[127] However, these complicated organisms were soon abandoned. This development may be attributed to an indirect influence of the physicist Friedrich entering the radiotherapeutic laboratory. Friedrich added not only technical knowledge but also introduced a special experimental style. After Friedrich's arrival the experiments

---

[122] Notgemeinschaft, Berichte 1926–1933. See footnote 2.
[123] For example Gaudillière, Biologists.
[124] Holfelder, Ergebnisse, p. 1168.
[125] Schinz, Jahre, pp. 86 ff.
[126] Gauß/Lembcke, Tiefentherapie, pp. 72–98.
[127] Ibid., p. 98.

were adjusted to statistical methods and based on large sample sizes with up to hundreds of cases. The physician Krönig made this quantitative rationale explicit when he explained why he had chosen beans, eggs and the tadpoles of frogs.[128] It was evident that the use of slow breeding mice was no longer practical.

In order to use experimental organisms appropriate, the medical researchers had to learn specific biological techniques for breeding and handling. In the words of a Freiburg physician they "had to become fishermen and gardeners".[129] But what did physicists do when they left the medical institutes to work on their own? Did they also become "fishermen and gardeners"? It is not possible here to offer a comprehensive overview of the biophysical experiments in the twenties. However, the example of Glocker shows that some proven model organisms such as beans were in use. Additionally, Glocker continued to rely on experienced researchers. First he cooperated with a radiologist experienced with experiments on beans who arrived in Stuttgart in 1926.[130] In fact, this physician was a pupil of the previously mentioned pioneering radiotherapist Perthes, a surgeon in the nearby town of Tübingen. Some years ago, Perthes had vehemently supported the foundation of Glocker's laboratory at Stuttgart. Now, the collaboration resulted in the construction of new laboratory for radiation biology at a municipal hospital.[131] Glocker and the physician decided to employ a biologist as the head of the new laboratory in order to have better expertise in the biology and breeding of model organisms.

This decision proved to be crucial for the direction research took at Stuttgart. In order to better meet the demands of the biophysical experiments, the biologist considered new model organisms.[132] One of these was the fruit fly, *Drosophila melanogaster*, a species that was originally used in biology for genetic research. Thus, it was not by chance that questions of the effects of radiation on chromosomes became more and more crucial in the experimental work of the radiation laboratory at Stuttgart in the years that followed. In short, they had been already introduced by the new model system. The new biophysical experimental system was disseminated by assistants of the Stuttgart laboratory when they moved to work with Walter Friedrich who then ran the Institute for Radiation Research in Berlin.[133] The X-ray laboratory in Stuttgart continued to reinforce connections with biology from the late 1920s.

With the biologist entering the medico-physical laboratory, it was acknowledged that the choice of the right model organism was a crucial part of the experimental system of medical physics. The responsibility of the biologist was to adjust the model organism to the requirements of the biophysical research interests. Thus, the hired biologist joined a laboratory of medical physics just as the

[128] Krönig/Friedrich, Grundlagen, pp. 139, also 152 and 154–155.

[129] Gauß, Geschichte, pp. 633–634.

[130] The surgeon and radiologist Otto Jüngling became chief of the surgical department of the nearby municipal Katharinenhospital. Jüngling, Abteilung, pp. 88 ff.

[131] Breitling, Glocker.

[132] Langendorff, Katharinenhospital.

[133] Grants Otto Risse, 1936–1939, BA Koblenz, R 73, 14000.

physicist Friedrich had once joined the laboratory of radiotherapists in Freiburg: for technical reasons. Obviously in this case, the bio-physical conjuncture was mediated by the technical configuration of the radiobiological experiment.

CONCLUSIONS

The history of biophysics began with medical physics in the 1910s. The scope of this paper ended with those physical-biological conjunctures that began to emerge in the late Twenties and marked the turn of medical physics towards biophysics in the term of new disciplinary alliances. These conjunctures have to be assessed more thoroughly in order to get a more comprehensive picture. It seems fruitful to do that, since the radio-therapeutic context of biophysics discussed here may revise some standard history of molecular biology. According to that history, theoretical physicists introduced X-rays to biology in order to analyze the structure of genes from the early thirties on.[134] Those stories highlight only the move of theoretical physicists towards the general reasoning about the fundamentals of life. Protagonists like Richard Glocker who has had a practical interest in radiotherapy are barely mentioned in these approaches.[135]

The history of the concrete and material conjunctures between medicine, physics and biology traces the origins of biophysics to the early years of radiology and, more precisely, of radiotherapy. When physicists joined the efforts of physicians, they entered the already established field of experimental radiobiological research. The biological effects of radiation on organisms were an epistemic object that linked physicians and physicists institutionally and discursively throughout the period described. However, as such the radiobiological research field was the product of the radio-therapeutic context. Physicists such as Walter Friedrich added their technical expertise by introducing and compelling the use of an exact dosimetry. Hence, a medical physicist was somebody who was convinced that the therapeutic use of X-rays was promising and that the understanding of biological mechanisms was dependent upon the exact technical control and the theoretical understanding of dosimetry.

Radiotherapists have often confirmed that the practice of radiotherapy changed with developments made in the technical facilities.[136] However, there was also a strong interdependence between the radiobiological problems and the configuration of the experiments that included the technical calibration in the laboratory, the use of model organisms – and industrial development. The radiobiological research field was changed and extended by the technical possibilities

---

[134] For an example see Fischer, Licht; a more comprehensive approach is Gausemeier, Ordnungen, pp. 150–186.

[135] This deficit may have been eased by the fact that Glocker's institute became part of the Kaiser Wilhelm Institute for Metal Research in 1934. Maier, Forschung, p. 304.

[136] Rajewsky, Theorie, p. 75; Willers/Hellmann/Beck-Bornholdt, Jahrhundert, p. 56; Frobenius, Röntgenstrahlen, pp. 95–97.

of modulating the characteristics and the performance of radiation in the human body and in organisms in general. For example, the invention of high-energy X-rays changed the conditions of the biological experiment because of the high secondary radiation they produced. Medical physicists had to distinguish artifacts from real biological changes; in this way they discovered by chance scatter radiation, which became a major problem for medical physics at the beginning of the 1920s. Thus, the practical context of radiotherapy generated a growing theoretical interest in the processes that were induced by radiation in living matter. These general interests paved the way for the emancipation of medical physicists from medicine after 1920.

Due to a lack of literature, it is difficult to compare the significance of radiology for biophysics in Germany and the United States.[137] In any event, the German history differs in research policy as it was implemented by the NGW. It has been shown that, in the US, a technical ideal of "human engineering" inspired the research policy.[138] This resulted in the comprehensive Rockefeller Foundation program that was intended to introduce physical methods centrifugation and spectroscopy into biological research.[139] All these techniques, including X-ray crystallography and spectroscopy, constituted the technical ensemble of molecular biology from the thirties on. But although the NGW supported the experimental research and the incorporation of technology into the natural sciences and medicine, it did not have a consistent program to combine physical techniques and biology.

BIBLIOGRAPHY

Blum, Martina: "Vom launischen Apparat zur präzisen Black Box". Die Einführung der Glüh-
    kathodenröhre in die Röntgentechnik, München 1999.
Breitling, Gerhard: Richard Glocker 21. 9. 1890–31. 1. 1978, in: Fortschritte auf dem Gebiete der
    Röntgenstrahlen und der Nuklearmedizin 129, 1978, pp. 149–150.
del Regato, Juan A.: The Unfolding of American Radiotherapy, in: International Journal of Radia-
    tion Oncology, Biology, Physics 35, 1996, pp. 5–14.
Dessauer, Friedrich: Über einige Wirkungen von Strahlen I., in: Zeitschrift für Physik 12, 1922,
    pp. 38–47.
Dessauer, Friedrich: Das Frankfurter Institut für physikalische Grundlagen der Medizin, in:
    Münchener Medizinische Wochenschrift 67, 1920, pp. 1524–1525.

[137] Rasmussen mentions the multiple origin of biophysics in (medical, electro- and general) physiology and radiology, but gives no deeper insights. Rasmussen, Biophysics, p. 253. Radiobiological research is only marginally mentioned in the literature about US radiology. Del Regato, Unfolding. This may be due to the focus of the studies. However, as similar theoretical approaches like Dessauer's were developed in France and Britain, but not in the US, this may suggest that radiology was less significant for the development of biophysics in the US.

[138] Rasmussen, Biophysics, p. 253; Kay, Vision, p. 49, in general for the program „Science for Man" see pp. 39–50.

[139] Kay, Vision, p. 49.

Dessauer, Friedrich: Das Problem der Röntgentiefentherapie vom Physikalischen Standpunkt, in: Klinische Wochenschrift 1, 1922, pp. 1395–1397.

Dessauer, Friedrich: Kontrapunkte eines Forscherlebens. Erinnerungen. Amerikanische Briefe, Frankfurt am Main 1962.

Dessauer, Friedrich: Zur Therapie des Karzinoms mit Röntgenstrahlen. Vorlesungen, Dresden 1923.

Dessauer, Friedrich: Zehn Jahre Forschung auf dem physikalisch-medizinischen Grenzgebiet. Bericht des Instituts für physikalische Grundlagen der Medizin an der Universität Frankfurt am Main, Leipzig 1931.

Dommann, Monika: Durchsicht, Einsicht, Vorsicht: Eine Geschichte der Röntgenstrahlen 1896–1963, Zürich 2003.

Fischer, Peter: Licht und Leben. Ein Bericht über Max Delbrück, den Wegbereiter der Molekularbiologie, Konstanz 1985.

Friedrich, Walter: Zur Geschichte des Institutes für Strahlenforschung der Humboldt-Universität zu Berlin, in: Friedrich, Walther and Hans Schreiber (eds.): Probleme und Ergebnisse aus Biophysik und Strahlenbiologie, Leipzig 1956, pp. 1–4.

Frobenius, Wolfgang: Röntgenstrahlen statt Skalpell. Die Universitäts-Frauenklinik Erlangen und die Geschichte der gynäkologischen Radiologie von 1914–1945, Erlangen 2003.

Gaudillière, Jean-Paul: Biologists at Work. Experimental Practices in the Twentieth-Century Life Sciences, in: Krige, John and Dominique Pestre (eds.): Science in the Twentieth Century, Amsterdam 1997, pp. 683–700.

Gausemeier, Bernd: Natürliche Ordnungen und politische Allianzen. Biologische und biochemische Forschung an Kaiser-Wilhelm-Instituten 1933–1945, Göttingen 2005.

Gauß, C. J. and H. Lembcke: Röntgentiefentherapie, ihre theoretischen Grundlagen, ihre praktische Anwendung und ihre klinischen Erfolge an der Freiburger Universitäts-Frauenklinik, Berlin 1912.

Gauß, Carl Joseph: Zur Geschichte der gynäkologischen Strahlentherapie, in: Strahlentherapie 100, 1956, pp. 633–648.

Glocker, Richard: Aus der Gründungsgeschichte des Röntgeninstituts der Technischen Hochschule Stuttgart, in: Fortschritte auf dem Gebiete der Röntgenstrahlen und Nuklearmedizin 95, 1961, pp. 2–3.

Goerke, Heinz: Fünfundsiebzig Jahre Deutsche Röntgengesellschaft, Stuttgart 1980.

Goes, Martin: Friedrich Dessauer (1881–1963). Röntgenpionier aus Aschaffenburg und seit 1934 im Exil. Würzburger medizinhistorische Mitteilungen 14, 1996, pp. 209–232.

Hammerstein, Notker: Die Deutsche Forschungsgemeinschaft in der Weimarer Republik und im Dritten Reich. Wissenschaftspolitik in Republik und Diktatur 1920–1945, München 1999.

Holfelder, Hans: Die derzeitigen Ergebnisse der experimentellen Strahlenforschung, in: Medizinische Klinik 19, 1923, pp. 1129–1131 and 1168–1170.

Holthusen, H.: Über die Dessauersche Punktwärmehypothese, in: Strahlentherapie 19, 1925, pp. 285–306.

Holthusen, Hermann: [Vortrag zur Geschichte], in: Fortschritte auf dem Gebiete der Röntgenstrahlen und der Nuklearmedizin 115, 1971, pp. 142–146.

Jüngling, Otto: Die chirurgische Abteilung, in: Stadtverwaltung Stuttgart (ed.): Festschrift zum hundertjährigen Bestehen des Katharinenhospitals in Stuttgart, Stuttgart 1928, pp. 77–93.

Kay, Lily E.: The Molecular Vision of Life. Caltech, the Rockefeller Foundation, and the Rise of the New Biology, Oxford 1993.

Krönig, Bernhard and Walter Friedrich: Physikalische und biologische Grundlagen der Strahlentherapie, Berlin 1918.

Langendorff, Hanns: Das Katharinenhospital als ein Ausgangspunkt strahlenbiologischer Forschung, in: 70 Jahre Radiologie am Katharinenhospital der Stadt Stuttgart, Stuttgart 1969, pp. 63–65.

Maier, Helmut: Forschung als Waffe. Rüstungsforschung der Kaiser-Wilhelm-Gesellschaft im NS-System am Beispiel des Kaiser-Wilhelm-Instituts für Metallforschung, Göttingen 2007.

Notgemeinschaft der Deutschen Wissenschaft: Bericht der Notgemeinschaft, Berlin 1922–1933 (annual reports).

Rajewsky, B.: Theorie der Strahlenwirkung und ihre Bedeutung für die Strahlentherapie, in: Kolle, W. (ed.): Wissenschaftliche Woche zu Frankfurt a. M., 2.–9. September 1934. Vol. 2: Carcinom, Leipzig 1935, pp. 75–91.

Rasmussen, Nicolas: The Mid-Century Biophysics Bubble: Hiroshima and the Biological Revolution in America, Revisited, in: History of Science 35, 1997, pp. 245–293.

Ratmoko, Christina: Hormone aus dem Industrielabor. Die Erforschung und Herstellung von Geschlechtshormonen bei der Ciba zwischen 1910 und 1940, in: Schweizerische Zeitschrift für Geschichte 55, 2005, pp. 84–94.

Schinz, Hans R.: 60 Jahre Medizinische Radiologie. Probleme und Empirie, Stuttgart 1959.

Schierhorn, Eike: Walter Friedrich, Leipzig 1983.

Schwerin, Alexander: Experimentalisierung des Menschen. Der Genetiker Hans Nachtsheim und die vergleichende Erbpathologie, 1920–1945, Göttingen 2004.

Willers, H., H.-P. Heilmann, and H.-P. Beck-Bornholdt: Ein Jahrhundert Strahlentherapie. Geschichtliche Ursprünge und Entwicklung der fraktionierten Bestrahlung im deutschsprachigen Raum, in: Strahlentherapie und Onkologie 174, 1998, pp. 53–63.

ALLIED CONTROL OF PHYSICS AND THE COLLEGIAL SELF-DENAZIFICATION OF THE PHYSICISTS

Gerhard Rammer

This article investigates how the end of war in 1945 and the occupation period affected physics in West Germany.[1] First of all, it analyzes denazification, which (consistent with its original purpose of ridding the German society of any remnants of National Socialism) also had an impact on scientific institutions and especially the academic staff. Second, it also examines science funding and science policy, thereby including the issues of demilitarization and the control of scientific research. To understand the changes regarding personnel and subjects, one has to consider more closely the changes in the Allied occupation policy. The first part of this essay intends to survey this briefly, focusing on the denazification process. This is followed by some remarks regarding research method, topic, and state of research concerning the denazification of science. Case studies on personnel policy and research funding in physics make up the body of this chapter.

HISTORICAL CONTEXT

The history of denazification is a complex matter. For one thing, denazification is part of a comprehensive Allied occupation policy (usually characterized by the 4 Ds: denazification, demilitarization, democratization, decentralization)[2] and cannot be regarded as an isolated component. For another, dividing Germany into four zones resulted in different policies for political purges. In spite of the consensus that existed initially among the occupation powers about the purposes of denazification – i. e. in short to abolish Nazism – and at first jointly issued di-

---

[1] On the history of physics during the post-war era, see Hoffmann (ed.), Physik; on denazification in particular Walker, Nazification; on the migration movement Schirrmacher, Wiederaufbau; on the mentality of German physicists immediately after the war, Hentschel, The Mental Aftermath.

[2] The four Ds refer for one thing to the main features of American occupation policy; for another to the statement made at the Potsdam Conference in August 1945. The original text of the Potsdam Agreement states the political objectives in rather general terms. In the literature the terms deindustrialization, dismantling und decartelization are sometimes also used as constituents of the "four" Ds. See Wolfgang Benz' handbook: Deutschland unter alliierter Besatzung, Berlin 1999, the articles: Amerikanische Besatzungspolitik (pp. 33–47), Demokratisierung (pp. 108–113), Entnazifizierung (pp. 114–117), Potsdamer Konferenz (pp. 214–217), Dekartellierung (pp. 336–338), Entmilitarisierung (pp. 342–346), JCS 1067 (pp. 349–351).

rectives, different views regarding the extent, approach and progress soon became apparent.

Initially the Allies were exclusively in charge of denazification. Prominent members of the political, military and economic leadership of National Socialism were arrested and put on trial. In January and October 1946 the Allied Control Council issued first directives to standardize its execution. Denazification panels were set up, which divided the people in question into five groups (ranging from Category I: major offenders to Category V: exonerated persons). To investigate the whole population, however, proved to be difficult. The professional perspectives of almost a third of the population were temporarily affected. There was a widespread belief on the German side that the purges largely hit the wrong people.

Before long the responsibility for the denazification process was transferred to the German administration, which turned denazification into one big rehabilitation procedure. The dwindling priority given to systematic political purges has to be seen in the context of the increasing East-West tensions. When denazification was ended in the Western zones in 1948, people previously identified as lesser offenders and fellow travelers were automatically downgraded to exonerated persons. In December 1951 legislation governing the liquidation of the denazification program (Gesetz zum Abschluss der Entnazifizierung) was passed.[3] Article 131 of the Basic Law (Grundgesetz) entitled former civil servants, who had not yet been rehabilitated, to return into their positions.[4]

Scientists were impacted by denazification in particular when they were tenured at a university, since political investigation there was implemented in a more thorough way than, for instance, in industry. In the American zone the removal quota in the public service amounted to more than 42%, and universities in Lower Saxony dismissed (at least temporarily, until mid-1947) about a third of their entire tenured faculty.[5] But politically unblemished scientists were affected by occupation policy as well, namely when their areas of research came into the scrutiny of Allied demilitarization policy. In the final stage of the war and during the first post-war year the Allies identified and extracted war-related research, arrested and evacuated selected scientists, and liquidated entire war research establishments. A systematic acquisition of German research work in innumerous reports was part of this extraction program.[6]

In April 1946 the Control Council issued Law No. 25 (Control of Scientific Research), introducing a new stage of research monitoring that led to a controlled reconstruction of German research. For the most part, the scientific research prohibited in Control Council Law No. 25 concerned fields of military importance.

---

[3] For a conspectus of denazification in the four zones see Vollnhals (ed.), Entnazifizierung; Niethammer, Schule; Königseder, Entnazifizierung.

[4] On the aftermath of denazification and social reintegration of former Nazi perpetrators, see Frei, Vergangenheitspolitik.

[5] Schneider, Entnazifizierung, p. 338.

[6] On the exploitation of German research institutions see Gimbel, Science.

With regard to physics that included, among others: applied nuclear physics, rocket propulsion, applied aero- and hydrodynamics, as well as different applications of electromagnetic, infrared and acoustic radiation especially exploitable for military purposes. The aim of research control was to guarantee the disarmament and demilitarization of science.

As Manfred Heinemann has remarked, the control of scientific research has to be regarded as an instrument of both the reconstruction of scientific research and its denazification. Thus according to Article 6 of Law No. 25, senior officials or scientists, who had been active members of the NSDAP (Nationalsozialistische Deutsche Arbeiterpartei, National Socialist German Workers Party) or other Nazi organizations "shall be removed and their replacement effected only by persons with suitable political records."[7] Especially the British insisted on a thorough purge of the scientists, since this allowed them to avoid a control of research programs, which they deemed hardly feasible. Politically reliable German scientists were then expected to steer scientific research into peaceful climes.[8] The last stage of research control started about 1951, characterized by a reduced control mode and the onset of Germany's remilitarization.[9]

The end of the war brought sharply modified conditions for research funding. Military employers like Army Ordnance (Heereswaffenamt), the Reich Aviation Ministry (Reichsluftfahrtministerium) or the Supreme Command of the Armed Forces (Oberkommando der Wehrmacht) that had a formative influence on the scientific landscape during the war, no longer existed. This development, in addition to the disruption of certain branches of industry like aviation, gave civilian establishments like the universities or the Kaiser Wilhelm/Max Planck Institutes a prominent position as research locations during the occupation. A large part of research funding was funneled through these institutions, given that other establishments, like the German Research Foundation (Deutsche Forschungsgemeinschaft, DFG), were not yet available. Hence the way to get funding for research programs during that era was above all a matter of being successful in getting a job in a civilian institution. Therefore this essay attaches special importance to the analysis of the denazification of the scientific community.

METHOD AND TOPIC

Both the directives for denazification and control of scientific research left a remarkable room for maneuver, thus making the results of both interventions subject to the local institutions in charge of them. Hence the historiography of science has not only to elaborate general characteristics of the stage of upheaval,

---

[7] See Control Council Law No. 25, published in: Neue Physikalische Blätter 2 (3), 1946, pp. 49–52, 50.

[8] See Cassidy, Controlling.

[9] See Heinemann, Überwachung, and on research organization in general Carson/Gubser, Science.

but must attach special attention to individual micro-histories. Moreover, a general development cannot be understood merely from formal regulations and directives. For instance, the small number of re-emigrants cannot be attributed to a corresponding regulation. On the contrary, re-emigrants enjoyed special attention of the military government and priority consideration in the (re-) appointment of positions.[10] That is, the actual development sometimes contradicted what was formally intended.

Thus it is difficult to derive a coherent picture of the change in the sciences around 1945. Certainly, the end of war represents a break in Germany's history of science. But obvious manifold changes in institutional, personnel and subject-related respects notwithstanding, one should not speak of a "zero hour" in science. The change on the different levels is marked by a coexistence of inconsistencies and continuities. Hence current research interprets the complex transformation process as an elaborately conducted "rearrangement of resource constellations", with no preset or predictable outcome.[11] Instead of merely asking, what was the result of this change, or what inconsistencies and continuities can be detected, the historiography of science should focus more on the processes of transformation per se, their conditions, courses and limitations. Hence Mitchell Ash, who advocates this position, talks about "constructed continuities", thus pointing to the considerable negotiating that was necessary to achieve a seemingly unbroken continuation of a biography, research program or institution.[12] Following this historiographical perspective the leading questions of this essay are: what were the conditions and terms for such constructed continuities? When were these successful and when did they fail?

The denazification of physics personnel in the Western zones can be considered as external interference into the cooptation traditions of academia, which had remained intact even under Nazi rule.[13] The scientific community tried, with increasing success, to insinuate its traditional rules of inclusion and exclusion into the denazification proceedings. In other words, scientists tried to defend their cooptation traditions, regarded by them as part of science's autonomy. Thus these efforts neither aimed to use denazification to restore the pre-1933 situation (too much had changed in the meantime in the individual biographies and

[10] See Szabo, Vertreibung.

[11] See Ash, Wissenschaftswandlungen, esp. p. 76. In this context the term resources implies physical, institutional, financial, cognitive, and/or rhetorical resources.

[12] On Mitchell Ash's methodical concept see also his works: Ash, Wissenschaftswandel; Ash, Umbrüche; Ash, Scientific Changes. Outside the history of science the phrase "Zero hour" has experienced a renaissance. For instance Uta Gerhardt uses it in a new study as a sociological characterization of the Umbruchsgesellschaft, see Gerhardt, Soziologie.

[13] That in fact cooptation traditions remained intact in spite of the officially introduced Führerprinzip is shown in the by now classical essay: Seier, Rektor. On the context of denazification in general and regarding the social elite and science, see Frei, Vergangenheitspolitik; Frei, Karrieren; Weisbrod, Vergangenheitspolitik; Loth/Rusinek, Verwandlungspolitik; Pehle/Sillem, Wissenschaft; Bruch, Nachkriegszeit; Szabó, Vertreibung; Schüring, Minervas. Defrance compares the denazification of universities in the four occupation zones in: Universitäten. See also: Oehler/Bradatsch, Hochschulentwicklung.

likewise the institutions), nor to save the entire personnel as of 1945. This article will examine the more precise question regarding the construction of personnel continuities, namely what were the criteria for renegotiating the 'membership' in academia?

Especially well-suited for an analysis of this nature are decisions of scientific societies – like for instance the German Physical Society (Deutsche Physikalische Gesellschaft, DPG) – that were determined by a policy of dealing with the Nazi Past (Vergangenheitspolitik), as well as the personnel investigations of the universities' denazification panels. Due to the fact that half of the members of these panels were professors – i.e. colleagues sat in judgment upon colleagues – the denazification of universities exhibited all the features of a self-denazification. On purely formal grounds this characterization might appear a bit too strong, since the judgment of the university panels (the other half of the panel members were public persons) had to be confirmed by the main denazification panel of each town. Practice proved, however, that the professors' votes carried a great deal of weight.

Although these internal panels were decisive, this should not belie the fact that both processes, denazification as well as self-denazification, were interlaced with each other, and cannot be separated into a 'correct' denazification under the leadership of the Allies and another one, somehow 'adulterated' by scientific institutions. The course of denazification was characterized by varying regulations, and emphases and aims that changed within the context of a hardening East-West confrontation.[14] In contrast to these changes the scientists tried to use the denazification proceedings to establish a permanent and solid new border of academe.

The further course of the personnel debate proves that in individual cases dismissals were certainly desired, and thus not just externally imposed. The "Act Regulating the Legal Status of Persons Coming within the Provisions of Article 131 of the Basic Law" from 1951, which entitled those faculty members who had been dismissed as a result of denazification to their former positions, caused a partially vehement defense of the previously hard-negotiated demarcation line. As a result, the university gates remained once and for all closed to some of the professors removed from office, despite their legal claim.[15] This "aftermath" of self-denazification, however, will not be subject matter of this essay, which focuses on the period up until the legal termination of denazification in 1951/52.

Due to the requirements of denazification procedures, political arguments were mostly used for both, the exclusion of "black sheep" from the scientific community as well as supporting the continuance in office of highly regarded

[14] On the dynamics of denazification shown at the example of Heidelberg University, see the exemplary analysis by Remy, Heidelberg.

[15] On this law and its development, see Frei, Vergangenheitspolitik. On dealing with the professors "removed from office", see Schael, Grenzen. On denazification as an instrument for redefining the academic community, which continued to welcome certain Nazis as members, see for instance Brynjølfsson's detailed study about the denazification of the Faculty of Arts at the University of Göttingen, Entnazifizierung. Also see Haude, Nachkriegssituation; Respondek, Wiederaufbau; Düwell, Entnazifizierungs- und Berufungsprobleme.

colleagues. Hence "whitewash certificates" (Persilscheine) often featured assertions such as: respectable scientific work in and of itself demonstrated keeping one's distance from National Socialism; or that the scientist in question had never inwardly supported National Socialism.[16] Yet the decision criteria were not merely, or not even in the first place political criteria – to stand aloof from National Socialism whatsoever was not the main point.[17] The maintenance of scientific excellence, for instance, was such a criterion. It often served as the motive for supporting politically incriminated scientists.[18] It does not provide, however, a complete picture of academe reorganized. This essay will show that this reorganization was determined to a considerable extent by collegiality. Although previous studies have occasionally attributed some significance to collegiality, its importance has yet not been specifically analyzed.[19]

It is the thesis of this essay that, both before and after 1945, adherence to the collegial code of behavior, i. e. loyal behavior towards colleagues as well as to the profession and scientific establishment, determined whether or not one could continue in the scientific community. Adherence to this code was reviewed and demanded over the course of self-denazification. Its purpose was to stabilize the community. In contrast, a revision according to political standards in particular would not have achieved stability, rather put that at risk. Similarly, merely requesting scientific excellence would not have been a sufficient criterion with regard to the community's desired cohesion. But though the following examples shall prove the great importance standards of loyalty possessed, this is not to argue that the other criteria played no part. In fact, case studies will be presented that demonstrate that the different factors for the demarcation in academia have to be reconsidered and reappraised.

---

[16] Beyler recently provided a detailed study about these topoi: Wissenschaft.

[17] The term political refers in this context to the general balance of power in society as negotiated by parties and other – in this sense – political bodies; it does not refer, however, to science and personnel political procedures in specific. That is, personal demarcations based on "standard" personnel political considerations without balancing National Socialist or, in more general terms, socio-political commitment will not be called political here, so as to obtain a conceptual distinction to distinguish between general scientific inclusion and exclusion standards as well as new, denazification-induced rules. This distinction only works due to the fact that socio-political commitment was neither a fixed nor generally acknowledged component of the standard inclusion and exclusion criteria.

[18] Weisbrod attaches special importance to this criterion in the academic Vergangenheitspolitik, see: Geist.

[19] In his study on the boundaries of political purges in Germany after 1945 Klaus-Dietmar Henke concluded that collegial bonds put an "almost impenetrable insulation layer around denazification." Henke, Grenzen, p. 130. Marc von Miquel mentions the great importance of collegiality for the self-denazification of justice, where, in a manner of speaking, judges acted for themselves. According to him, even decisions by the Supreme Court showed "actually open collegiality to the accused." Miquel, Juristen, p. 197.

COLLEGIAL SELF-DENAZIFICATION OF PHYSICISTS

The behavior of Max von Laue is revealing for the issues outlined above, because he was regarded as the best-known adversary of the Nazis among the German physicists. Indeed this was one of the main reasons why he was allotted such a prominent role in the post-war era.[20] He was very committed to the physicists' self-denazification, but instead of categorical antagonism to Nazis, he showed a remarkable willingness to support the case of incriminated colleagues. At first sight this conduct may seem astonishing, but such amazement is arguably caused by a false picture of Laue's activities in the National Socialist era. Some observers have been pleased to generalize Laue's clearly visible resistance to a certain kind of National Socialist transformation of science (especially noted is his courageous and consequent opposition to Johannes Stark and the "Aryan Physics", "Deutsche Physik", movement) as if he had put up resistance against "any form of National Socialist influence on science".[21]

In fact Laue fought for the maintenance of traditional power relations within the physics community, which, for example, Stark's plans had jeopardized. The defense of tradition was not achieved by fighting National Socialism, but instead by making a pact with those Nazis who accepted the established hierarchy in science as well as the concept of autonomy coupled with this, which they left untouched. When Laue's behavior is interpreted in this way, his 1948 judgment of a physicist, who had been a member of the SS, becomes comprehensible: "Kohler had belonged to the S.S. and unfortunately is not yet qualified for a faculty position. [...] He is by no means heavily implicated."[22]

This passage in Laue's letter to his Berlin colleague Wilhelm Westphal tells the reader that Max Kohler belongs to the group of desirable professors, who had not yet returned to the academic arena due to their formal political incrimination. Like some other Nazi physicists Kohler too remained a permanent feature in the physicists community, because apart from professionalism he distinguished himself with his loyal behavior towards the elite of physicists – namely disregarding political orientations. Laue was well suited to judge Kohler's adherence to the loyal code of behavior, given that for years the latter had been his assistant at the Berlin University. Such close collegial relations were characteristic for academe and determined the content of alleged political judgments. In most cases these close collegial relationships outlived the political change of system in 1945 unharmed; for one reason, the political homogeneity of the people involved was

---

[20] For the following example see Rammer, Sauberkeit. On Laue's role in the post-war era see Oexle, Hahn, as well as the new Laue biography by Katharina Zeitz.

[21] Zeitz, Laue, p. 42.

[22] "Kohler hat der S.S. angehört und kommt deshalb bisher leider für eine Hochschulstellung noch nicht infrage. [...] Besonders schwer belastet ist er keinesfalls", Laue to Westphal, 6. March 1948, Archive of the Max Planck Society (henceforth MPG-A), Laue papers, III, 50, 2125. Kohler had been Laue's assistant at the University of Berlin from 1933 to 1943, then became associate professor at the University of Greifswald. On Kohler's vita see Goenner/Klein, Nachruf.

simply no prerequisite for their constitution and continuity. The next case studies provide a more nuanced view of the effect of collegial commitment.

Over the course of denazification at the University of Göttingen, the only physics instructor to be dismissed was Karl-Heinz Hellwege, member of both the NSDAP and National Socialist Faculty League (NS-Dozentenbund).[23] Born in 1910, Hellwege climbed his first rungs on the university ladder during the National Socialist era. Since memberships in the Nazi Party or its subsidiary organizations were rather the order of the day for the scientists of his generation, this is not very much help in making a retrospective political classification. Thus it is of vital importance for a proper understanding of the Hellwege case to know that his colleagues knew him as a convinced National Socialist. After his dismissal, due to the office he held in the Storm Troopers' riding squad (Reiter-SA), he busily collected political references in order to bring about his reappointment.

At least 57 people supported his cause with whitewash certificates. Even those physicists who had avoided him for political reasons became involved. This included, for example, a research group from his institute that issued a whitewash certificate signed by everyone. This certificate, however, merely attested that Hellwege would not have taken advantage of political differences in order to denounce them. Despite critical comments regarding his political attitude, the still existing bond between colleagues caused them to support his rehabilitation. Even more valuable for Hellwege proved to be his close relationship, not to say friendship, with the full professor for theoretical physics, Richard Becker. The latter one attested that, along with academic expertise, Hellwege was a sincere character. Hellwege not only enjoyed Becker's high esteem as a scientist, but also as a person.[24]

Dean Arnold Eucken used Becker's whitewash certificate as the basis for a formal faculty statement:

> I consider Professor Becker's remark that Dr. Hellwege's character is distinguished by great sincerity and honesty to be of great significance for the assessment of this case. Hence in his appeal he [Hellwege] avoids the otherwise often used phrases like "I had no real understanding of National Socialism" and such. He admits implicitly that he approved of National Socialism per se, which is sympathetic and should not be held against him, but rather be appreciated. Yet by no means does this attitude of his preclude that he was *primarily* a scientist, who unambiguously defended the position of his science once Nazi policies conflicted with the former. Given that furthermore he was neither politically active as a Nazi nor had he thereby caused damage to the university, the faculty altogether feels rather obliged to support his appeal of his dismissal, the more so as he belongs without question to the elite of our younger physicists.[25]

---

[23] On the following two case studies see Hentschel/Rammer, Neuanfang; Rammer, Wirkung; Rammer, Nazifizierung, chap. 1.

[24] Certification by Richard Becker, 27. July 1945, HStAH, Nds. 171 Hildesheim 11847.

[25] Faculty statement on Hellwege's objection, signed by Dean Eucken, 4. February 1946, ibid. (Emphasis in the original.)

In doing so, the faculty expressed its interest in continuing to employ a Nazi with three arguments: he was an excellent scientist, a man of good character, and was capable of distinguishing politics from science. The rhetoric of this statement deviates from typical patterns because it admitted that a physicist was a convinced Nazi, but nonetheless should remain employed.[26] As compensation for this political stigma, his noble mind was emphasized. Moreover, the faculty attached great importance to human qualities in many contexts, so that it was not a distinctive feature of the denazification procedure. Thus it was not a matter of sheer rhetoric, but a serious argument.

In reference letters for the habilitation (essentially a second, more demanding doctoral thesis, which was one of the prerequisites for university teaching) and the selection of new tenured professors, character assessments were also common – both before and after 1945. Even in the comprehensive entry exams of the post-war university the character of applicants was scrutinized and preference given to former officers with the argumentation that they had developed a particularly mature character during the war. Their quota among the post-war students in Göttingen amounted to 40 per cent. As the debates on denazification practices between the occupation and German authorities show, the concept of character served as substitute for and circumvention of the political purge. But this was also not limited to academics, as demonstrated by the general debates about denazification and the decency of office-holders in National Socialism. German emigrants like Carl Zuckmayer also used a character classification as the basis for an alleged political purge. Thus Zuckmayer's secret report on creative artists in Nazi Germany from 1943 begins with a general examination of character.[27]

Let us return to the Hellwege case, which had more going for it than testimony during the denazification process about Hellwege's decency and honesty. The topos of decency formed part of an exculpation strategy, which in the sciences also included a discourse on scientific autonomy. One of the key elements of this strategy was the separation of individual, science and politics in order to construct a scientific concept of intellectual "cleanliness" – pure task orientation, so to speak – as well as the continuity of personal integrity. Eucken pursued exactly this strategy when he emphasized that Hellwege had been "*primarily* a scientist". His remark suggesting an occasionally occurring difference between Nazi customs and his [Hellwege's] science similarly implied that political practice and science were separated on principle, since in fact Eucken asserted explicitly that these two were distinguishable from each other.

According to this version Hellwege had not practiced Nazi physics, but clean physics. Since according to this, he could not have damaged the university's reputation, the faculty supported him and reserved his position as an untenured lecturer for him, although they could not have counted upon a positive outcome of his denazification procedure. In late 1948 Hellwege was able to return to his position. In 1952 he received a tenured professorship at the TH Darmstadt. The

---

[26] On the rhetoric of Persilscheine, see Beyler, Wissenschaft; Sachse, Persilscheinkultur.
[27] Zuckmayer, Geheimreport.

year before his colleague from Göttingen, Hans König, had also been appointed to the chair of experimental physics there.

Born in the same year, 1910, König was just like Hellwege a dyed-in-the-wool Nazi. That held, by his own admission, at least for the 1930s; according to contemporary witnesses, however, it was also true for the period that followed. Following a job as a lecturer at the TH Danzig, König arrived at Göttingen in 1941, where he conducted research on behalf of the Reich Air Ministry (Reichs-luftfahrtministerium). He avoided dismissal in 1945 due to the fact that he was not employed by the university. In 1947 he initiated his denazification within the course of his efforts to relocate his teaching position (Umhabilitierung). This all went smoothly because König denied his early membership in the NSDAP, which he had in fact joined in 1929, before the National Socialists had come to power. Two years later on he was appointed tenured associate professor (außer-planmäßiger Professor).

This post-war career was successful because König was so well-embedded into the network of his Göttingen colleagues. Though his "juvenile enthusiasm" for National Socialism was a known fact (something he did not conceal from the denazification panel, and which was turned into proof of his decent character), his colleagues regarded him above all as an earnest scientist, conscientious experimental physicist and specialist for electron microscopy, likewise suitable for a successful application for the above-explained exculpation strategy. The next step he took on the ladder in 1951, full professor (Ordinarius), depended due to academic cooptation traditions on the support of other established tenured professors.

Along with the close Göttingen and former Danzig colleagues, Laue was also included in the list of supporters. Laue had already responded to the appointment as associate professor in 1949 with a very warm letter of congratulations: "Dear colleague! With great joy I learned from the rector's bulletin that you have been appointed professor. Allow me to congratulate you whole-heartedly, although I don't believe that you will remain for long at this stage, but will soon receive a tenured [full] professorship."[28] This example shows that favoritism towards Nazis was not solely responsible for the post-war careers of former Nazis. The physicists' elite, which determined personnel policy, was too inhomogeneous a conglomerate for that. Rather than the candidate's political views, his strong loyal bonds to the community's elite were the determinant for the career opportunities in academia, with all its specific and traditional advancement mechanisms made of schooled rites and individual methods of quality assessment.

These successful post-war careers of two Nazis stand in contrast to the failed restitution (Wiedergutmachung) in the case of Kurt Hohenemser.[29] In this case study too the setting is the University of Göttingen. In the summer of 1945 Hohenemser returned after a 12-year absence to the University and requested his

---

[28] Laue to König, 3.3.1950, MPG-A, Laue papers III, 50, 1079.
[29] On the Hohenemser case see Szabó, Vertreibung, pp. 214–232; Hentschel/Rammer, Neu-anfang; Rammer, Wirkung; Rammer, Nazifizierung, chap. 5.

reinstatement as assistant at the Institute for Applied Mechanics – the same position from which he had been dismissed in 1933 for "racial" reasons. Following that, he had worked as a scientist mainly in the helicopter industry until the end of war. His (not retrospectively acquired) interpretation of Nazi rule as a criminal and thus on principle unacceptable regime was anything but standard at that time, and had in this case to do with Hohenemser's early commitment to left-wing policies and his later personal experiences. The NS regime was responsible for the death of the majority of his family. This personal matrix should be kept in mind throughout his dispute with the university.

In 1945 a passage in his application for reinstatement turned into the bone of contention. In this he had mentioned that former institute director Ludwig Prandtl had been divested of his office in 1934 for political reasons, and in his place Max Schuler had been appointed. Though in accordance with the facts, this allusion was defying the carefully practiced discretion and consideration for colleagues in political matters during the post-war era. Although Hohenemser regarded touching upon the so far condoned political interference with the institute as a moral rehabilitation of his admired teacher Prandtl, the faculty considered it primarily as a sharp, unpleasant and inappropriate attack on the well-regarded colleague Schuler.

This alleged attack was the real offence, the "indecency" Hohenemser committed. Dean Eucken wrote a clearly dismissive faculty statement defeating the reinstatement of Hohenemser on the grounds that "fertile collegial cooperation" with him seemed to be out of question. Moreover Eucken maintained that Hohenemser's account of the directorate take-over was not correct.[30] Yet the particular investigation of the facts, the university had announced, never occurred. As a result, however, the faculty could allege to the argument of incorrectness, which served to implement institutional autonomy. In fact, it was the university's concern to construe personnel matters of the past as independent of party politics, thus creating a line of argumentation that was based on a largely intact academic autonomy during the National Socialist era, and which – given that the core of the university resisted being inducted by politics – was to remain being politically untouched; i.e. by the denazification process.[31] Hohenemser was a threat to these efforts of constructing a presentable university identity; in this sense he behaved disloyal to the institution.

What is interesting in this conflict between Hohenemser and the university is Prandtl's behavior. Not only was he on friendly terms with Hohenemser, but also entertained loyal relations to his colleague Schuler, in spite of the fact that the disempowerment of Prandtl at the institute in 1934 opened the way for Schuler. Prandtl's behavior in 1945 was marked by his effort to keep both relationships as unaffected as possible. He confirmed that his resignation from directorship had been solely due to the political pressure on part of the student body. But then again he also defended Schuler's position at the institute, and pronounced

---

[30] Dean Eucken to Rector Smend, 7.7.1945, UAG, Math.-nat. Fak. 19 a.
[31] See Schildt, Kern.

against Hohenemser's reinstatement, since he deemed it inappropriate to require Schuler to take on a hostile assistant. He gave Hohenemser the well-meant piece of advice to apologize to Schuler.

As a solution for the conflict-laden situation Prandtl suggested (and the faculty acceded to this proposal) to place Hohenemser at another university. In doing so Prandtl intended to defuse the difficult situation arising from the presence of the adversaries Schuler and Hohenemser in the same place. Hohenemser, however, asserted his right to rehabilitation in Göttingen and tried to gain the support of the ministry of education as well. But the ministry not only refused to interfere with the universities' intended practice of self-recruitment, but instead supported it. After two years of fruitless and sometimes unfair negotiations, the resigned Hohenemser emigrated to the USA. The Hohenemser case shows that the demarcation line within academia will remain incomprehensible, if one only pays attention to political and scientific criteria. As the faculty had emphasized repeatedly, it had been "the great danger that collegial cooperation cannot be expected" which prevented his reintegration.[32]

Part of the requested loyalty towards the colleagues was not to cheat on the exculpation strategy, which had begun to establish itself in delicately conducted discourses, part in public, part in private; and which strived to whitewash both the academic institutions as well as its learned members. Hohenemser's remarks indicated that he was not willing to play this discursive "game" according to the rules of the university. His refusal had to do with his biography. Being someone previously segregated from the Nazi so-called "people's community" (Volksgemeinschaft), and who had lost almost all his relatives to the Nazis, he demanded political consequences from the past. Sure enough, the "threat" averted by the faculty was real, yet not alone due to Hohenemser's biography as a victim and an outsider, but above all due to his behavior. The example of Fritz Houtermans, a physicist with a quite similar biographical background, who was rehabilitated in the post-war era in Göttingen, makes this clear. Houtermans succeeded where Hohenemser failed not least thanks to his willingness to collegially cooperate with (former) Nazis like Hellwege and König.

A debate dealing with the politics of the past within the DPG that was touched off during an important physics meeting in Göttingen in 1947, provides an impressive example for the self-denazification of the physicists.[33] The annual meeting in 1947 was of great importance for the German physicists. It represented one of the very few opportunities where physicists of all four occupation zones were able to convene and discuss the many pressing issues concerning the future of research as well as one's own professional life. How strong the desire for communication was, can be deduced from letters of that time. There is, for instance, a letter written by Hermann Senftleben, former Obertruppführer in the SA and tenured professor in Münster, who in the summer of 1946 complained to Robert Pohl about his dismissal: "So far I have no idea what will become of

---

[32] Dean Eucken to Rector Smend, 14. 7. 1945, UAG, Math.-nat. Fak. 19 a.
[33] For this case see Rammer, Sauberkeit.

me, but for now I have not quite lost courage. Göttingen has now become the scientific center at least of the British zone, and I would like to visit and talk to the colleagues there from time to time. I hope the occasion to do that will arise anytime soon."[34]

The meeting was also an occasion to share information about the state of denazification, to organize support for colleagues concerned, but also for taking measures to keep undesirables away. An example for the latter is the following application submitted by Richard Becker on the DPG board meeting, which was directed against the Berlin physicist and science administrator Erich Schumann:

> Recently there has been disturbing news to the effect that alleged physicists like this, who acted during the war in the worst kind of way by causing damage to the careers of colleagues with the help of the Party's instruments of power, attempt to reconnect and try to use their proven savoir faire to gain influence with the present governments. I request that the Society formally decide to oppose such attempts. Though we are not thrilled about the denazification law, we regard it our duty to provide for cleanliness in the circle of colleagues.[35]

The sources allow reconstructing beyond any doubt that this motion was exclusively directed against Erich Schumann. But instead of individualizing personnel issues, it was Becker's strategy (expressed in the anonymous mode of speaking) to create general provisions, which were to serve as the basis for future interferences by the DPG board. The aim of these rules was to keep away from academic positions those physicists, who had behaved in a particular wrong way during the National Socialist era. The standard used to determine this misdemeanor was clearly a collegial one. The serious offense Becker addressed in his application was damaging a colleague. No other kind of misconduct, not to mention of a political nature, is included. The object of this kind of demarcation was, in Becker's words, "to provide for cleanliness in the circle of colleagues." Likewise his disassociation from denazification, that the physicists had "not been thrilled by the denazification law", makes plain that this was not a matter of a political purge.

The DPG board was indeed subsequently commissioned "to prevent those physicists, of whom it was known that their behavior during the era of the Nazi rule conflicted in scientific or human terms with the conception of decency and morality generally accepted among scientists, from acquiring official positions."[36] That is, this decision defines explicitly that the boundaries of the community were to be defined consistent with moral criteria. What is more, the term "decency" points here to the fact that, in the view of the physicist elite, right or at least acceptable behavior in the Nazi era had something to do with character and personality. The term "decency" thus ran counter to the term "politics", which is precisely the reason why it has been used with such frequency in the whitewash certificates. This implies that in both the discursive practice of formal denazifica-

---

[34] Senftleben to Pohl, 2.7.1946 [Archive of the DPG, henceforth] DPG-A, Nr. 40028.
[35] Minutes of the Board Meeting on 6. September1947, DPG-A, Nr. 40043.
[36] Ibid. Also see the announcement of this decision in Physikalische Blätter 3, 1947, p. 281.

tion, as well as in the physicists' internal attempt at self-denazification, the political purge was to be transferred to another, non-political level.

Another reason why this strategy was so successfully established may have been that both the Allies, as well as the German denazification panels, depended to a considerable degree on the statements made in the affidavits for their assessment of the human aspects, hence revaluating the importance of these and in doing so providing the faculties with an instrument of self-denazification in their whitewash certificates. The scientists thereby succeeded in transferring the negotiations over personnel to a level that was largely determined by their own assessments. But it remained a discursive conflict due to the lack of a uniform understanding of moral criteria necessary for the assessment of decent behavior.

In 1947 the elite of German physicists achieved consensus within the course of their self-denazification to exclude Erich Schumann. Obviously the remaining more than 300 participants of the annual meeting 1947 gave no offence. Yet some of the participants were politically incriminated and outside the elite their presence and career plans met with disapproval. Two months after the physicists' meeting a graduate student in physics, Ursula Martius (subsequently Professor Ursula Franklin), published an article in the Deutsche Rundschau in which she named, along with Schumann, another four physicists and expressed her displeasure with the fact that they had been reaccepted and were aspiring again to university positions. In order to justify her concern, she also indicated that the criticized colleagues had supported National Socialism in word and deed.[37] Three of the scientists named were regarded as superb physicists and honorable colleagues: Pascual Jordan, Herbert Stuart, and Hans Kneser. Martius mentioned in particular those scientists she knew from her Berlin environment. She knew from her own experience how they had conducted themselves and regarded it as a threat for youth to be educated by scientists who were politically incriminated in such a manner. Naming names, however, meant breaking a taboo and incurred displeasure among the leading heads of the DPG.

This case of public critique was absolutely unique. One reason for Martius to choose this path lies in her biography. For one thing, since she had been excluded from the National Socialist Volksgemeinschaft, she held a different perspective on what had happened during the Third Reich. For another, as a graduate student she was not yet fully mainstreamed into the codex of loyalty among colleagues – she still stood outside the circle of colleagues. Martius expected a personal rebuke in reaction to her essay. In fact she received a personal letter from DPG president Max von Laue, in which he criticized the method of her article: *"The accused are entitled to remain unmolested after the official settlement of their matters. And said right is violated by articles such as the one written by you* [...]. What matters in establishing order are less personnel issues but the reintroduction of the *rule* of law."[38] This kind of reprimand meant no great threat for Martius' career, since at that time she was already seriously considering the idea of emigration. In 1949, one year after

---

[37] See Martius, Videant.
[38] Laue to Martius, 26. 12. 1947, MPG-A, Laue papers, III, 50, 2395. (Emphasis in the original.)

successfully completing her Ph. D., she emigrated to Canada, where she became full professor and one of the most-honored Canadian scientists.

The DPG tried to neutralize Martius' public criticism. The executive board of the society considered writing a reply. In the end, however, it decided against this, but the correspondence of the board members that arose in this connection is revealing in several respects. For one thing, there was no unity regarding whether the DPG should respond to this article or not; and if so, what stance it should adopt towards the colleagues criticized by name. For another, the letters, no matter how they commented on the individual cases, provided no arguments that referred to precise deeds of the physicists. It remained unclear what the deeds in question were.

Nonetheless one letter claimed, "that it was basically a matter of trivialities that did not matter compared to the scientific importance of the gentlemen in question." What is typical about this argumentation is the attempt to reduce political incrimination because of scientific achievements. If deeds were mentioned at all, then only in a belittling way: some individuals, it said for instance, had been "fooled into talking drivel."[39] On this level a discussion of the deeds was impossible; and there was no discussion whatsoever, and it seems as if there had not been much of a need for one. According to these sources, the prevailing attitude seems to have been: in dubio pro "reo" (When in doubt, in favor of the accused). Furthermore, they suggest that ignorance of the deeds of the accused was a comfortable argument for the support of them. Removing the doubts would have threatened the stability of the community.

As in the Hohenemser case, a desired picture of events was projected, but not verified. In concrete terms, this meant that decent physicists had to be regarded as decent, even if critics provided references to their "indecency". Indications of that kind were suppressed and instead the "doubts" preserved. Sometimes a personal experience, not even necessarily dating from the Nazi era, was proof enough of decency. For instance Clemens Schaefer passed the following judgment on the criticized Kneser: "In contrast I consider Kneser, whom I know from his youth, as a man beyond any reproach."[40]

To illustrate this attitude we will take a closer look at the Stuart case. Stuart was a renowned physicist, whose scientific achievements were acknowledged. In 1925 he completed his Ph. D. in Göttingen (with James Franck as his supervisor) with a study about resonance fluorescence in mercury-vapor. He then worked as an assistant to Otto Stern and Richard Gans. The assessment of Stuart may have been affected by the fact that he received his education mainly from Jewish professors. In 1934 his first monograph on the determination of molecule structures was published at Springer. Stuart was appointed as a tenured associate professor (Extraordinarius) in Königsberg (now Kaliningrad) in the following year. In 1936 he temporarily held the chair for theoretical physics at the University of Berlin that Erwin Schrödinger had vacated in 1933. During this (rather beneficial to his

---

[39] Unsöld to Laue, 2. 12. 1947, DPG-A, Nr. 40048.
[40] Schaefer to Laue, 8. 12. 1947, ibid.

career) period in Berlin he became acquainted with the German physicist elite. In 1938, on the occasion of Planck's 80th birthday, he performed a comic play together with Arnold Sommerfeld, Peter Debye, Ernst Ruska, Werner Heisenberg and Walther Gerlach.[41] In 1939 he was offered a chair at the TH Dresden. In 1942 Springer published his Short Physics Textbook (Kurzes Lehrbuch der Physik), which proved him to be a good pedagogue, well-grounded in every field of physics.

Thus Stuart was well integrated into the physicists' community in 1947 when, Martius not only criticized him in her article, but also, at the physicists' meeting mentioned above, the Amsterdam physicist Teun Michels accused Stuart of denouncing him as a saboteur. Michels had only been able to escape the arrest by the Gestapo by fleeing. As a result, the military government informed the rector of the TH Hannover that Stuart was not to be employed any longer. In this situation Laue stood up for Stuart without even inquiring about the actual deeds that had caused the comments. In his correspondence with Stuart he switched from the hitherto employed "Sehr geehrter Herr Kollege!"[42] to the rather amicable "Lieber Kollege!"[43]. Since Stuart could not have a university position, Laue found him a job with the German chemical company Farbenwerke Bayer. Laue's support of Stuart was probably a result of the former's respect of the latter's scientific achievements, which caused him to suppress, at least partially, any qualms.

When Stuart tried to resume his academic career in 1951, he suddenly met with strong opposition from Laue, whom Samuel Goudsmit had given incriminating material about Stuart during Laue's journey to the USA in 1948. The emigrant Goudsmit told Laue that in 1939 Stuart had tried to place the DPG more firmly under Nazi control. With a past like that, Stuart now became intolerable as a university teacher for Laue and some other physicists. In fact, he would have been already disqualified in 1947/48 for the same reason, but at that time (when denazification kept all incriminated physicists in suspense) questions about political misdemeanor had been a taboo. For the same reason existing doubts were not followed up.

Laue himself had admitted before embarking to the USA that Stuart's "behavior in Berlin had given rise to a lot of suspicion".[44] Nonetheless he took the view that Stuart could and should teach at a German university due to his proven skills. In cases like that, the majority of the relevant physicists argued with Laue that scientific quality outweighed political incrimination. Just like the emphasis placed on the decency of the individual physicist, diverting the discursive conflict from the political level formed part of the exculpation strategy. This was the main objective, especially during the early stage of denazification. As Stuart's case illustrates, political considerations gained increasingly in importance later on. The suspicion acknowledged by Laue in 1948 does not seem to have carried enough

[41] See Hermann, Max, p. 97.
[42] Laue to Stuart, 30.8.1946, 11.6.1947, MPG-A, Laue papers, III, 50, 1951.
[43] Laue to Stuart, 16.1.1948, MPG-A, Laue papers, III, 50, 1951.
[44] Laue to Bartels, 10.2.1948, MPG-A, Laue papers, III, 50, 210.

weight. But the balance had changed three years later, when a few concrete acts committed by Stuart were on the table, and due to the approaching end of the denazification process it was no longer as important to preserve or construct general doubts about physicists' involvement with National Socialism. Now resistance against Stuart's return to faculty could be offered in word and deed, which then proved to be effective to that extent that in 1952 Stuart was only employed as a visiting professor and his appointment as full professor (Ordinarius) postponed until 1955.

In late 1947 Laue outlined the politics of the past (Vergangenheitspolitik) practiced in a letter as follows: "We try to realize a policy here in physics, which unfortunately has not succeeded on the national level, that is to grant a general amnesty for all Nazi followers after having passed harsh sentences on the real evildoers."[45] Yet Laue's statement is misleading inasmuch as there was neither a reliable identification of the "real evildoers", nor an agreement regarding the measures to be applied. The climate that would have been necessary to advance a discussion of past behavior did not exist. Hence it was unclear how to proceed. However, the physicists reached a consensus on excluding those, who had clearly violated the collegial code of behavior.

RESEARCH FUNDING – A CASE STUDY

The III. Physical Institute at the University of Göttingen founded in 1947 serves as an impressive example for studying the problems of research control and research funding as well as its very problematic connection to denazification. It is also a case of general importance, since it is the first example of the British successfully funding a research project that was carried out in Germany during the post-war era. It is a rather intricate story and in this sense typical for that time. It is subdivided into three narrative strands: the starting position of the Göttingen institutions, the research team that arrived there in 1947, and the British plans and objectives.

The University of Göttingen had two institutes of applied physics, whose history can be traced back at least until 1905: an institute for applied electricity and one for applied mechanics. In 1946 the directors of both institutes retired: one because of his age, the other one, because the military government had suggested early retirement due to his political past. The faculty tried until the autumn of 1946 to find new directors for both institutes, but this was hampered by the fact that applied research was severely restricted by Control Council Law No. 25 as well as the uncertain future of the institutes. At the same time the faculty was fully occupied with fending off Hohenemser, who tried to force his reinstatement at the Institute for Applied Mechanics. There was the "risk" that the military government would support Hohenemser's rehabilitation actively, thus interfer-

---

[45] Laue to H. Pechel, 11. 11. 1947, DPG-A 40048. Quoted according to Hentschel, Mentalität, pp. 48 f.

ing with the carefully practiced self-denazification, which could have been easily thrown off balance by "disturbances" of that kind. In this situation it was rather convenient for the faculty that a new candidate arose, someone who promised to solve all the unsettled problems in one go. This windfall for Göttingen arrived in the person of Erwin Meyer.

Meyer had been professor for technical physics at the TH Berlin until the end of war, where, as specialist for acoustics, he had been head of department at the Heinrich Hertz Institute for Oscillation Research (Institut für Schwingungs-forschung). During the war he researched for the navy, for instance on sound absorbers in water. He developed a sophisticated rubber fuselage for submarines, which drastically reduced the possibility to locate them acoustically. This camouflage skin called "Alberich" was not deployed during the war, but after 1945 knowledge about this technique still remained valuable, since the Allies were now interested in it. Meyer's institute fell into British hands since it had been transferred to Pelzerhaken, Holstein, during the war to escape the Berlin air raid. Being aware of Meyer's military importance the British kept him as "frozen scientist" in their zone. Instead of returning to Berlin, he was obliged to be at the disposal of the Royal Navy. In this time Meyer and his collaborators wrote a voluminous report on "Sound Absorption and Sound Absorbers in Water (Dynamic properties of rubber and rubber-like substances in the acoustic frequency region)" which was based on their working reports from wartime. It was finished in 1947; three years later an English translation was published by the Department of the Navy, Bureau of Ships in Washington D. C.

In England, officials now discussed how to make best use of the German research for their own purposes. The Admiralty had great interest in continuing to employ Meyer and his research group. This required removing some formal obstacles and the involvement of numerous institutions in doing so. To overcome these problems the Department of Scientific and Industrial Research (DSIR) put in charge of the coordination of all parties involved. Two options were considered. In the first plan, the DSIR offered Meyer employment at the Building Research Station in England. The second plan would locate the research in Germany at the University of Göttingen, where Meyer should be offered a chair. In Göttingen, the university supported this plan by abandoning its former plans. In October 1946 the faculty merged both institutes of applied physics into a so-called III. Physical Institute and nominated Meyer as new director. Since university officials had not previously considered establishing an acoustics department at Göttingen, nor had any special need for such a department been evident, it seems to be a safe bet that the faculty's procedure was an accommodation to decisions made somewhere else.[46] The Research Branch as well as the German Ministry of Culture (Kultusministerium) approved the faculty's new proposal. In

---

[46] Making acoustics the new field of interest in Göttingen provided the faculty with a matter-of-fact argument not to re-employ Hohenemser at their university: the new institute of acoustics was no longer a place for his research of applied mechanics.

April 1947 Meyer was appointed full professor and director of the III. Physical Institute.

The open question on the British side was how to place a research contract in Germany. In January 1947 several institutions met to discuss this matter: Ministry of Supply, Admiralty, Ministry of Defence, Control Office for Germany and Austria, Colonial Office, Foreign Office, Treasury, Board of Trade, and DSIR.[47] The policy of placing research contracts in Germany was laid down by ministers as a means of relieving unemployment among German scientists and as a counter-measure to "enticement" (for going elsewhere). An additional advantage of the placement in Germany was that it enabled the Treasury to declare the expenses as German exports. As a result, 50 per cent of the export income would be reimbursed to the exchequer. This proved to be lucrative in particular for those research programs that British institutions had already approved, because transfer of the research to Germany cut the expenses by half.

Yet the transfer had problems. The large number of institutions involved caused problems in communication and coordination. Given the kind of acoustics research that the Admiralty desired, a conflict with Control Council Law No. 25 was to be expected. Therefore it was considered useful "to provide some form of civilian cover for contracts the service departments wished to place, which, although peaceful and innocent in themselves, and in conformity with Control Council Law No. 25, might give occasion for misinterpretation in the press or elsewhere." This consideration was, moreover, one of the reasons not to leave the coordination of the negotiations to the Admiralty, but instead to place it in the hands of the civilian institution DSIR.[48] In addition to that, the research program was focused on architectural acoustics, and thus on civilian applications.

The military relevant water acoustics remained, however, an explicit component of the program. The scope of research included the transmission of vibration through solid structures, partitions with granular fillings, measurement of attenuation of sound in liquids, absolute calibration of hydrophones by various methods, work on the general extension of acoustical standards of measurement in air to the supersonic range of frequency, and the development and calibration of a microphone for the measurement of sound in air at frequencies up to 100 kc/s.[49] This program shows the typical discrepancies that emerge when military based research is placed into a civilian context. Though some of the work was the continuation of old research, the entire venture had to be regarded as inherent "peaceful and innocent" and be rewritten accordingly.

[47] The meeting was held at the Economic Department of the Control Office for Germany and Austria on 15 January 1947. See the notes of the meeting, Public Record Office, FO 943/300.

[48] See the notes of a meeting held on 15 January 1947, Public Record Office, FO 943/300.

[49] See the list of "Proposed Subjects of Research", undated, probably December 1947, UAG, Kur. XVI. V. C. hh. 6.

Given the conditions of the early post-war era funding was splendid: the contract covered five Ph. D. and four junior scientists, plus four mechanics and a secretary, that is 60,000 RM in personnel expenditure annually. Another 80,000 RM were provided for equipment. It took two solid years to reach that agreement. Unexpectedly difficult proved to be the fixing of the exchange rate between the Mark and the pound Sterling. The DSIR tried to assert the highest exchange rate possible so as to minimize the British expenses, and suggested 20 M for one £. The Joint Export-Import Agency (JEIA), however, insisted on 13 M/£; probably to keep conditions favorable for British exports. As basis of negotiation served that the conversion rate "should be based upon an assessment of the cost of similar work in the United Kingdom."[50] In August 1947 Meyer traveled to England to discuss the contract's details with the Admiralty, the Building Research Station and the National Physical Laboratory. But the financial problems had not yet been solved with that. Following extensive negotiations especially among JEIA, Foreign Office, and DSIR the contract was signed in May 1948, including a exchange rate of 20 M/£. The DSIR possessed ownership on results, publications and patents. Obviously co-operation proved to be lucrative for both sides; the contract was extended each year even beyond Meyer's tenure, who retired in 1968. In addition to the contract with DSIR, in 1953 the III. Physical Institute received a research contract called "Microwaves" with the European Office Air Research and Development Command United States in Brussels, with a contract sum of $ 7,000 for the first year and $ 10,000 for the second.[51]

SUMMARY

This last example illustrated that, with Allied support, a new Centre for Oscillation Physics, in particular in the field of acoustics, could be established during the occupation period in Göttingen. Funded by the British, a formally purely civilian research program was established, which, however, originated from a military context. Yet this example is the exception that proves the rule regarding research funding in occupied Germany. Research during this time was usually state-funded and located at universities. In order to be able to do research, a scientist first of all had to be employed at a university. This meant that a scientist had to survive the political purge, which (as demonstrated in the main part of this essay) should be understood more than anything else as a process of collegial self-denazification.

The circumstances under which personnel continuity was achieved were one issue in this context. These requirements included the successful effort the physicists' elite made to establish academic cooptation traditions even within the process of denazification. In other words, an existing elite of physicists decided about the future fate of the physics community. This elite now decided about the

---

[50] K. H. Lauder (director Research Branch, Economic Sub-Commission) to the rector of the university of Göttingen, 25.2.1948, UAG, Kur. XVI. V. C. hh. 6.

[51] See UAG, Kur. XVI. V. C. hh. 6.

exclusion of undesired, or inclusion of welcome colleagues, whatever the case may be. The community's demarcation lines were drawn according to traditional criteria (proof of scientific excellence, practiced loyalty towards one's colleagues, and character qualities), the decisions, however, were politically embellished due to the framework given by the denazification procedure. Thus scientists claimed that esteemed colleagues had not really been that incriminated, whereas others had (allegedly) become politically intolerable.

But this line of argumentation is highly misleading. During the period of denazification severely incriminated physicists could also count on their colleagues' support, without being questioned about their deeds. It seems that political aspects have been largely suppressed, and not just by former Nazis. Particular emphasis has been placed on the attitude of the first post-war president of the DPG and renowned adversary to National Socialism, Max von Laue, in order to demonstrate this. For the reintegration of National Socialists and other former supporters of the regime is not primarily a result of ideological continuity or still-operating right-wing networks, but rather has to be seen as a defense reaction of a loyal community against intrusion.

This was a situation that produced strong internal solidarity, which became useful for the incriminated physicists, yet sometimes proved to be an obstacle for the victims of National Socialism. The latter scientists' interest in learning a political lesson and drawing corresponding personnel policy conclusions from the past, collided with the need for stability of the physicists community, most of whose members were politically compromised to some degree. The "cleanliness" rhetoric used by the physics elite meant not the exclusion of National Socialists, but the demarcation lines drawn by the community according to collegial criteria. Along with some scattered perpetrators, this kind of exclusion affected victims of National Socialism as well. This bitter side of self-denazification has, however, been rather neglected so far.

BIBLIOGRAPHY

Ash, Mitchell G.: Wissenschaftswandel in Zeiten politischer Umwälzungen: Entwicklungen, Verwicklungen, Abwicklungen, in: Internationale Zeitschrift für Geschichte und Ethik der Naturwissenschaften, Technik und Medizin 3, 1995, pp. 1–21.

Ash, Mitchell G.: Verordnete Umbrüche – Konstruierte Kontinuitäten: Zur Entnazifizierung von Wissenschaftlern und Wissenschaften nach 1945, in: Zeitschrift für Geschichtswissenschaft 43, 1995, pp. 903–923.

Ash, Mitchell G.: Scientific Changes in Germany 1933, 1945, 1990: Towards a Comparison, in: Minerva 37, 1999, pp. 329–354.

Ash, Mitchell G.: Wissenschaftswandlungen in politischen Umbruchszeiten – 1933, 1945 und 1990 im Vergleich, in: Acta Historica Leopoldina 39, 2004, pp. 75–95.

Beyler, Richard H.: "Reine" Wissenschaft und personelle "Säuberungen". Die Kaiser-Wilhelm-/Max-Planck-Gesellschaft 1933 und 1945 (= Ergebnisse 16, preprint from the research program "History of the Kaiser Wilhelm Society in the National Socialist era"), ed. by Carola Sachse on behalf of the Presidential Commission of the Max Planck Society for the Advancement of Sciences, München 2004.

Bruch, Rüdiger vom: Nachkriegszeit: Einführung, in: Bruch, Rüdiger vom and Brigitte Kaderas (eds.): Wissenschaften und Wissenschaftspolitik. Bestandsaufnahmen zu Formationen, Brüchen und Kontinuitäten im Deutschland des 20. Jahrhunderts, Stuttgart 2002,
pp. 369–372.
Bruch, Rüdiger vom and Brigitte Kaderas (eds.): Wissenschaften und Wissenschaftspolitik. Bestandsaufnahmen zu Formationen, Brüchen und Kontinuitäten im Deutschland des 20. Jahrhunderts, Stuttgart 2002.
Brynjølfsson, Einar: Die Entnazifizierung der Universität Göttingen am Beispiel der Philoso-
phischen Fakultät, Thesis Göttingen 1996.
Carson, Cathryn and Michael Gubser: Science advising and science policy in postwar West Germany: The example of the Deutscher Forschungsrat, in: Minerva 40, 2002, pp. 147–179.
Cassidy, David C.: Controlling German science, I: U. S. and Allied forces in Germany, 1945–1947,
in: Historical Studies in the Physical and Biological Sciences 24, 1994, pp. 197–235.
Defrance, Corine: Deutsche Universitäten in der Besatzungszeit zwischen Brüchen und Traditio-
nen 1945–1949, in: Pappenfuß, Dietrich and Wolfgang Schreder (eds.): Deutsche Umbrüche
im 20. Jahrhundert, Köln u. a. 2000, pp. 409–428.
Düwell, Kurt: Zwischen Entnazifizierungs- und Berufungsproblemen. Die RWTH im Kontext
der deutschen Universitätsgeschichte nach 1945, in: Loth, Wilfried and Bernd-A. Rusinek
(ed.): Verwandlungspolitik. NS-Eliten in der westdeutschen Nachkriegsgesellschaft, Frankfurt am Main 1998, pp. 313–331.
Frei, Norbert: Vergangenheitspolitik. Die Anfänge der Bundesrepublik und die NS-Vergangenheit, München 1999.
Frei, Norbert (ed.): Karrieren im Zwielicht. Hitlers Eliten nach 1945, Frankfurt am Main, 2001.
Gerhardt, Uta: Soziologie der Stunde Null. Zur Gesellschaftskonzeption des amerikanischen Be-
satzungsregimes in Deutschland 1944–1945/1946, Frankfurt am Main, 2005.
Gimbel, John: Science, Technology, and Reparations. Exploitation and Plunder in Postwar Ger-
many, Stanford 1990.
Goenner, Hubert and R. Klein: Obituary for Max Kohler, in: Physikalische Blätter 38, 1982,
pp. 298–299.
Haude, Rüdiger: Die Nachkriegssituation an deutschen Hochschulen – Wer machte Karriere?, in:
Ungeahntes Erbe. Der Fall Schneider/Schwerte: Persilschein für eine Lebenslüge, hrsg. vom
Antirassismus-Referat der Studentischen Versammlung an der Friedrich-Alexander-Univer-
sität Erlangen-Nürnberg, Aschaffenburg 1998, pp. 48–60.
Heinemann, Manfred: Überwachung und "Inventur" der deutschen Forschung. Das Kontrollrats-
gesetz Nr. 25 und die alliierte Forschungskontrolle im Bereich der Kaiser-Wilhelm-/Max-
Planck-Gesellschaft (KWG/MPG) 1945–1955, in: Mertens, Lothar (ed.): Politischer System-
umbruch als irreversibler Faktor von Modernisierung in der Wissenschaft?, Berlin 2001,
pp. 167–199.
Henke, Klaus-Dietmar: Die Grenzen der politischen Säuberung in Deutschland nach 1945, in:
Herbst, Ludolf (ed.): Westdeutschland 1945–1955. Unterwerfung, Kontrolle, Integration,
München 1986, pp. 127–133.
Hentschel, Klaus: The Mental Aftermath. The Mentality of German Physicists 1945–1949, Ox-
ford 2007.
Hentschel, Klaus and Gerhard Rammer: Kein Neuanfang: Physiker an der Universität Göttingen
1945–1955, in: Zeitschrift für Geschichtswissenschaft 48, 2000, pp. 718–741.
Hermann, Armin: Max Planck in Selbstzeugnissen und Bilddokumenten, Reinbek 1973.
Hoffmann, Dieter (ed.): Physik im Nachkriegsdeutschland, Frankfurt am Main 2003.
Königseder, Angelika: Entnazifizierung, in: Benz, Wolfgang (ed.): Deutschland unter alliierter
Besatzung 1945–1949/55, Berlin 1999, pp. 114–117.
Loth, Wilfried and Bernd-A. Rusinek (ed.): Verwandlungspolitik. NS-Eliten in der westdeutschen
Nachkriegsgesellschaft, Frankfurt am Main 1998.
Martius, Ursula Maria: Videant consules …, in: Deutsche Rundschau 70 (11), 1947, pp. 99–102.

Miquel, Marc von: Juristen: Richter in eigener Sache, in: Frei, Norbert (ed.): Karrieren im Zwielicht. Hitlers Eliten nach 1945, Frankfurt am Main 2001.

Gerhardt, Uta: Soziologie der Stunde Null. Zur Gesellschaftskonzeption des amerikanischen Besatzungsregimes in Deutschland 1944–1945/1946, Frankfurt am Main, 2005, pp. 181–237.

Oehler, Christoph and Christiane Bradatsch: Die Hochschulentwicklung nach 1945, in: Führ, Christoph and Carl-Ludwig Furck (ed.): Handbuch der deutschen Bildungsgeschichte, vol. VI: 1945 bis zur Gegenwart. 1.Teilband: Bundesrepublik Deutschland, München 1998, pp. 412–446.

Niethammer, Lutz: Schule der Anpassung. Die Entnazifizierung in den vier Besatzungszonen, in: Niethammer, Lutz: Deutschland danach. Postfaschistische Gesellschaft und nationales Gedächtnis, ed. by Ulrich Herbert and Dirk van Laak, Bonn 1999, pp. 53–58.

Oexle, Otto G.: Hahn, Heisenberg und die anderen. Anmerkungen zu "Kopenhagen", "Farm Hall" und "Göttingen" (= Ergebnisse 9, preprint from the research program "History of the Kaiser Wilhelm Society in the National Socialist era"), ed. by Carola Sachse on behalf of the Presidential Commission of the Max Planck Society for the Advancement of Sciences, München 2004.

Pehle, Walter H. and Peter Sillem (eds.): Wissenschaft im geteilten Deutschland. Restauration oder Neubeginn nach 1945?, Frankfurt am Main 1992.

Rammer, Gerhard: Göttinger Physiker nach 1945. Über die Wirkung kollegialer Netze, in: Göttinger Jahrbuch 2003, pp. 83–104.

Rammer, Gerhard: Die Nazifizierung und Entnazifizierung der Physik an der Universität Göttingen, Diss. Göttingen 2004 (forthcoming).

Rammer, Gerhard: "Sauberkeit im Kreise der Kollegen": die Vergangenheitspolitik der DPG, in: Hoffmann, Dieter and Mark Walker (eds.): Physiker zwischen Autonomie und Anpassung. Die Deutsche Physikalische Gesellschaft im Dritten Reich, Weinheim 2007, pp. 359–420.

Remy, Steven P.: The Heidelberg Myth. The Nazification and Denazification of a German University, Cambridge 2002.

Respondek, Peter: Der Wiederaufbau der Universität Münster in den Jahren 1945–1952 auf dem Hintergrund der britischen Besatzungspolitik, Diss. Münster 1992.

Sachse, Carola: "Persilscheinkultur". Zum Umgang mit der NS-Vergangenheit in der Kaiser-Wilhelm/Max-Planck-Gesellschaft', in: Weisbrod, Bernd (ed.): Akademische Vergangenheitspolitik. Beiträge zur Wissenschaftskultur der Nachkriegszeit, Göttingen 2002, pp. 217–246.

Schael, Oliver: Die Grenzen der akademischen Vergangenheitspolitik: Der Verband der nicht-amtierenden (amtsverdrängten) Hochschullehrer und die Göttinger Universität, in: Weisbrod, Vergangenheitspolitik, pp. 53–72.

Schildt, Axel: Im Kern gesund? Die deutschen Hochschulen 1945, in: König, Helmut, Wolfgang Kuhlmann and Klaus Schwabe (eds.): Vertuschte Vergangenheit. Der Fall Schwerte und die NS-Vergangenheit der deutschen Hochschulen, München 1997, pp. 223–240.

Schirrmacher, Arne: Wiederaufbau ohne Wiedergutmachung. Die Physik in Deutschland in den Jahren nach 1945 und die historiographische Problematik des Remigrationskonzepts, in: Münchner Zentrum für Wissenschafts- und Technikgeschichte, working paper 2005.

Schneider, Ullrich: Zur Entnazifizierung der Hochschullehrer in Niedersachsen, in: Niedersächsisches Jahrbuch für Landesgeschichte 61, 1989, pp. 325–346.

Schüring, Michael: Minervas verstoßene Kinder. Vertriebene Wissenschaftler und die Vergangenheitspolitik der Max-Planck-Gesellschaft, Göttingen 2006.

Seier, Hellmut: Der Rektor als Führer, in: Vierteljahrshefte für Zeitgeschichte 12, 1964, pp. 105–146.

Szabó, Anikó: Vertreibung, Rückkehr, Wiedergutmachung. Göttinger Hochschullehrer im Schatten des Nationalsozialismus, Göttingen 2000.

Vollnhals, Clemens (ed.): Entnazifizierung. Politische Säuberung und Rehabilitierung in den vier Besatzungszonen 1945–1949, München 1991.

Walker, Mark: The Nazification and Denazification of Physics, in: Judt, Matthias and Burghard Ciesla (ed.): Technology Transfer out of Germany after 1945, Amsterdam et al. 1996, pp. 49–59.

Weisbrod, Bernd: Dem wandelbaren Geist. Akademisches Ideal und wissenschaftliche Transformation in der Nachkriegszeit, in: Weisbrod, Bernd (ed.): Akademische Vergangenheitspolitik. Beiträge zur Wissenschaftskultur der Nachkriegszeit, Göttingen 2002, pp. 11–35.

Weisbrod, Bernd (ed.): Akademische Vergangenheitspolitik. Beiträge zur Wissenschaftskultur der Nachkriegszeit, Göttingen 2002.

Zeitz, Katharina: Max von Laue (1879–1960). Seine Bedeutung für den Wiederaufbau der deutschen Wissenschaft nach dem Zweiten Weltkrieg, Stuttgart 2006.

Zuckmayer, Carl: Geheimreport, hrsg. von Gunther Nickel und Johanna Schrön, Göttingen 2002.

PHYSICS AND THE IDEOLOGY OF NON-IDEOLOGY: RE-CONSTRUCTING THE CULTURAL ROLE OF SCIENCE IN WEST GERMANY

Richard Beyler

The re-construction of a viable political culture was a high on the agenda of both the victorious Allies and German authorities after the end of the Second World War. It was easier to say, however, what this political culture should not be than to say what it should be. The political culture of the nascent Federal Republic defined itself largely through differentiation from other systems, past and present. Most important at the outset was establishing a line of demarcation from the Nazi regime; with the development of the Cold War, mainstream contributors to West German political discourse were also keen to differentiate themselves from the Communist bloc.

Both the erstwhile National Socialist regime and the contemporary Communist regimes were held to be predominantly *ideological* in a pejorative meaning of that term. That is, they were held to be founded on tendentious, unreasonable, perhaps even deliberately deceptive principles. The deceptions of ideology led, in all probability, to autocratic or totalitarian government; such dictatorial regimes depended in turn on the obfuscatory power of ideology – so went the proposition. The new West German polity required some other foundation. What should this something else be? Resurrection of a German tradition predating 1933 or 1918? Wholesale emulation of the American, British, or French victors? Integration into a new pan-European sensibility? Free market principles? Christian principles? Social-democratic principles? On these questions, there was no unity of opinion, but mainstream political discourse held the consensus that, somehow, a rejection of, or at least skepticism against, ideology as such was required for a healthy political future. Thus, thirty years into the history of the Federal Republic, historian Karl Dietrich Bracher asserted that its political culture was, at its outset, "marked by a need for de-ideologization."[1] This course was marked out, for example, in the move of the Christian Democratic Union (CDU) from a party based (largely) on a program of Christian, non-materialistic communalism to a party that presented itself primarily as the advocate of free-market recipes for pragmatic economic success.[2]

---

[1] Bracher, Parteienstaat, p. 34; for a comprehensive survey of the cultural scene, see Glaser, Kulturgeschichte.
[2] See Mitchell, Materialism.

But as the CDU's strategy suggests, there was a broader meaning of the term ideology – not necessarily pejorative – as, roughly speaking, the set of shared, usually tacit assumptions and principles of a given discourse. In this broader meaning, no political discourse can function without an ideology. The notion of political culture free from ideology was, itself, then an ideology – a political move made in rivalry with other political moves. This paper examines the efforts made by spokesman of the scientific community, as well as a few of its patrons and critics from the broader society, to define a role for science within this ideology of non-ideology. The task for the leaders of the scientific community was to convince their constituencies (and themselves) that science was non-ideological – indeed, to make that case that science was perhaps more than any other field of human endeavor intrinsically anti-ideological – and hence able to make contributions to and deserving of support from the new polity. This discursive task was far from straightforward, for (at least) three main reasons:

1) From 1933 to 1945, the leaders of German science had faced cooptation (Gleichschaltung) by the Nazi regime, and in many cases had adopted policies of self-cooptation (Selbstgleichschaltung). Whereas during the twelve years prior to 1945 the scientific community had undertaken establish its ideological compatibility with National Socialism, the task after 1945 for the self-same institutions and leaders was now exactly the opposite. Could a denazified reading of the history of science during the Third Reich be attained?

2) Differentiation from the East bloc was also no trivial task, given that Communism – at any rate in its Leninist-Stalinist interpretation – billed itself as a supremely scientific social philosophy. Was the proper response a claim that science transcended the Cold War divide altogether, or a claim that science was intrinsically linked to anti-Communism?

3) Particularly relevant for physics was the widespread sense that this branch of science had – in the epoch of rockets, radar, and nuclear weapons – become inevitably part of politics and strategy on the grandest scale. Could such a strategically important science then sustain a claim of ideological innocence?

This paper will analyze these three themes in the re-construction of the cultural role of science in West Germany in the late 1940s, 1950s, and early 1960s. The concluding section will present (very briefly) a comparison with contemporary discussions of science and ideology in other national contexts. Although there were some commonalities, the de-ideologization discourse in West Germany had distinctive features. Before undertaking the thematic analysis, however, it is worthwhile to examine a specific example of how the ideology of non-ideology informed West German science policy debates.[3]

---

[3] Related aspects of the re-constuction of a role for science in post-war Germany are considered in Beyler/Low, Science; Beyler, Demon.

THE SELF-CONCEPTION OF THE DFG IN THE 1960s

The re-established Emergency Association for German Science (Notgemeinschaft für deutsche Wissenschaft, NGW) and its successor organization (from 1951 on), the German Research Foundation (Deutsche Forschungsgemeinschaft, DFG), sought to mediate between the state and the scientific community. For general director Kurt Zierold, the keyword to describe the optimal position of the DFG was self-administration (Selbstverwaltung), a conception that he developed in several articles in the early 1960s and in his massive history of the DFG published in 1968.[4] Alongside the historical narrative, the book was, simultaneously, a piece of advocacy for Zierold's vision of public science policy and a defense of the strategy and tactics deployed by the DFG up to that time. It was, to be more specific, a response to ongoing transformations of state-science relations – most importantly, the creation of the Federal Ministry for Scientific Research (Bundesministerium für wissenschaftliche Forschung, BMwF) in 1962 – and to pointed accusations that the German scientific establishment was insufficiently democratic in orientation and practice – critiques raised, above all, in programmatic writings produced by a group of students affiliated with the Social Democratic Party and informed by Frankfurt School critical theory.[5]

As depicted in Zierold's historical narrative, since the re-establishment of the NGW in 1949, and the creation of the DFG through merger of the NGW with the German Research Council (Deutscher Forschungsrat, DFR) in 1951, the DFG had undertaken a delicate balancing act, or rather several simultaneous balancing acts.[6] Zierold and other leaders of the organization had attempted – and largely succeeded – to accommodate several conflicting agendas in both the political and scientific spheres. In the early years of its renewed existence, the NGW/DFG faced sometimes tense relationships with other key German scientific organizations such as the Max Planck Society (Max-Planck-Gesellschaft, MPG). Eventually, an accommodation was reached by which the MPG became one of the key constituent members of the DFG, but not subordinate to it and thus able to retain its primacy in the arena of long-term, elite research in the natural sciences. Conflicts with the nascent DFR envisioned by Werner Heisenberg et al. as an

---

[4] Zierold, Deutsche Forschungsgemeinschaft; Zierold, Selbstverwaltungsorganisationen; Zierold, Wie soll man; Zierold, Forschungsförderung. The final chapter of the latter book, pp. 528–542, is: Die Forschungsgemeinschaft als Selbstverwaltungsorganisation der Wissenschaft. Ihre Stellung in Staat und Gesellschaft. This chapter includes material closely parallel to or borrowed from the articles cited.

[5] Nitsch et al., Hochschule; see esp. the foreword by Jürgen Habermas, pp. v–vi.

[6] Historical overviews of science policy in the Federal Republic of Germany include: Stamm, Staat; Osietzki, Wissenschaftsorganisation; Stamm, New Opportunities; Eckert/Osietzki, Wissenschaft; Osietzki, Reform; Cassidy, Controlling. Some illustrative examinations of particular figures and organizations include: Nipperdey/Schmugge, 50 Jahre; Eckert, Primacy; Heinemann, Hochschulkonferenzen; Vierhaus/vom Brocke, Forschung; Trischler, Luft; Stucke, Institutionalisierung; Oexle, Wie in Göttingen; Carson, New Models; Trischler/vom Bruch, Forschung; Carson, Nuclear; Carson/Gubser, Science; Schüring, Minervas; Zeitz, Max von Laue.

elite scientific steering committee for the German government, were resolved largely in favor of the NGW. Although ostensibly the new DFG emerged out of a merger between the NGW and the DFR, in fact the functions of the latter were largely subsumed into those of the former.[7]

As these scientific organizations occasionally competed and occasionally co-operated with each other, a protracted struggle was also underway between the Länder (individual German states) and the federal government (Bund) for authority in science policy.[8] The several Länder sought to assert their constitutionally guaranteed sovereignty over cultural affairs, while the federal government (and, increasingly, many scientists) pressed the point that many projects of modern science were of a scale which required funding and coordination at a national or even international level. Both out of a sense of pragmatism and due to personal connections–before joining the NGW administration Zierold had been the deputy secretary for higher education in the Lower Saxony Cultural Ministry – the NGW was, initially, largely the creature of the Länder within this polarized field. This did not mean, however, that the DFG was later averse to increasing support from the Bund; by 1954, then DFG president Ludwig Raiser was protesting against a "false federalism" based on a stubborn insistence on Land prerogatives in the cultural sphere.[9] Less agreeable, however, would be any indications of centralized governmental direction or planning of science. Precisely the autonomy of science from supposed governmental control, and hence the danger of political or ideological interference, was the crucial element of science policy as envisaged by most spokesmen of German science after 1945; we will return to this point in the next section.

The problematic but increasingly close liaison between the central government and the DFG manifested itself in several ways throughout the 1950s, but became more acute with the creation of the BMwF out of the more narrowly delimited Federal Ministry for Atomic Questions (Bundesministerium für Atomfragen) in 1962. The appearance on the scene of this federal ministry seemed many ways a seemingly unavoidable product of the increasing scope and scale of science. In other words, it seemed to be a necessary conduit for the financial and political resources necessary to do (especially) "big science." However, as Andreas Stucke (among others) has described, the creation of the BMwF also raised anxieties for the DFG and other organizations of scientific self-administration: Would this new bureaucratic entity impinge on their prerogatives and autonomy? Once the decision became a fait accompli despite a brief campaign of protest, the leadership of the scientific bodies fixed upon a policy of cooperation rather than continued intransigence, in the service of assuring friendly interest and of gaining a degree of influence within the ministry itself. Moreover, it soon became clear

---

[7] See Eckert, Primacy; Carson/Gubsers Science.
[8] See, e.g., Stamm, Staat, pp. 141–150.
[9] Raiser, Föderalismus; see also Zierold, Forschungförderung, p. 532.

that fostering relations with a strong Science Ministry could counterbalance the growing role of the Defense Ministry in science funding.[10]

Negotiating a working relationship between new governmental institutions and the traditional institutions of the scientific community was one important part of the context for Zierold's historical apologia for the DFG. The key for Zierold was self-administration: with this principle as a guide, science could exercise both its intrinsic freedom (Freiheit) and, simultaneously, maintain its vital connectedness (Bindung) to the broader society.[11] As Zierold noted, the term had a rather technical meaning in the prior history of political and economic organization, having been articulated in some detail by the Prussian political reformer Karl vom Stein and, later, associated with the liberal ideals of 1848.[12] Zierold now deployed it a broader sense. The self-administered organization "stands between the state administration on the one hand and a private interest group on the other."[13] It therefore played a mediating role between the expectations of society, broadly conceived, and the necessarily more narrow concerns of the specific community it represented. A self-administered organization received the confidence and endorsement of the state to control a particular sector of social activity – in this case, scientific research – ultimately on behalf of the commonweal, but proximately according to standards and policies intrinsic to that sector. As Zierold expressed it, such institutions "require recognition from the state, but act self-responsibly. Recognition by the state does not mean that duties for the state must be [explicitly] delegated; state sanction suffices."[14] In short, the relationship required that the government place a considerable degree of trust in the self-administrative organization. Conversely, it was implicit in this arrangement that the self-administrative organization maintained the unique qualities of the sub-community to which it related: otherwise its mediating function would be moot. In short, the self-administrative organization vouched for political disinterestedness on both sides of the relationship. Such, at least, was the theory.

Devotion to the concept of self-administration was by no means unique to Zierold among the leaders of the DFG. In a 1963 article, president Gerhard Hess argued "that exactly the natural polarity [between] research governing itself and the interests of the state is a particularly favorable condition to allow research to unfold itself freely for the augmentation of knowledge and for the good of society."[15] Or again, in his valedictory address upon leaving the presidential post in 1964, Hess lauded the "insight of the re-founders of the Research Society into the necessary freedom of science and into the co-responsibility of government in

---

[10] Stucke, Institutionalisierung, pp. 64–66.

[11] Zierold, Deutsche Forschungsgemeinschaft, p. 482; Zierold, Forschungsförderung, p. 528. Zierold in these passages relies strongly on Huber, Selbstverwaltung.

[12] Zierold, Deutsche Forschungsgemeinschaft, p. 482.

[13] Ibid., p. 484.

[14] Zierold, Selbstverwaltungsorganisationen, p. 687.

[15] Hess, Gesellschaft, p. 249.

a self-administrative organization as the precondition of the work and success of our institution ...".[16]

In such an arrangement, ideology other than the values of the professional community itself could play no role. This required a considerable amount of self-restraint on the side of the state, and on this point the advocates of self-administration came close to equating the success of this model of scientific administration with the success of democracy as it was practiced in the Federal Republic and its allies. Zierold, for example, asserted that "only in mature democracies" were politicians able to countermand their "natural feeling" that "whoever pays, rules," and thus to commit to funding self-administrative organizations while at the same time "renouncing the right of state decision-making" with respect to their affairs. Fortunately, he continued, public opinion in the Federal Republic was sympathetic to the general idea of self-administration, and so therefore "many politicians were induced to praise [it]."[17] Hess, for his part, indicated that the states "with a living, not [merely] formal democratic tradition" such as Great Britain, Switzerland, and the United States had based their science policy on the self-administration model.[18] Elsewhere he asserted that the freedom of science "excludes despotism as well as a form of life approximating 'laisser aller'" – in short, neither authoritarianism nor unrestrained capitalistic individualism, but the middle ground of the social welfare state.[19] And in his valedictory address he directly rejected a role for ideology in settling upon this model:

> It is certainly not ideological fundamentalism ... which leads me to think this way. I believe I am judging thoroughly realistically, i.e., on grounds of principles [derived from] experience valid for every fruitful science, namely freedom and self-determination, guarded and bounded by the cooperation of state and society.[20]

This rosy picture of the democratic virtues of self-administration was not universally accepted. In this same period, voices demanding reform in the social structures of German academia were becoming more and more voluble. A prominent source of such critique was a study group of the Socialist German Student League (Sozialistische Deutsche Studentenbund, SDS), organized in 1961 to consider possible reforms in the German university system; its studies culminated in a 1965 volume entitled Hochschule in der Demokratie. Strongly influenced by critical theorists such as Max Horkheimer, Herbert Marcuse, and Jürgen Habermas, the authors presented a detailed analysis of what they saw as a number of ills in German higher education and research institutions which rendered them far from democratic. Above all, what they described as the overwhelmingly hierarchical structure of these institutions, in which power was concentrated in the

---

[16] Hess, Gesellschaft, p. 321.
[17] Zierold, Selbstverwaltungsorganisationen, pp. 693–694.
[18] Hess, Gesellschaft, p. 315.
[19] Ibid., p. 308.
[20] Ibid., p. 323.

hands of relatively few professors and patrons in industry and government, called for reform.[21]

Insofar as this critique was accurate, the notion of self-administration became quite dubious. The SDS group's picture of the "freedom" of science as it existed in 1960s Germany was more differentiated – and more pessimistic – than the one offered by (for example) Zierold or Hess. Zierold had argued that decision making in a self-administrative organization (such as science possessed) was "cooperative" (genossenschaftlich); hence the "leadership principle" (he used here the Nazi jargon Führerprinzip) was incompatible with it.[22] From the SDS perspective, however, the "corporative and institutional" aspects of scientific work formed a kind of self-reproducing "feedback" loop in which individual aspirations had to take a subordinate place. It was far from a fully cooperative process:

> the liberal promise of *individual* freedom of research and teaching for anyone who identifies with science and [who is scientifically] active is taken back ... Freedom is polarized in science, as everywhere in our society, into the 'positive' freedom of the leaders [Führer] to act in a 'formative' way upon the behavior of others, and the 'negative' [merely] apparent freedom of action of those who are led, to whom remains only the free choice of the relations of domination and subordination.[23]

The notion of a science which was both "free" to act according to its own rules and "connected" to broader social currents thus only reproduced the monopolizing tendencies of the latter in the former: the supposed freedom of science was reserved for a select few.

While this critique was broadly applied to German academia, the authors also directed some specific criticisms at the DFG. They were skeptical of the degree to which such an organization really guaranteed the openness of the scientific process to the public that it supposedly served. They pointed out that, in contrast to the universities, the DFG had no "public parliamentary or ministerial supervision." The DFG was, moreover, "no independent self-administrative body of West German science (as it claims) but rather an interest consortium, organized according to private law, of public and private scientific establishments, of commercial industry, of governmental agencies with special research budgets, and of state and federal ministries *responsible for science policy*."[24] In other words, public transparency was lost in favor of increasing dominance of research by "calculations according to commercial or power-political interests" such as the military – a pattern they saw exemplified in contemporary developments in the United States.[25] A more genuinely public supervision of the scientific process was necessary.

Not surprisingly, Zierold reacted with umbrage to these criticisms. He, himself, had warned that "the degeneration of self-administration is called 'represen-

---

[21] Nitsch et al., Hochschule.
[22] Zierold, Selbstverwaltungsorganisationen, p. 687.
[23] Nitsch et al., Hochschule, p. 89; emphasis in the original.
[24] Ibid., pp. 221–222; emphasis in the original.
[25] Ibid., p. 223.

tation of interests,'" so the accusation that the DFG was merely a special interest group, without an overarching social perspective, must have hit close to home.[26] He professed "astonishment and consternation ... that from circles of the ... SDS a stronger state control over the Research Foundation is being demanded."[27] Telling is Zierold's interpretation of the origins of these criticisms of the self-administration concept: justifiable discussions about the need for university reform in the face of growing student numbers had been extended and affected opinions about other scientific institutions. According to Zierold, the most negative statements sprang from "circles of academic youth who no longer had experience with National Socialism and, within them, among a politically ultra-left oriented minority [who] toy with the other variety of totalitarianism, Communism ...".[28] Zierold thus counterposed his vision of the politically connected but disinterested self-administration of science with alternatives rooted, ultimately, in totalitarianism. In so doing, he appropriated several themes in the discourse of non-ideology which had grown up around science since World War II: science, once cleared of Nazi abuses, was intrinsically compatible with the new, ideologically restrained West German polity, while at the same time it should and could bear an elevated sense of social responsibility. I now turn to explore some aspects of these themes as they developed after 1945.

DENAZIFICATION AND THE ASSERTION
OF POLITICAL INNOCENCE

In science as in other fields after the collapse of the Third Reich, former Nazis were to be purged from positions of responsibility in both the public and private sector.[29] The details of the process varied from zone to zone, and also modified over time, but the core of the process was assessment of the individual's involvement in the Nazi power structure though a lengthy questionnaire (Fragebogen) and, often, testimony from colleagues, neighbors, employers or employees, etc. While denunciations did occur, more typical were the laudatory character references that came to be known colloquially as "whitewash certificates" (Persilscheine). At the institutional level, as Allied occupation authorities, the German states, and the new federal government all jockeyed for control over educational and research affairs, it was critical to make the case that the scientific community had maintained its autonomy from the Nazi state. The existence of collaborative relations (in the terminology of historian Herbert Mehrtens) was hereby not

---

[26] Zierold, Deutsche Forschungsgemeinschaft, p. 484; Zierold:, Forschungsförderung, p. 540.

[27] Zierold, Forschungsförderung, p. 540. The SDS document had not stated specifically "state" supervision but referred rather to "public" supervision.

[28] Zierold, Forschungsförderung, p. 540.

[29] On Allied military government science policy and denazification in the sciences, see, e. g., Beyerchen, German Scientists; Gimbel, U.S. Policy; Schneider, Entnazifizierung; Heinemann, Hochschuloffiziere; Cassidy, Controlling; Ash, Umbrüche; Ash, Denazifying; Hentschel/Rammer, Physicists; Beyler, Wissenchaft, pp. 16–43; Schüring, Minervas; Hentschel, Mental Aftermath.

necessarily denied, but frequently obscured. For scientists, again and again in these exculpatory testimonials the argument appeared that participation in science constituted, in itself, a kind of non-conformity to the Nazi regime.[30]

For the physics profession, the sorry history of "Aryan physics" figured centrally in the discourse of denazification. As several recent historical studies have demonstrated, the Aryan physicists hardly constituted a unified movement in the 1930s.[31] Although they did experience some successes in garnering or preventing appointments, and although they created problems for several lightning-rod cases such as that of Heisenberg, both their influence among physicists as a whole, as well as the concern of physicists as a whole, actually remained rather constrained. Moreover, the eventual opposition to the Aryan physicists developed precisely through selective cooperation with particular centers of power within the Nazi state – above all, the military. It was only after the war that there developed the narrative of a coherent campaign against "Nazi science" rooted in a pure, untrammeled science removed from entanglements with the regime. This narrative developed precisely in the context of the attempt to rehabilitate the reputation of physics as a discipline and the reputation of several particular physicists.

Thus Werner Heisenberg in his questionnaire described several of his publications as contributing to a struggle against the anti-Semitic physics of Johannes Stark and Philipp Lenard.[32] This was true enough, but implied that Stark and Lenard spoke for all National Socialists, which was not the case. The picture was painted of a war within the physics community in which even non-activists were part of a network of moral support, even if this resistant activism was not obvious to outsiders. For the proceedings of one of his institute members, Karl Wirtz, Heisenberg wrote in a testimonial that he (Wirtz) "had always diligently pursued reasonable scientific inquiry, and had sought to fend off Nazi Party attacks on fellow scientists as axiomatic."[33] Taken for granted, again, was the idea that the Party was somehow intrinsically opposed to reasonableness and scientific diligence. For the denazification proceedings of Kaiser Wilhelm Society (Kaiser-Wilhelm-Gesellschaft, KWG) general director Ernst Telschow, Heisenberg wrote that he (Telschow) had "represented the KWG in its dealings with Nazi authorities in a way that always relied upon purely factual, scientific arguments and avoided reference to party politics (Parteipolitik)."[34] Similar themes of the purity of the scientific endeavor were echoed by atomic physicist Walther Gerlach, who attested that he "had maintained close connection [with Telschow] in questions of the promotion of pure scientific research, which threatened to succumb to the higher education policy of those years." Telschow's Party membership and

[30] Mehrtens, Kollaborationsverhältnisse; see also Mehrtens, Missbrauch; Sachse, Persilscheinkultur.

[31] This point is explored in several essays in Hoffmann/Walker (eds.), Deutsche Physikalische Gesellschaft; see also Walker, Nazi Science; Litten, Mechanik.

[32] Heisenberg, questionnaire, 8 Nov. 1946, Nds. 171 Hildesheim, Nr. 9793.

[33] Heisenberg, testimonial letter, 9 Jun. 1948, Niedersächsisches Hauptstaatsarchiv-Außenstelle Pattensen (hereafter NsHsta-AP) Nds. 171 Hildesheim, no. 13940, NsHsta-AP.

[34] Heisenberg, testimonial letter, 20 Jul. 1946, NsHsta-AP, Nds. 171 Hildesheim, no. 12773.

interaction with other Party members had, Gerlach attested, merely served the cause of moderating the "evil consequences of politics."[35] Gerlach thus alludes to Telschow's adeptness at maneuvering through the power structures of the NS state, but attributes this not to political skill but rather the avoidance of politics. Moreover, it is perhaps worth noting that the opposition presented in these statements is not between science and National Socialism but rather between science and "politics" or "party politics" generically. In other words, responding to the Nazi situation after the fact, the spokesmen for the rehabilitation of German science implicitly equated "party politics," whatsoever, with the danger of ideological interference with scientific freedom. Perhaps this was a euphemism in a legally sensitive situation, but perhaps also it reflects a desire to present science not as the antagonist of a specific political philosophy but rather as transcending political philosophies altogether.

Neither Heisenberg's nor Wirtz's cases were particularly problematic – neither had been Party members and both had been quasi vetted in advance by the British authorities. Telschow's position was less obviously secure, but a massive and well-organized campaign of whitewash certificates contributed to a positive conclusion. However, similar arguments applied in more difficult cases. Theoretical physicist Pascual Jordan, who gained something of a reputation as a pro-Nazi propagandist due to several inflammatory books and articles from the 1930s and early '40s, faced a protracted and vexatious process of denazification. Jordan in his own defense, and various colleagues including Heisenberg, pointed strenuously to his continued contributions to the modern quantum physics despised by the Aryan physicists. In the end, these arguments, combined with Jordan's prestige – and the prospect of his taking his talents to East Germany – contributed to his formal rehabilitation.[36]

In any specific case, the outcome of denazification frequently depended on very local and personal circumstances and connections. Nevertheless, an overarching discourse of a pure science as the natural opponent of Nazi ideology also began to emerge. One example was a series of articles in the Physikalische Blätter that rehearsed the Aryan physics controversy and recounted efforts of leaders of mainstream leaders of the physics community, such as those of Carl Ramsauer (president of the German Physical Society 1941–1945) to protect and promote its interests against or within the National Socialist state.[37] But the choice of preposition is exactly the problem: the post-1945 articles cast these actions as actions *against* the Nazi regime, whereas at the time Ramsauer's appeals for support of mainstream physics were made precisely on terms that would be valued *within* the Nazi state – above all, on terms of the usefulness of physics for a modern military infrastructure. But in the post-war context, the pragmatic alliances between the physics community and powerful state constituencies were obscured. Instead,

---

[35] Gerlach, testimonial letter, 2 Jul. 1946, NsHsta-AP, Nds. 171 Hildesheim, no. 12773.

[36] On Jordan's post-war situation, see Wise, Pascual Jordan, pp. 250–254; Schirrmacher, Physik, pp. 1–5.

[37] E. g., Ramsauer, Geschichte; Planck, Besuch; Laue, Bemerkung.

the spokesmen of physics presented itself as maintaining its own values and interests, allegedly indifferent to the desires of the state that had, indeed, sought to displace these values with its own ideological agenda.[38]

Sympathetic non-scientists made similar arguments. Thus Munich philosopher Aloys Wenzl saw a lesson of the Third Reich in the conflict between ideologically motivated "Aryan physicists" and those physicists who continued, overtly if possible but covertly if necessary, solid work in modern physics.[39] For Wenzl, therefore, there was a self-evidence opposition between Nazi ideology and modern science: the two were mutually exclusive. The message was that if scientists had persevered in ignoring the ideological pressure of Nazism, they could certainly be trusted within the nascent Federal Republic.

It may have been that the scientists were protesting too much out of a sense of insecurity. But the Allied occupation authorities soon came to prioritize a prosperous, stable Germany over protracted scrutiny of the past; this approach likewise became congenial to a large proportion of the German population who had become tired of, or cynical about, the process of denazification.

DEMOCRATIC SCIENCE IN THE COLD WAR CONTEXT

Some humanistically oriented intellectuals warned of potential alliances between ideology and instrumental rationality; eminent historian Friedrich Meinecke blamed the "German catastrophe" on inadequate cultural education (Bildung); and several authors, led by Friedrich Georg Jünger, portrayed a "demon of technology" that threatened to overwhelm humanity in modern mass culture; Martin Heidegger also philosophized against the productionist materialism of modern society.[40] These authors saw a humanistic-classical orientation as the best bulwark against totalitarianism. The dominance of scientific-technical thinking, on the other hand, correlated with the rise of mass society – an imprecisely defined but resonant term. Totalitarianism was, in turn, almost by definition the outgrowth of mass society. Many humanists were alarmed by the rise of the scientific-technical expert to a prominent social position. The scientific-technical expert, now rising to a position of social prominence, was lacking in breath of vision and therefore particularly susceptible to the blandishments of ideological power. In this representation of the social role of science, then, the emergence of mass society and the dominance of the specialist, both phenomena driven by the advance of scientific-technical reason, were at least partially responsible for the rise of Nazism. They were also perhaps responsible for the rise of Communism. In any case, they posed a threat to *Western* culture. In the wake of Hiroshima, physics

---

[38] See Schmithals, Carl Ramsauer; Hoffmann, Carl Ramsauer; Hoffmann, Autonomy; Hoffmann/Walker, Deutsche Physikalische Gesellschaft.

[39] Wenzl, Strömungen, p. 41.

[40] Meinecke, Katastrophe; Jünger, Perfektion; Heidegger, Technik; on Heidegger's views on technology, see Zimmermann, Heidegger's.

was especially implicated in these critiques as the most powerful and potentially terrifying exemplar of scientific-technical expertise.[41]

To some degree these reflections simply continued a prior humanistic-philosophical tradition. For this reason, in the view of one historian, the intellectual climate in post-war Germany can be described as "stale."[42] But the very prominence of apparently tired tropes is worthy of attention. In a climate of defeat, of mutual recrimination for the rise of Nazism, and of deep anxiety about the future, bold solutions were apparently not in demand. Rather, many German intellectuals relied on verities that gave a sense of continuity and stability, even if to some observers these verities seemed exhausted.

Science might serve as a model for democratic education. For example, Adolf Grimme, the culture minister of Lower Saxony – and at that time Zierold's superior – sought a middle ground between tradition and reform. Grimme pleaded strongly for a renewal of the Gymnasium – the traditional university-track secondary school – as a site for the rigorous training of talented students. Standards had slipped under Nazism and during the war, Grimme noted, inter alia because the curriculum had become ideologized. The new curriculum should still be centered on the classics, Grimme believed, but he envisioned a major place for mathematics and the natural sciences.[43] He argued that mathematics and sciences were especially valuable because they were an "antidote to tumors, including those of democracy, namely that [kind of] 'democracy' one can call smart-alecky adolescence [Schwatzbubenhaftigkeit] ... It is an ethical viewpoint, where the young person sees that he must subordinate himself to law."[44] Hence, for Grimme, science was a rigorous mental discipline that would be valuable to the new social and political order. He envisioned the ideal democracy as run according to rational principles by people with demonstrated competence. Similar arguments were made by Carl Friedrich v. Weizsäcker in support of the much discussed (though ultimately unsuccessful) Blaues Gutachten for university reform in 1947 in the British zone.[45]

Conversely, science might be billed as the enemy of the enemy: namely, East bloc Communism. One of the most persistent and vocal advocates of this line was Jordan whose politics remained on the far right in the post-war context. In numerous books, and dozens of articles and public lectures, he made the argument that modern physics, above all through the introduction of indeterminism, had revealed materialism to be a metaphysical dogma that had to be discarded and thereby removed the philosophical underpinnings of Marxism.

---

[41] I discuss the "demon" of technology in relation to mass culture further in Beyler, Demon; post-war cultural-political debates are surveyed in Laurien, Zeitschriften; on technoscientific pessimism, see Herf, Belated Pessimism; Beyler, Hostile Environmental Intellectuals.

[42] Muller, How vital was the Geist, p. 198.

[43] Heinemann, Hochschulkonferenzen, pp. 65–67, 131, 177–179.

[44] Minutes, Kulturpolitischer Ausschuss beim Länderrat, 27–28 Sep. 1946, Bundesarchiv Koblenz, Z1/1009.

[45] Weizsäcker, Geschichte, pp. 5–7; Studienausschuss, Gutachten; for discussion of Weizsäcker's role, see Stamm, Staat, pp. 65–67; Phillips, Intitiative, p. 187.

More broadly, both socialism and liberalism were part of a "failed revolt" – so the title of his 1956 book – in modern European culture.[46] Ideology was, according to Jordan, the attempt to re-structure society according to a set of preconceived theories; such attempts rested upon a deterministic understanding of human life and society and were, given the findings of modern science, doomed to failure. The alternative was adherence to the "concrete values" of conservatism.[47]

Jordan's line of argument was rather idiosyncratic, but more mainstream efforts to recruit science to the cause of anti-Communism can be found. Prominent forums for this cause were provided by the congresses and publications of the Congress for Cultural Freedom (CCF), an international organization of intellectuals – or rather, an umbrella organization for a series of national organizations – presenting a cultural front against Soviet Communism. The CCF organized a series of prominent international conferences, the first in 1950 in West Berlin, followed by meetings in Paris in 1951, Hamburg in 1953, and Milan in 1955. The CCF also supported the publication of several cultural-political magazines in various countries, such as Encounter in Great Britain. The journal Der Monat, though founded independently of the CCF, was de facto its organ in West Germany. Since the 1960s, rumors circulated that the CCF and its various conferences and journals were paid for by the CIA. In more recent years, with the opening up of various archival records, it has become indisputable that the CCF was, indeed, largely funded by the CIA through various "front" foundations, and that (at least one) CIA officer was actively involved in planning many of its activities. The consensus among historians – though it is far from unanimous – is that most of the occasional conference participants or authors probably weren't aware of this connection, but that the conference organizers and journal editors probably were; Melvin Lasky, editor of Der Monat, certainly was. There is an argument to be made that the CIA simply seized upon sentiment that was already there; its involvement may therefore have affected the quantity but not necessarily the quality of intellectuals' anti-Communist discourse. Historian Giles Scott-Smith has argued that the CIA's motive was not so much promoting anti-Communism, which was going on anyway, but rather promoting pro-American, or rather combating anti-American, attitudes among West European intellectuals.[48]

The CCF's 1950 international conference was held in West Berlin, a site deliberately chosen for its highly charged symbolism of East-West confrontation. Even more provocatively, just days before the opening of the conference, Communist forces invaded South Korea. Austrian physicist Hans Thirring demonstratively withdrew his paper, which had planned to strike a conciliatory tone, because its premise that the East bloc was not actively threatening the West had been violated. Other speakers pointed to censorship of various forms of intel-

---

[46] Jordan, Aufstand.

[47] See Wise, Pascual Jordan, pp. 250–254; Beyler, Positivism, pp. 480–541; Beyler, Demon, pp. 232–237; Schirrmacher, Physik, pp. 6–17.

[48] Scott-Smith, Politics, esp. pp. 82–84; see also, for the history of the CCF, Coleman, Liberal Conspiracy; Saunders, Cultural Cold War.

lectual activity, including science, in the Soviet Union and suggested an intrinsic incompatibility between the Communist system and free inquiry.[49]

The theme of science's affinity to Western-style democracy was even more prominent at the 1953 meeting in Hamburg, organized around the theme of "Science and Freedom." The twenty-one members of the honorary committee included seven physicists: Percy Bridgman, A. H. Compton, James Franck, Max v. Laue, Lise Meitner, Robert Oppenheimer, and George Thomson. Other scientists on the committee included Otto Hahn, Henry Dale, Bertrand Russell, and H. J. Muller. The Lysenko affair provided abundant material for speakers to argue that free scientific inquiry and totalitarian ideology were fundamentally incompatible – so Hans Nachtsheim, Theodosius Dobzhansky, Max Hartmann, Alexander Weissberg-Cybulski, Michael Polanyi, Muller, and others. James Franck drew explicit analogies between Communist and National Socialist approaches to science, but also deplored what he saw as a censorship of science taking place in the name of security in the United States.[50]

THE SOCIAL RESPONSIBILITY OF SCIENCE
IN A POST-ATOMIC AGE

A third strand of argument was closely related to the second, but rested (I would suggest) on a different set of premises: namely, that with the sudden power revealed by physics (and above all physicists) it was not merely possible but rather incumbent upon scientists to participate in the political process – not as representatives of a particular ideology, but rather, representatives of a rational viewpoint that transcended ideological differences. Zierold adumbrated this new and awesome responsibility in a 1948 essay:

> For the first time in history humanity has the technological means to bring everything to an end in a great self-annihilation …. In such times of radical transformation … it is necessary to preserve traditional values, but in the first instance it is required to reach the new shore and to entrust leadership to men who are looking towards tomorrow.[51]

Tradition should not be discarded, but likewise new ways of thinking had to be explored. In this new world, science – above all, physics – presented both dangers and opportunities.

At the immediate end of the war, Allied policy seemed to be directed at stringent restriction of science and technology, in the name of demilitarization. Allied Control Law 25 forbade or heavily restricted research with potential military applications, including many fields of pure and applied physics: atomic energy,

---

[49] The conference proceedings appeared as Lasky, Kongress; on Thirring's withdrawal, see Coleman, Liberal Conspiracy, p. 28.

[50] The conference proceedings appeared in German as Kongress, Wissenschaft, and in English as Congress, Science.

[51] Zierold, Hochschulprobleme, p. 12.

aerodynamics, electronics, etc. This approach was soon modified in the cause of economic recovery and development. One set of arguments used to advance the cause of science focused on the economic advantages to be gained from scientific research and development, especially for the export-based economy widely assumed to be West Germany's only hope for long-term survival.[52] This line of argument was implicitly de-ideological: decisions had to be based on what worked, not on doctrinaire prejudices, and it fit well alongside the economic pragmatism characteristic of the Adenauer era.[53]

But, as noted above, the awesome power revealed though nuclear physics entailed deeper reflections. Should the scientist simply offer the products of his work as a tool to be used any which way? Physicist Erich Regener, reflecting on the implications of the atomic bomb for the future of his profession, called for an end to the indifference German scientists had traditionally taken towards the practical application of their work. "Political activity is not under discussion here," he wrote in the Physikalische Blätter, "but rather the cultivation of a sense which grows out of the free and honest scientific activity and which is itself the presupposition for such activity."[54] In other words, scientists were now to take a more active role in public discourse about the results of their work, but this was cast as a direct extension of scientific activity and habits of mind that entailed "freedom" and "honesty." Regener seemed to be searching, then, for a new way to intervene in the civic arena, but this would require inculcating a greater willingness among scientists to speak and work beyond the walls of their laboratories, to use their skills, aptitudes, and prestige to influence debates that were no longer about "pure" science – to function, though Regener did not use this word, as public intellectuals. But this was, as far as he was concerned, precisely *not* political discourse or activity. Conventional politics, by implication, suffered from dishonesty and a lack of freedom. Regener's understanding of "politics" was confined to "party politics," so that his plea becomes one for a "non-partisan" participation in political discourse.

Regener's article was a small sign of a broader shift towards a professional philosophy of public responsibility taken by many German scientists in the post-war era, but not without some ambivalence and unease. Cathryn Carson has analyzed in some detail Werner Heisenberg's trajectory of rethinking the role of scientist as a political and public intellectual in the post-war period, a role which involved for Heisenberg striking a careful balance between quiet work with policy makers behind the scenes and carefully selected public advocacy for important causes.[55] A landmark on this trajectory, for Heisenberg as well as a number of his disciplinary colleagues, was the publication of the "Göttingen Eighteen" mani-

---

[52] See Beyerchen, German Scientists; Gimbel, U.S. Policy; Cassidy, Controlling; Beyler/Low, Science Policy.

[53] See Mitchell, Materialism.

[54] Regener, Mitverantwortlichkeit, p. 170.

[55] Carson, New Models.

festo against nuclear armaments in 1957.[56] This episode is sometimes taken as a sign of a new willingness on the part of German physicists to participate in political debate. But indeed the organizers of the manifesto, above all Weizsäcker, agonized over just that possibility, and attempted to construe the campaign as a non-political or at the very least non-partisan move. Arguably this was either somewhat disingenuous or somewhat naive; in any even, the manifesto soon became embroiled in the general election campaign of that year. Critics of the manifesto were quick to condemn it as an attempt by scientists to arrogate to themselves, on the basis of their prestige, authority about matters – strategies of war and peace, and international relations – about which they had no more right to speak than anyone else in the polity. Moreover, some critics quickly labeled the manifesto – confirming Weizsäcker's trepidations – a contribution to Social Democratic opposition to Konrad Adenauer's administration. Jordan went even further and essentially accused the Göttingen Eighteen of being dupes of the Soviet Union: they had let their objectivity as scientists be swayed or tainted by the global ideological conflict. Jordan himself, of course, billed his arch-conservative, vigorously anti-Communist stance as an intrinsic outgrowth of the development of modern physics. Patently, sustaining the distinction between partisan and non-partisan participation in the public arena was no longer so easy, but all sides relied on the comforting notion that their side, at least, had not been swayed by "merely" political or ideological concerns, but was rather expressing something more authentically grounded in scientific truth.[57]

Literary reflections on the changing public responsibility of science also reflected this tension between self-conception and de facto effect. Robert Jungk's Brighter than a Thousand Suns (Heller als tausend Sonnen, 1956, English translation 1958) showed two sides of scientists' interactions with the state in the atomic age: Oppenheimer, seduced by the prospect of power, acceded to the state's interest to build and use the bomb, but this compliance didn't do him any good when his own political reliability became suspect. The German scientists in Jungk's tale, on the other hand, stepped back from cooperation with the state, conceding power in the short run but retaining it in the long run, along with their political innocence.[58] Jungk later asserted that this account was based on (charitably) misunderstood or (less charitably) deceptive interviews with Weizsäcker and other physicists.[59] A few years later, Friedrich Dürrenmatt's play The Physicists (Die Physiker), which drew on Jungk's account as well as the Oppenheimer affair for inspiration, had the title characters withdraw into an insane asylum in order to protect the world from a terrifying discovery, only to learn that the asylum's director has stolen the secret, thus rendering moot their strategy of self-sacrificial

---

[56] Bopp et al., Erklärung.

[57] On the Göttingen Eighteen manifesto and response thereto, see Cioc, Pax, pp. 72–91; Carson, New Models, pp. 155–165; Beyler, Demon, pp. 232–237; Schirrmacher, Physik, pp. 17–38.

[58] Jungk, Heller; cf. Carson, New Models, pp. 134–135.

[59] Jungk, Trotzdem, pp. 297–299.

but individualistic refusal to participate.[60] The moral of Dürrenmatt's tale seemed to be that only self-conscious and cooperative participation among the physicist held out any hope of their being able to influence the course of world history (though even this was no guarantee) – echoes of the plans for international cooperation on and supervision over nuclear weapons proposed by Bohr and others around the end of World War II.

Claiming a special responsibility for science to contribute to the public good thus seemed necessary in order to respond to the terrific fears aroused by its power. However, it became all the more difficult to claim that this could happen in a non-ideological or non-political way, or to avoid the impression that scientists were simply being advocates for their own parochial interests.

A BRIEF INTERNATIONAL COMPARISON BY WAY OF CONCLUSION

The shadow of the Third Reich was, uncomfortably, still quite long in post-1945 Germany. This was perhaps the most salient factor distinguishing the ideology of non-ideology in German from some apparently analogous discussions about science and ideology underway elsewhere. In other Western countries, and especially in the United States, the "end of ideology" discourse arose primarily out of intramural debates among disillusioned Marxists and ex-Marxists, and methodological and theoretical debates among social scientists.

The notion of science as a model of democracy carried strong echoes to the science planning debate in Great Britain between Michael Polanyi and the "visible college" of leftist scientists such as J. D. Bernal, J. G. Crowther, P. M. S. Blackett, and J. B. S. Haldane. Polanyi promoted to the notion that "pure science" functioned best, or functioned at all, as a self-governing "republic" in which everyone was committed to the same methodology of the pursuit of truth – perhaps not entirely unproblematic as a model of democracy. In any event, despite the rather overarching rhetoric, the contents of this debate, on both sides, clearly arose out of concerns quite specific to the state of affairs in British science policy.[61]

There also seem to be parallels between the German situation and the "end of ideology" debate in the United States. This refers to a flurry of books and articles in the late 1950s and early 1960s promoting, in various ways, the idea that "ideology" per se (whatever that meant) had (or should) come to an "end." After appearance of the concept in work by Albert Camus and by Raymond Aron, Edward Shils entitled his report in Encounter on the 1955 CCF conference in Milan "The End of Ideology?" (with a question mark). [62] But probably the most prominent example of the genre was a book of essays published by Daniel Bell

---

[60] Dürrenmatt, Physiker.
[61] Representative texts include Bernal, Social Function; Polanyi, Logic; for discussion, see Werskey, Visible; Nye, Blackett.
[62] Camus, Neither Victims; Aron, Opium, pp. 305–324; Shils, End.

in 1960 under the title The End of Ideology, without a question mark.[63] Various other critics, political scientists, sociologists, and philosophers chimed in pro and con, resulting in a rather extensive literature.[64] The American "end of ideology" debate was primarily an intramural methodological disagreement among American social scientists and, simultaneously, a running battle between leftists and former leftists who were on their way from various forms of anti-Stalinist Marxism to various forms of neoconservatism. Proponents of the "end of ideology" were often accused, with some justification, of being technocrats, but the technical expertise in question was largely social-scientific or managerial in nature: there was relatively mention of the changing character of the natural sciences or of technology in the more literal meaning of that word. The American "end of ideology" discourse thus represented its participants' attempts to re-assess a now suspect past, but it was the authors' flirtation with various Marxist heterodoxies, rather than the problematic heritage of Nazism.

In Germany, there was a related but distinct agenda: advocates of the "end of ideology" proclaimed it in the first instance as defense against, or in some cases a shelter against the National Socialist past. The bracketing off of the Eastern bloc was, so to speak, a bonus, with the added benefit of tighter integration into the north Atlantic alliance. In this specifically German context, natural science served for its spokesmen as a model of freedom of thought and political innocence; for critics of science, the converse side of these same virtues was the danger of materialism, hyper-specialization, and thoughtless conformity. In short, it appears that even the ostensible end of ideology, which would seem to be by definition a nationally and culturally transcendent phenomenon, took on a nationally and culturally dependent coloration.

BIBLIOGRAPHY

Aron, Raymond: The Opium of the Intellectuals, trans. Terence Kilmartin, Westport, Conn. 1977 (French orig. 1955).

Ash, Mitchell G.: Verordnete Umbrüche – konstruierte Kontinuitäten: Zur Entnazifizierung von Wissenschaftlern und Wissenschaften nach 1945, in: Zeitschrift für Geschichtswissenschaft 43 (10), 1995, pp. 903–923.

Ash, Mitchell G.: Denazifying Science – and Scientists, in: Judt, Matthias and Burghard Ciesla (eds.): Technology Transfer Out of Germany After 1945, Emmaplein 1996, pp. 61–80.

Bell, Daniel: The End of Ideology. On the Exhaustion of Political Ideas in the Fifties, Glencoe, Ill. 1960.

Bernal, J. D.: The Social Function of Science, London 1939.

Beyerchen, Alan D.: German Scientists and Research Institutes in Allied Occupation Policy, in: History of Education Quarterly 22 (3), 1982, pp. 289–299.

[63] Bell, End of Ideology.

[64] Examples include Lipset, Political Man; Waxman, End of Ideology; Rejai, Decline; for discussion, see Dittberner, End of Ideology; Brick, Daniel Bell; Dorrien, Neoconservative Mind; Scott-Smith, Politics, pp. 138–159.

Beyler, Richard H.: The Demon of Technology, Mass Society, and Atomic Physics in West Germany, 1945–1957, in: History and Technology 19 (3), 2003, pp. 227–239.

Beyler, Richard H.: Hostile Environmental Intellectuals? Critiques and Counter-Critiques of Science and Technology in West Germany after 1945, in: Berichte zur Wissenschaftsgeschichte 31 (4), 2008, pp. 393–406.

Beyler, Richard H.: "Reine" Wissenschaft und personelle "Säuberungen". Die Kaiser-Wilhelm-/Max-Planck-Gesellschaft 1933 und 1945 (Forschungsprogramm "Geschichte der Kaiser-Wilhelm-Gesellschaft im Nationalsozialismus," Ergebnisse 16), Berlin 2004.

Beyler, Richard H. and Morris F. Low: Science Policy in Post-1945 West Germany and Japan. Between Ideology and Economics, in: Walker, Mark (ed.): Science and Ideology. A Comparative History, London, New York 2003, pp. 97–123.

Bopp, Fritz et al.: Die Göttinger Erklärung, in: Physikalische Blätter 13 (5), 1957, pp. 193–194.

Bracher, Karl Dietrich: Der parlamentarische Parteienstaat zwischen Bewahrung und Anfechtung, in: Scheel, Walter (ed.): Nach dreissig Jahren. Die Bundesrepublik Deutschland. Vergangenheit, Gegenwart, Zukunft, Stuttgart 1979, pp. 29–46.

Brick, Howard: Daniel Bell and the Decline of Intellectual Radicalism. Social Theory and Political Reconciliation in the 1940s, Madison 1986.

Camus, Albert: Neither Victims Nor Executioners, trans. Dwight MacDonald, New York 1960 (French orig. 1946).

Carson, Cathryn: New Models for Science in Politics. Heisenberg in West Germany, in: Historical Studies in the Physical and Biological Sciences 30 (1), 1999, pp. 115–171.

Carson, Cathryn: Nuclear Energy Development in Postwar West Germany. Struggles over Cooperation in the Federal Republic's First Reactor Station, in: History and Technology 18 (3), 2002, pp. 233–270.

Carson, Cathryn and Michael Gubser: Science Advising and Science Policy in Post-War West Germany. The Example of the Deutscher Forschungsrat, in: Minerva 40 (2), 2002, pp. 147–179.

Cassidy, David: Controlling German Science, in: Historical Studies in the Physical and Biological Sciences 24 (2), 1994, pp. 197–235, and 26 (2), 1996, pp. 197–239.

Cioc, Mark: Pax Atomica. The Nuclear Defense Debate in West Germany During the Adenauer Era, New York 1988.

Coleman, Peter: The Liberal Conspiracy. The Congress for Cultural Freedom and the Struggle for the Mind of Postwar Europe, New York 1989.

Congress for Cultural Freedom: Science and Freedom, Boston, London 1955.

Dittberner, Job L.: The End of Ideology and American Social Thought, 1930–1960 (= Studies in American History and Culture 1), Ann Arbor 1979.

Dorrien, Gary: The Neoconservative Mind. Politics, Culture, and the War of Ideology, Philadelphia 1993.

Dürrenmatt, Friedrich: Die Physiker. Eine Komödie in zwei Akten, Zürich 1962.

Eckert, Michael: Primacy Doomed to Failure. Heisenberg's Role as Scientific Adviser for Nuclear Policy in the FRG, in: Historical Studies in the Physical and Biological Sciences 21 (1), 1990, pp. 29–58.

Eckert, Michael and Maria Osietzki: Wissenschaft für Macht und Markt. Kernforschung und Mikroelektronik in der Bundesrepublik Deutschland, München 1989.

Gimbel, John: U.S. Policy and German Scientists. The Early Cold War, in: Political Science Quarterly 101 (3), 1986, pp. 433–451.

Glaser, Hermann: Die Kulturgeschichte der Bundesrepublik Deutschland, 3 vols., Frankfurt am Main 1990.

Heidegger, Martin: Die Technik und die Kehre, Pfullingen 1962.

Heinemann, Manfred (ed.): Hochschuloffiziere und Wiederaufbau des Hochschulwesens in Westdeutschland 1945–1952, 3 vols., Hildesheim 1990–1991.

Heinemann, Manfred (ed.): Nordwestdeutsche Hochschulkonferenzen 1945–1948 (Geschichte von Bildung und Wissenschaft, ser. C., 1), Hildesheim 1990.

Hentschel, Klaus: The Mental Aftermath: The Mentality of German Physicists 1945–1949, Oxford 2007.

Hentschel, Klaus and Gerhard Rammer: Physicists at the University of Göttingen, 1945–1955, in: Physics in Perspective 3 (2), 2001, pp. 189–209.

Herf, Jeffrey: Belated Pessimism. Technology and Twentieth Century German Conservative Intellectuals, in: Ezrahi, Yaron, Everett Mendelsohn, and Howard P. Segal (eds.): Technology, Pessimism, and Postmodernism, Amherst 1994, pp. 115–136.

Hess, Gerhard: Gesellschaft, Literatur, Wissenschaft. Gesammelte Schriften 1938–1966, ed. by Hans Robert Jauss and Claus Müller-Daehn, München 1967.

Hoffmann, Dieter: Between Autonomy and Accomodation. The German Physical Society during the Third Reich, in: Physics in Perspective 7 (3), 2005, pp. 293–329.

Hoffmann, Dieter: Carl Ramsauer, die Deutsche Physikalische Gesellschaft und die Selbstmobilisierung der Physikerschaft im "Dritten Reich", in: Maier, Helmut (ed.): Rüstungsforschung im Nationalsozialismus. Organisation, Mobilisierung und Entgrenzung der Technikwissenschaften, Göttingen 2002, pp. 273–304.

Hoffmann, Dieter and Mark Walker (eds.): Die Deutsche Physikalische Gesellschaft im Dritten Reich, Weinheim 2007.

Huber, Ernst Rudolf: Selbstverwaltung der Wirtschaft, Stuttgart 1958.

Jordan, Pascual: Der gescheiterte Aufstand. Betrachtungen zur Gegenwart, Frankfurt am Main 1956.

Jünger, Friedrich Georg: Die Perfektion der Technik, Frankfurt am Main 1946.

Jungk, Robert: Heller als tausend Sonnen, Bern, Stuttgart 1956.

Jungk, Robert: Trotzdem. Mein Leben für die Zukunft, München 1993.

Kongress für kulturelle Freiheit: Wissenschaft und Freiheit. Internationale Tagung Hamburg, 23.–26. Juli 1953, Berlin 1954.

Lasky, Melvin (ed.): Der Kongress für kulturelle Freiheit in Berlin, in: Der Monat 2 (22–23), Jul.–Aug. 1950, pp. 381–487.

Laue, Max v.: Bemerkung zu der vorstehenden Veröffentlichung von J. Stark, in: Physikalische Blätter 3 (8), 1947, pp. 272–273.

Laurien, Ingrid: Politisch-kulturelle Zeitschriften in den Westzonen 1945–1949. Ein Beitrag zur politischen Kultur der Nachkriegszeit, Frankfurt am Main, New York 1991.

Lipset, Seymour M.: Political Man. The Social Basis of Politics, New York 1960.

Litten, Freddy: Mechanik und Antisemitismus. Wilhelm Müller (1880–1968), München 2000.

Mehrtens, Herbert: Kollaborationsverhältnisse: Natur- und Technikwissenschaften im NS-Staat und ihre Historie, in: Meinel, Christoph and Peter Voswinckel (eds.): Medizin, Naturwissenschaft, Technik und Nationalsozialismus. Kontinuitäten und Diskontinuitäten, Stuttgart 1994, pp. 13–32.

Mehrtens, Herbert: "Missbrauch". Die rhetorische Konstruktion der Technik in Deutschland nach 1945, in: Kertz, Walter (ed.): Hochschule und Nationalsozialismus, Braunschweig 1994, pp. 33–50.

Meinecke, Friedrich: Die deutsche Katastrophe: Betrachtungen und Erinnerungen, Wiesbaden 1946.

Mitchell, Maria: Materialism and Secularism: CDU Politicians and National Socialism, 1945–1949, in: Journal of Modern History 67 (2), 1995, pp. 278–308.

Muller, Jerry Z.: How Vital Was the Geist in Heidelberg in 1945? Some Skeptical Reflections, in: Hess, Jürgen C. et al. (eds.): Heidelberg 1945 (Transatlantische historische Studien 5), Stuttgart 1996, pp. 197–200.

Nipperdey, Thomas and Ludwig Schmugge: 50 Jahre Forschungsförderung in Deutschland. Ein Abriß der Geschichte der Deutschen Forschungsgemeinschaft 1920–1970, Berlin 1970.

Nitsch, Wolfgang et al.: Hochschule in der Demokratie, Berlin 1965, repr. New York 1977.

Oexle, Otto Gerhard: Wie in Göttingen die Max-Planck-Gesellschaft entstand, in: Max-Planck-Gesellschaft Jahrbuch, 1994, pp. 43–60.

Osietzki, Maria: Reform oder Modernisierung – Impulse zu neuartigen Organisationsstrukturen der Wissenschaft nach 1945, in: Fischer, Wolfram et al. (eds.): Exodus von Wissenschaften aus Berlin. Fragestellung – Ergebnisse – Desiderate (Akademie der Wissenschaften zu Berlin Forschungsbericht 7), Berlin, New York 1994, pp. 284–295.

Osietzki, Maria: Wissenschaftsorganisation und Restauration. Der Aufbau außeruniversitärer Forschungseinrichtungen und die Gründung des westdeutschen Staates 1945–1952, Köln 1984.

Planck, Max: Mein Besuch bei Adolf Hitler, in: Physikalische Blätter 3 (5), 1947, p. 143.

Phillips, David: Britische Initiative zur Hochschulreform in Deutschland. Zur Vorgeschichte und Entstehung des "Gutachtens zur Hochschulreform" von 1948, in: Heinemann, Manfred (ed.): Umerziehung und Wiederaufbau: Die Bildungspolitik der Besatzungsmächte in Deutschland und Österreich (Veröffentlichungen der Historischen Kommission der Deutschen Gesellschaft für Erziehungswissenschaft 5), Stuttgart 1981, pp. 172–189.

Polanyi, Michael: The Logic of Liberty. Reflections and Rejoinders, London, Chicago 1951.

Raiser, Ludwig: Falscher Föderalismus, in: Deutsche Universitätszeitung 9 (17), 6 Sep. 1954, pp. 3–5.

Ramsauer, C[arl].: Zur Geschichte der Deutschen Physikalischen Gesellschaft in der Hitlerzeit, in: Physikalische Blätter 3 (4), 1947, pp. 110–114.

Regener, Erich: Mitverantwortlichkeit der wissenschaftlich Tätigen, in: Physikalische Blätter 3 (6), 1947, pp. 169–170.

Rejai, Mostafa (ed.): Decline of Ideology?, New York 1971.

Sachse, Carola: "Persilscheinkultur". Zum Umgang mit der NS-Vergangenheit in der Kaiser-Wilhelm-Gesellschaft, in: Weidbrod, Bernd (ed.): Akademische Vergangenheitspolitik. Beiträge zur Wissenschaftskultur der Nachkriegszeit, Göttingen 2002, pp. 217–246.

Saunders, Frances Stonor: The Cultural Cold War. The CIA and the World of Arts and Letters, New York 1999.

Schirrmacher, Arne: Physik und Politik in der frühen Bundesrepublik. Max Born, Werner Heisenberg and Pascual Jordan als politische Grenzgänger (Max-Planck-Institut für Wissenschaftsgeschichte, Preprint 296), in: Berichte zur Wissenschaftsgeschichte 30 (1), 2007, pp. 13–31.

Schmithals, F.: Carl Ramsauer und das Dritte Reich, in: Physikalische Blätter 36 (11), 1980, p. 345.

Schneider, Ullrich: Zur Entnazifizierung der Hochschullehrer in Niedersachsen 1945–1949, in: Niedersächsisches Jahrbuch für Landesgeschichte 56, 1989, pp. 325–346.

Schüring, Michael: Minervas verstoßene Kinder. Vertriebene Wissenschaftler und die Vergangenheitspolitik der Kaiser-Wilhelm-Gesellschaft (= Geschichte der Kaiser-Wilhelm-Gesellschaft im Nationalsozialismus 13), Göttingen 2006.

Scott-Smith, Giles: The Politics of Apolitical Culture. The Congress for Cultural Freedom, the CIA and Post-War American Hegemony, London, New York 2002.

Shils, Edward: The End of Ideology?, in: Encounter 5, Nov. 1955, pp. 52–58.

Stamm, Thomas: New Opportunities – Old Traditions. The Struggle for Autonomy of Post-War German Science, in: Michelangelo De Maria, Mario Grilli, and Fabio Sebastiani (eds.): The Restructuring of Physical Sciences in Europe and the United States 1945–1960, Singapore et al. 1989, pp. 228–246.

Stamm, Thomas: Zwischen Staat und Selbstverwaltung: Die deutsche Forschung im Wiederaufbau 1945–1965, Köln 1981.

Stucke, Andreas: Institutionalisierung der Forschungspolitik. Entstehung, Entwicklung und Steuerungsprobleme des Bundesforschungsministeriums, Frankfurt am Main 1993.

Studienausschuss für Hochschulreform: Gutachten zur Hochschulreform, Hamburg 1948.

Trischler, Helmuth: Luft- und Raumfahrtforschung in Deutschland 1900–1970. Politische Geschichte einer Wissenschaft, Frankfurt am Main, New York 1992.

Trischler, Helmuth and Rüdiger vom Bruch: Forschung für den Markt. Geschichte der Fraunhofer-Gesellschaft, München 1989.

Vierhaus, Rudolf, and Bernhard vom Brocke (eds.): Forschung im Spannungsfeld von Politik und Gesellschaft. Geschichte und Struktur der Kaiser-Wilhelm-/Max-Planck-Gesellschaft, Stuttgart 1990.

Walker, Mark: Nazi Science. Myth, Truth, and the German Atomic Bomb, New York 1995.

Waxman, Chaim (ed.): The End of Ideology Debate, New York 1968.

Weizsäcker, Carl F. v.: Die Geschichte der Natur, Stuttgart 1948.

Wenzl, Aloys: Die geistigen Strömungen unseres Jahrhunderts, München 1948.

Werskey, Gary: The Visible College. The Collective Biography of British Scientific Socialists of the 1930s, New York 1978.

Zeitz, Katharina: Max von Laue (1879–1960). Seine Bedeutung für den Wiederaufbau der deutschen Wissenschaft nach dem Zweiten Weltkrieg, Stuttgart 2004.

Zierold, Kurt: Die Deutsche Forschungsgemeinschaft als Selbstverwaltungsorganisation der Wissenschaft, in: Die öffentliche Verwaltung 13 (13), 1960, pp. 481–485.

Zierold, Kurt: Forschungsförderung in drei Epochen. Deutsche Forschungsgemeinschaft, Geschichte, Arbeitsweise, Kommentar, Wiesbaden 1968.

Zierold, Kurt: Selbstverwaltungsorganisationen der Wissenschaft in der Bundesrepublik, in: Die öffentliche Verwaltung 14 (17–18), 1961, pp. 686–695.

Zierold, Kurt: Wie soll man die Wissenschaft verwalten?, in: Neue Sammlung 4, 1964, pp. 176–182.

Zimmermann, Michael E.: Heidegger's Confrontation with Modernity. Technology, Politics, and Art, Bloomington 1999.

BEYOND RECONSTRUCTION: CERN'S SECOND-GENERATION ACCELERATOR PROGRAM AS AN INDICATOR OF SHIFTS IN WEST GERMAN SCIENCE

Cathryn Carson

In late 1959, the European Center for Nuclear Research (CERN) in Geneva marked a milestone when its 28-GeV (billion electron volt) proton synchrotron went into action. Built by physicists and their governments working together, this particle accelerator concretely realized the postwar period's enthusiasm for European collaboration and nuclear science. The era of postwar reconstruction, its efforts coming to fruition around 1960, has been the subject of much historical attenion. Indeed, when the PS began operating, it opened up unmatched experimental possibilities. At the same moment, it opened up an as yet indeterminate future. And shaping this *post*-postwar future would be no simple task.

Looking at the decade-long process (1960–1970) by which the Federal Republic committed itself to a second round of CERN accelerator construction, this paper aims to recount a controverted story in a nuanced fashion, using its details over a decade to give insight into larger forces at work. The West German decision to participate in CERN's second-generation program was a drawn-out episode of wrangling, politicking, and personal conflict. Given the stories still told about it by veteran physicists with partisan gusto, this messy history needs clarification on its own, relatively confined terms. At the same time, the episode can be used to illuminate something larger and clearer. For the second-generation commitment to CERN, this paper argues, marked a culmination of the same postwar system it ultimately helped usher out. It not only secured the Federal Republic's role in CERN, that model European scientific organization, but also underwrote domestic expansion of research and training in high-energy physics. Paired with the construction of DESY, the Deutsches Elektronen-Synchrotron in Hamburg, it signaled West Germany's recovery of competitive status on the European high-energy physics scene. And yet, though the outcome satisfied all parties, the struggles leading to it were intense – precisely because much was at stake in an emerging new order.

The debate over CERN's post-PS program in fact delineates the trajectory of West German science out of the reconstruction period. To replace the postwar science policy and funding regime, which ruled the 1950s, new procedures were instituted in the decade that followed that were more deliberate, bureaucratic, and financially cautious. From this story we get a different picture of the 1960s than either high-level debates about funding and policy competence, or else the new institutions of "big science" (Großforschung). The transition to the post-

postwar was visible across science as a whole, but it set in more sharply and early in high-energy and particle physics, for two reasons. First, no other branch of basic science was remotely as expensive; and the skyrocketing costs of research were, this paper suggests, a primary driver in instituting the new mode of policymaking. Second, the decisionmaking structures in high-energy physics were the very exemplar of the postwar regime. The field practically had its own ministry, the Atomic Ministry (the origin of the Research Ministry), with a critically important advisory body, the Deutsche Atomkommission's Arbeitskreis Kernphysik (Nuclear Physics Commission), initially set up by the Deutsche Forschungsgemeinschaft in order to disburse federal monies for that 1950s Leitwissenschaft, nuclear physics. Until the early 1960s, the Bund's funding decisions lay almost entirely in the hands of this tiny group of researchers, who were simultaneously the beneficiaries of the policies they helped decide. In real life, the actors who initially set the terms for the second-generation CERN debate in West Germany were the dozen or so self-assured institute heads who made postwar careers building machines. The personalized character of science funding was entirely characteristic of the postwar order. Though it would not entirely vanish – remnants of it are still visible – it would never be so strong again as it was after the war.

That has had consequences, especially, for how scientists have retold the CERN conflict. Signally, the conventional account of the debate pits a rising younger group led by the accelerator-builder Wolfgang Gentner (1906–1980), against the commission's chair, Werner Heisenberg (1901–1976), its éminence grise. Either told as a generational conflict or a theorist-experimentalist battle, the personalized narrative makes for a good story. But while this head-butting picture captures some of the dynamics quite well, it collapses the participants' complications. More importantly, it occludes larger forces at work, which were gradually pushing in the direction of de-personalization. What happened in the 1960s was not really a victory of Gentner over Heisenberg. In fact, it was much less a personal victory of any sort than the scientists involved typically took it to be. Nor did the outcome reinforce the special position that Gentner took for granted, and that Heisenberg was beginning to question – the special position of nuclear and particle physics within physics as a whole, or physics among other branches of science. Rather, as the reconstruction period came to an end, that special position was beginning to erode. Physics was coming to seem a little more like one field among others, with interests it had to find new ways to defend.

THE FUNDING SYSTEM: SHIFTS ON THE HORIZON

By end of the 1950s, in nuclear and particle physics a modern research and funding system was largely in place. For historical reasons relating to Allied concerns and Konrad Adenauer's interest in nuclear energy, the system had crystallized around nuclear technology. However, its financial gestures towards pure science were generous. In its procedures it also showed respect for the principle of free space for research to follow its own lead. The postwar settlement gave immense power to elite scientists to set the agenda in self-governing institutions like the revived Deutsche Forschungsgemeinschaft and Max-Planck-Gesellschaft. It also created a climate in which the Wissenschaftsrat, in existence since 1957 and coming into its own in the early 1960s, could powerfully advocate for massive spending increases for education and science, meeting with broad parliamentary, party, and public support that sometimes even exceeded its recommendations.[1]

In the space of a decade, counting from 1962, science would nearly double as a fraction of government spending. Even if part of the reason was that it started out low, public funding was starting on a track of impressive expansion. [Fig. 1] Significantly, this growth occurred at a time when government spending at all levels was mounting in its own right. Yet it was exactly because science was securing attention and resources that the postwar settlement was opened up for debate. For once science had entered into partnership with the state, that relationship would alter science's own dynamic. A source of good things, science was also expensive, abstruse, and hard to direct. If the government funded it, how could the government exercise oversight?

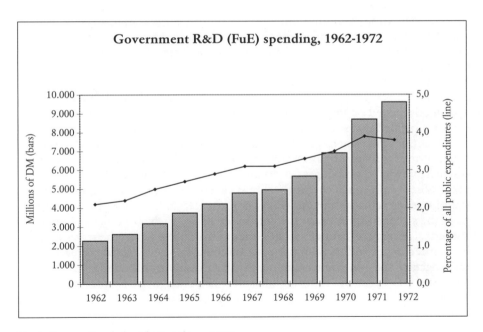

Fig. 1. Source: Bundesbericht Forschung 2000.

[1] For an introduction see Bartz, Wissenschaftsrat; Eckert/Osietzki, Wissenschaft; Fischer, Atomenergie; Hohn/Schimank, Konflikte; Stamm, Zwischen Staat; Stucke, Institutionalisierung, Weingart/Taubert, Wissensministerium.

The financial forces shaping the funding system in the 1960s are complicated enough to merit a few words of their own.[2] Research funding fit into the Federal Republic's uneasy, unsteady, but seemingly unstoppable shift toward a more expansive construal of the role of the state. Yet counterintuitively, because of West Germany's federal structure and strict fiscal policy, the Bund's mounting expenditures were accompanied by ever sharper demands for budgetary restraint. This combination played out in grand public gestures of Erhardian appeal, but equally in small-scale but potentially draconian negotiations about individual budget items. And to scientists' dismay, despite a broad political consensus in favor of science, federal research funding frequently ended up as a target.

Research was an easy target precisely because its year-to-year totals were growing so fast. However much politicians favored spending money on science, few or none of those financial commitments were secured by legislation. And when demands for belt-tightening came, they really hit the fifteen percent of the budgetary volume in that insecure position. With budgeting still done on a year-to-year basis, the government's initial tools were to defer obligations and impose a freeze (Mittelsperre) that often turned into a cut. Of course, pushing expenditures into the next fiscal year did not solve the problem, while across-the-board reductions (the dreaded Globalkürzung) undermined ministerial planning. Unless science were exempted, it could be hit hard exactly because it had begun planning for expansion.[3] Moreover, some independent-minded members of the Bundestag's budget committee, spearheaded by the young finance expert Gerhard Stoltenberg, were determined to impose fiscal discipline and scrutinize the government's proposals for science. In committee, global cuts in nonobligated programs were actually deepened by targeted reductions in certain scientific budget lines. Here fiscal problems became a wedge for delicate questions about control.[4]

Scientific spokesmen were not pleased. In the early 1960s the economy was still growing, perhaps not as impressively as before, but soundly. Revenues were still increasing, and yet budget cuts loomed. But if one thing was constant through the 1960s, even through the first real postwar recession of 1966–67, public monies for science actually kept growing. They just were not growing fast enough to meet the scientists' proposals: the cuts were relative to extraordinarily expansive requests. Clearly, the attitude towards science within the government was not fully captured by a climbing number of Deutschmarks, nor by the repeated flurries of budgetary cuts. As science entrenched itself in the regular budget, it became less a special or separate thing. While still small, it might be spared means-ends scrutiny. As it grew it became more commensurable with ordinary political concerns.

---

[2] Lessman, Politics; Adami, Haushaltspolitik. Carson, Heisenberg, ch. 10, gives a fuller account relative to science.

[3] Haushaltsreden, pp. 187 n. 7, 165, 257.

[4] Starting with Drucksache IV/325, Mündlicher Bericht des Haushaltsausschusses zum [Haushaltsgesetz 1962], hier: Einzelplan 31, 30.3.1962, in: Verhandlungen, 4. Wahlperiode. Cf. Boenke, Entstehung, pp. 208–215.

ACCELERATOR PLANNING: NEW CHALLENGES

Well into the 1960s, while budgetary turmoil preoccupied the scientific community's leaders, working researchers were still largely sheltered. In a few exposed domains, however, the nature of the emerging regime of governance would become plainer, earlier, than elsewhere. Along with sensitive fields such as aerospace, nuclear and particle physics would feel the effects: no other field of basic research had so visibly profited from the old order or made remotely comparable claims. Building particle accelerators, the tools of the trade, made resource demands in excess of anything previously imagined; and once they were built, running and administering them created other sorts of strains. The 1960s were the years when German physicists really learned to work with their machines: to work with them technically, but also organizationally and politically. They ended up pushing the boundaries of policy notions and ushering in new ways of dealing with science, mostly against their own wishes.

To understand the physicists' situation, we need to recover a sense of those "golden years" of the mid- to late 1950s. Through that period the task of the atomic ministry's Arbeitskreis Kernphysik had been fairly easy. Its initial membership had been selected by assembling everyone presumed to have a say in nuclear and particle physics. The government had proved willing to follow its advice in funding new construction (and personnel and equipment) on a scale previously unseen. Most of the money naturally went to experimental research: fusion and plasma physics, nuclear structure and forces, CERN in Geneva, larger and smaller accelerators at home. The commission had elected Werner Heisenberg as its chair; he was known as a good discussion leader, a consensus-builder rather than a table-pounder, and a man whose name recognition and facility in the give-and-take of negotiating gave him some access to political actors. The latter attributes had raised him into the main scientific statesman of West German physics, with greatest success perhaps in relation to CERN. He was also a theorist, though that did not mean he had no stake in the commission's decisions. The Max Planck Institute for Physics, which he headed, ran experimental programs in cosmic rays, detector construction for accelerators, data analysis, plasma physics, and space research. Heisenberg also served as the Arbeitskreis's unofficial liaison to the Max Planck Gesellschaft, where he was one of the powers behind the throne.[5]

Fiscal abundance and start-up enthusiasm meant the commission's work in the 1950s was easy. In the following decade everyone's job grew less pleasant. A major building program was underway, one almost without parallel in modern German science, and its complications became clearer only once it was underway. Physicists were irked by emergent operational kinks, like troubles following up one-time funding for construction with long-term funding for staff. A requirement that spending in key federal budget lines be matched by Länder monies left scientists frustrated by financial constraints, to say nothing of administrative

---

[5] Carson, Wissenschaftsorganisator, and references therein; see also Hermann, Germany's; Eckert, Primacy; Carson, Energy.

burdens.[6] Above all, they faced trouble from those watching over federal spending, namely, the finance ministry and the Bundestag budget committee. The watershed was their experience in early 1962, when the all-important Budget Line Item 950, for modernization and expansion of nuclear research institutes, was threatened with sharp cuts after institute heads had already begun obligating the money. That caught their attention as each scrambled to cope.[7]

Along with complaints to the ministers and short-term attempts to fill gaps, the physicists and their advocates in the ministry began to look for a better way to do business. Ministry staffers drew the scientists' attention to the political uses of proactive planning. "In nuclear technology," one noted, "the Deutsche Atomprogramm has done good service for years in getting claims past the government departments and the Bundestag." To do their work, such documents had to be, the same staffer noted, "as impressive and concrete as possible."[8] So as a result of the 1962 crisis, the physicists began planning. And with the express approval of the ministry, they, not the staffers, took the lead. The plans that emerged in the early 1960s had few roots in an ideology of preprogrammed technical progress. Rather, in the research field they took their origins in part from the state's fiscal predicament, following the same logic as the government's venture into multiyear budgeting (mittelfristige Finanzplanung).[9]

An initial draft of a plan, quickly drawn up that fall, was followed by formation of committees for specific fields of research. These committees sent out questionnaires to ascertain what ventures were proposed, assembling the answers into lists.[10] The results were folded into the next Atomprogramm, which, while not binding, was meant to serve as a touchstone. Even if these tools were not instantly effective in stabilizing long-term funding, the scientists swiftly made the procedures and instruments of planning their own. And perhaps unsurprisingly, the result of the exercise was less a plan in any strict sense than a "catalog of current research themes," an "inventory," an "overview" – plus the conclusion that "the need for accelerators can hardly be seen as nearly satisfied."[11]

---

[6] [Heisenberg,] Stand der Tätigkeit im Arbeitskreis II/3, [10–11.1963], Nachlaß Werner Heisenberg, Max-Planck-Institut für Physik und Astrophysik, München (WHM) under BMAt AKKP II/3c. AKKP records are in many participants' papers, though not in the Bundesarchiv Koblenz (BAK), B 138, until 1966.

[7] Balke, Zur Frage, p. 3; minutes of AKKP, 18.6.1962, WHM under BMAt AKKP II/3c; Heisenberg to Balke, 27.2.1962, WHM under BMAt Bundesminister – Schriftwechsel a); Heisenberg to Starke, 22.6.1962, WHM under Ministerien, Bundes- (verschied.) auch Bundeskanzler.

[8] Hesse to Heisenberg, 14.6.1962, WHM under BMAt AKKP II/3c. Cf. Braun, Steuerung, p. 59.

[9] If sometimes more for the phenomenon than the argument, Trischler, Planungseuphorie; Ruck, Sommer; Metzler, Konzeptionen; also Frese/Paulus/Teppe, Demokratisierung, sect. III.

[10] Minutes of AKKP, 26.9.1962, WHM under BMAt AKKP II/3c; Gentner, Haxel, and Maier-Leibnitz, Merkpunkte für eine gezielte Grundlagenforschung, 20.10.1962, under BMAt Bundesminister – Schriftwechsel b); minutes of AKKP, 26.11.1962, WHM under BMAt AKKP II/3d.

[11] Minutes of AKKP, 26.9.1962 (Katalog der aktuellen Forschungsthemen); 18.6.1962 (Bestandsaufnahme); minutes of HEP committee, 10.12.1962 (Überblick); minutes of AKKP, 4.2.1963 (need for more accelerators), WHM under BMAt AKKP II/3c and II/3d.

It could hardly have been otherwise in a system of elite self-governance of personal fiefdoms, which is what the system of funding without constraints had become. The listing exercise forced the Arbeitskreis to confront for the first time the welter of projects it proposed to support. But order or overview was hard to achieve by this opportunistic approach – especially in a setting that one participant characterized dead-on as a "conflict of interested parties."[12] In the background, finally, lurked the concern that the mess of individual proposals would cut the ground out from underneath the biggest going concern, namely DESY. That high-energy electron accelerator (6.5 GeV), the payoff from commitments from the mid-1950s onward, would go on line in 1964.[13] DESY needed to show results quickly and prove its value before competition clogged the field. And like nearly all projects in this novel domain, its cost estimates were proving far too low.[14]

THINKING BIG: INTERNATIONAL, NATIONAL, INSTITUTIONAL, PERSONAL

Where foresight became urgent was in preparations for the next generation of Europe's high-energy machines. As mentioned above, these officially came on the agenda after CERN's 28-GeV accelerator went into operation in 1959. Co-ordinated by the scientists of the European Committee for Future Accelerators (ECFA), the planning process initially was led by scientific and machine-building interests.[15] In the heady spirit of expansion so typical of the dawning 1960s, the ECFA plan included not one, but two major international projects: a colliding-beam machine with intersecting storage rings (ISR) at the existing Geneva laboratory, plus a huge new proton synchrotron reaching 300 GeV at another site yet to be named. These two accelerators did not stand on their own, either. Rather, they were meant to be the apex of a European pyramid of (comparatively) smaller new national machines; for making fullest use of European offerings depended on training and experience at home.

ECFA was the projection into the indefinite future of the gains and the rules of the postwar years. Yet at the beginning, not all West German particle physicists were clearly behind CERN's expansion. Questions regarding its wisdom were

[12] Schmidt-Rohr, Teilchenbeschleuniger; cf. Müller, Geschichte, v. 1, app. 8; Bundesbericht Forschung II, pp. 237–238; Memorandum des Arbeitskreises Kernphysik zur Errichtung weiterer Niederenergiebeschleuniger, [May 1965], WHM under BMAt AKKP II/3d. On the Streit der Interessenten in the HEP committee, Gentner to Paul, 2.3.1963, Dienstzimmernachlaß Wolfgang Gentner, Archiv zur Geschichte der Max-Planck-Gesellschaft, Berlin (MPGB), III/68A (WGB), 20.

[13] Minutes of AKKP, 1.3.1958, 2.12.1962, WHM under BMAt AKKP II/3a and II/3d; minutes of MPG Verwaltungsrat, 3.12.1964, MPGB, II/1A.

[14] Habfast, Großforschung, pp. 60–66. This came at exactly the wrong moment: minutes of AKKP, 26.2.1962, WHM under BMAt AKKP II/3c; Heisenberg to Gambke, 22.6.1962, WHM under Stiftung Volkswagenwerk. Cf. Krige, Finance Policy, pp. 574–576.

[15] Pestre, Generation.

expressed early on by some of the Federal Republic's leading institute-builders, including Wolfgang Paul (Bonn), Willibald Jentschke (DESY), and Otto Haxel (Heidelberg), as well as Heisenberg, the Arbeitskreis chair.[16] However, as CERN plans were more explicitly built into the European pyramid, the appeal of the overall scheme grew. Its demand for domestic training and education of particle physicists meshed nicely with the accelerator princes' desires to justify machines of their own. However, preliminary responses from member states were called in for mid-fall 1963, just when the uncertainty of West German budgetary planning was reaching its high point.[17]

From the start, CERN had been a school for West Germany's accelerator designers, builders, and users in an era when the Federal Republic was still in reconstruction, digging its way out from Allied restrictions on research. As ambitious physicists looked to Geneva, or in some cases returned home from there, they tended to take CERN-style budgets and programs for granted.[18] This younger generation – most of them in their late thirties in 1960, like Herwig Schopper (b. 1924) and Anselm Citron (b. 1923) – found an advocate in Wolfgang Gentner, an outstanding experimentalist somewhat older than they. Starting after the war as professor at Freiburg, in 1955 Gentner had taken on the directorship of the first accelerator group at CERN, the smaller synchrocyclotron. In 1958 he returned to take the reins of the Max Planck Institute for Nuclear Physics, newly created from Walther Bothe's former division in the MPI for Medical Research.[19] And Gentner knew how to build an institute. Coming into the MPG position just as funding took off, he always pushed the envelope of what was doable, financially and politically as well. His plans for the new MPI started out from two low-energy accelerators (12-MeV deuterons in the inherited cyclotron, plus a new 12-MeV tandem van de Graaff) and covered a range of topics in cutting-edge nuclear physics.[20] For the next half-decade, the new director would put through a rapid succession of proposals for expansion. Even as colleagues and administrators passed on the requests, they noted that sketchy planning made his proposals hard to evaluate. The costs increased with each change and also with time, at a moment when administrators were growing more sensitive.[21]

[16] Minutes of AKKP, 15.6.1961, WHM under BMAt AKKP II/3c.

[17] Heisenberg to Lenz, 10.9.1963, WHM under BMAt Bundesminister – Schriftwechsel b). On the national-international dynamic see also Hars, Wenn Forschung.

[18] Heisenberg to Walcher, 20.12.1963; emphatically agreeing, Walcher to Heisenberg, 12.2.1964, WHM.

[19] Trischler, Wolfgang Gentner; Rechenberg, Gentner; Schmidt-Rohr, Teilchenbeschleuniger.

[20] Draft minutes of commissions of CPT Sektion, 8.5.1960, WHM under MPG CPT/WR 1959–1963; on the MPI's research see Schmidt-Rohr, Erinnerungen. The early version of the research program can be read from Gentner to Hahn, 20.5.1959, WGB, 139. There is no mention yet of a synchrocyclotron.

[21] Minutes of AKKP throughout this period, E.g., 17.12.1959, 6.2.1960, 26.2.1962, WHM under BMAt AKKP II/3b and II/3c; Ballreich to Gentner, 27.2.1964, and Gentner to Ballreich, 25.4.1964, WGB, 140; minutes of Verwaltungsrat, 16.4.1962 and 11.3.1965, MPGB, II/1A. For Gentner's tendency to start construction before getting information and permits lined up, Schmidt-Rohr, Erinnerungen, pp. 118, 136; cf. Ballreich to Gentner, 30.5.1961, WGB, 139.

In West Germany Gentner was from the start the most emphatic proponent of CERN's expansion. Many of his colleagues gradually adopted his view.
However, regarding the formal ECFA proposal of 1963, Heisenberg, for one,
remained considerably more cautious. While describing German participation
in CERN's future construction (particularly the storage rings) as "urgently desirable," he thought that other obligations, like DESY, came first. In this he did
not stand alone, but he made the point more deliberately, and at this stage, as
chair of the advisory commission, his word was more likely to carry weight with
ministry staffers.[22] Vis-à-vis Gentner, occasional programmatic friction between
Gentner's and Heisenberg's MPIs in Heidelberg and Munich – for instance, over
their bubble chamber groups – fortified a certain feeling of conflict.[23] Thus two
positions, representing clusters of scientific, institutional, and political strategies,
came to be embodied in two leading figures. Again, in the face-to-face world of
West German atomic physics, this was nearly foreordained.

The conflict showed up in more than one realm, as money was getting tighter
across the board. For instance, Heisenberg's critique in late 1963 of West Germany's supposed surplus of low-energy accelerators – better, he said, to invest
in visitor experiments, data processing, or even solid state physics – crossed
Gentner's plans to put another tandem van de Graaff in his MPI.[24] However, it
was CERN's second-generation proposal that brought matters to a head. The
ECFA strategy, to state it baldly, was to avoid setting financial priorities. But even
with the most expansionist scientists in the Arbeitskreis Kernphysik, the need to
make choices gradually settled in.[25] And more than anyone else, except ministerial staffers, Heisenberg raised concerns that a lack of selectivity could damage the
venture and take money from other branches of research.[26] The research ministry
staffers in fact valued Heisenberg's leadership because he had the ability "to mod-

[22] Heisenberg to Cartellieri, 12.6.1963, WHM under BMAt Bundesministerium – Schriftwechsel div. a). For a similar assessment see Schmelzer, Bemerkungen zu dem europäischen und
dem deutschen Hochenergieprogramm, 3.6.1963, WHM.

[23] On Filthuth, Heisenberg to Gottstein, 19.7.1963, WHM, and Gottstein to Gentner,
23.10.1963, WHM under Gentner; cf. Schmidt-Rohr, Aufbaujahre, pp. 20–24. A Heidelberg-
Munich front, with positions dug in as far back as the war years, seems to have been perceived
more saliently in Heidelberg than in Munich. Still, the history as viewed in Gentner's circle shaped
understandings of the present. Rechenberg, Gentner, sees harmonious cooperation through 1958;
Carson, Heisenberg, ch. 8–9, has more instances of conflict. Schmidt-Rohr, Erinnerungen, esp.
p. 113, reports the view from within Gentner's MPI.

[24] Heisenberg, Zum Stand der experimentellen Kernphysik in der Bundesrepublik,
[17.12.1963], WGB, 92, sent to Wagner 17.12.1963, to Walcher 20.12.1963, to Cartellieri, Ballreich, and Schmezler 10.1.1964, to Gentner 17.1.1964, and to Stoltenberg 27.1.1964. Heisenberg was targeting Lauterjung, not Gentner, of whose tandem plans he was unaware. Gentner,
Emperor-Antrag, [14.2.1964], WGB, 140; as it played out, minutes of AKKP, 24.2.1964, WHM
under BMAt AKKP II/3a; minutes of Verwaltungsrat, 12.3.1964, MPGB, II/1A.

[25] E.g., Gentner to Walcher, 29.7.1963, WGB, 87 (with the caveat that Gentner's proposal
was certain to work to his advantage).

[26] Heisenberg to Gentner, 26.7.1962, WHM; cf. Hesse to Heisenberg, 14.6.1962, WHM
under BMAt AKKP II/3c; Heisenberg, Probleme, p. 187.

erate," as one put it, "the strong expansive tendencies present in the Arbeitskreis, without letting the ministry neglect something it ought to support."[27] Heisenberg would have agreed; that was about how he saw it himself. To the expansive forces, however, he seemed overly cautious, potentially restraining younger colleagues' dynamism.

PLANS FOR THE FUTURE

First up for discussion was the base of the ECFA pyramid. This proposed second national high-energy machine, alongside DESY, was supposed to be a proton accelerator of several GeV. The project was an unconcealed desire of the CERN generation. The reconstruction-style argumentation was laid out even before ECFA: the French and British had similar machines underway, so the Germans found themselves "distinctly behind."[28] In spring 1963 Gentner was chair of the ministry's new high-energy committee. Bringing its emphatic recommendation to the Arbeitskreis Kernphysik, he proposed a study group of Citron, Schopper, Joachim Heintze, and Volker Soergel under a neutral leader, Christoph Schmelzer, to coordinate potentially competing plans.[29] In some ways, nonetheless, the study group's choices were likely to preshape the design and the siting. In particular, Gentner was persuaded that his local area was ideal, with its double poles of Heidelberg (university plus MPI) and Karlsruhe (Kernforschungszentrum, West Germany's initial nuclear reactor center). And from the circle of younger experimentalists he invited Citron to help launch the technical planning.[30]

What institutional arrangement could house the accelerator? Gentner proposed to bring his younger colleague Citron into his MPI as coequal director. This made the ECFA plan an issue for the Max Planck Gesellschaft as well. Here especially Gentner would need to deal with Heisenberg, the MPG's power broker in matters of physics. Heisenberg, too, had long favored bringing high-energy accelerators into the MPG – for instance, with DESY, which he had been interested in incorporating for a long time.[31] However, about this proposed proton accelerator Heisenberg was skeptical, and he backed siting it in Munich if anywhere.[32]

---

[27] [Prior], draft of Vermerk, Betr.: Frage des Ausscheidens von Prof. Heisenberg, [mid-1965], BAK, B138/3923. Cf. Cartellieri to Heisenberg, 14.6.1963, WHM under BMAt Bundesminister – Schriftwechsel b).

[28] Gentner, Haxel, and Maier-Leibnitz, Merkpunkte für eine gezielte Grundlagenforschung, 20.10.1962, WHM under BMAt Bundesminister – Schriftwechsel b). I am skimming over the friction between advocates of proton and electron accelerators.

[29] Gentner to Heisenberg, 5.4.1963, WGB, 92; cf. Gentner to Jentschke, 4.5.1963, WGB, 86.

[30] Materialien für die Sitzung des Verwaltungsrats, 16.3.1963, WHM, MPG Verwaltungsrat. For plans in Karlsruhe see Schopper to Gentner, 20.3.1963, WGB, 92.

[31] Habfast, Großforschung, pp. 69–72, 185–200; Carson, Heisenberg, ch. 9.

[32] Heisenberg to Uebelacker, 14.6.1963, WHM, Bayer. staatl. Kommission zur friedl. Nutzung der Atomkräfte; see also von Elmenau to Heisenberg, 25.3.1963, WHM under Bayerisches Kultusministerium. To complicate matters, Citron, Begründung des Vorschlages, in München einen Protonenbeschleuniger zu bauen, 14.10.1962, WGB, 17 under Bopp.

More than that, the MPG's planning procedures were at issue, a problem that Heisenberg had made his own.

Gentner contended that his MPI was compelled to expand from nuclear studies into particle physics. Initially he suggested that his codirector would be occupied with visitor experiments at DESY or CERN. However, when asked, the candidate himself indicated that he wanted to build a machine. Indeed Gentner soon made clear that the move was some sort of step towards building the national proton accelerator.[33] The new accelerator had to be built in the MPG, he argued, or it would get built outside it.[34] Heisenberg certainly agreed. Yet even the machine's planning (to say nothing of its construction) would mean a massive expansion of the Heidelberg institute. In Citron's private negotiations for a hoped-for "big accelerator" (8–10 GeV) he mooted staff figures exceeding 300 and annual operating costs of 40 million marks.[35] Within the MPG it all raised questions: Would Gentner cut back somewhere else? Where would the money come from? Was there even a plan?[36]

Gentner interpreted the MPG's questions as evidence of perilous disinterest. His concern was to make room for high-energy physics, and good institutional solutions were few.[37] He could hardly avoid noticing that Heisenberg's plans for plasma physics and space research were no less ambitious, expensive, or plastic. The difference, of course, was that Heisenberg was expanding into fields where he competed directly with no one, while Gentner picked a domain where resource conflicts were sharpening. For the first time, indeed, in December 1963, a ministerial staffer said categorically that a new large-scale machine would force cuts somewhere else. For the moment, the Arbeitskreis declined to support a proposed 10-GeV proton accelerator for studies with kaons, speaking instead for a

[33] Gentner to Butenandt, 10.5.1963 and 27.5.1963, WGB, 140; minutes of CPT Sektion, 14.5.1963, MPGB, II/1A Az. I A 5/6 Protokolle; Ballreich, Vermerk, 25.7.1963, WHM under MPG CPT/WR b); Gentner to Heisenberg, 7.10.1963, WHM.

[34] Ballreich, Vermerk, 20.1.1964, MPGB, II/1A Az. I A 3/- Organisation der MPG, Beiakte Neue Forschungsgebiete.

[35] Citron to Kammerer, 30.9.1963; cf. Citron to Gentner, 21.4.1964, WGB, 47. Schmidt-Rohr, Erinnerungen, p. 152, recalls that Citron was brought in to head a synchrocylotron that he reports as drawn onto institute plans (p. 138). Such a machine was not specified at the start of the deliberations.

[36] Heisenberg to Wagner, 26.9.1963, WHM under MPG CPT/WR b); Heisenberg to Gentner, unsent draft, [mid-October 1963], WHM (quotation). The draft uses Filthuth to illustrate.

[37] Gentner to Butenandt, 5.12.1963 and 3.12.1963 (not sent until 5 December), MPGB, II/1A Az. I A 5/6 Wagner VIII; Gentner to Citron, 23.6.1964, WGB, 47. On the Kaiser Wilhelm Gesellschaft's leadership in accelerators see Weiss, Harnack-Prinzip.

smaller machine (1.5–2 GeV).[38] Gentner put his institute proposal on hold until the MPG could reach its own clarification.[39]

Gentner could hardly help interpreting this as a personal conflict. When Heisenberg had earlier raised questions about the Heidelberg proposal, his colleague had read this as turning against high-energy physics tout court. Heisenberg had always, Gentner suggested, shown "a certain dislike for accelerators"; Gentner even believed that Heisenberg had written letters to block DESY's founding.[40] Confoundingly, the theorist now seemed open to some high-energy plans, but not others. Gentner had no explanation except for Heisenberg's pique.[41] But equally important, Gentner was concerned to keep opportunities open for the younger generation, at just the age when their hopes for independent standing might finally be realized. He was also inclined to give the openness of experimental research more weight than to credit budget constraints. In part this attitude was anchored in his own casual approach to planning. But as he was probably right to suggest, it might also mirror a theorist-experimentalist split. Experimenters of the early 1960s were impressed by phenomena, starting with parity violation, that had caught most theorists by surprise. In the specific terms that politicians wanted, however, theoretical physicists were more articulate in explaining what new accelerators might bring. That meant – at least Gentner saw it this way – that "the scientific program increasingly ends up in the hands of the theorists."[42] In Heisenberg's case, Gentner thought this was magnified by a conviction that his own theory (about which more below) had the future of physics wrapped up.[43]

The study group Gentner proposed did not begin meeting until 1965. Since the group started from the assumption that a national proton accelerator was necessary, the outcome could have been foreseen. Within months its advocates were arguing that further studies only made sense with a statement of intent to

---

[38] Minutes of AKKP, 2.12.1963, WHM under BMAt AKKP II/3d.

[39] Gentner went into the MPG meetings intending to withdraw the nomination. He was persuaded to hold off: minutes of commission and CPT Sektion, 4.12.1963, WHM under MPG CPT/WR b) and c); see also WGB, 47. Given his intent, the moment of deep awkwardness described in Schmidt-Rohr, Erinnerungen, pp. 153–154, seems partly tactical. Citron took a double post in Karlsruhe in 1965.

[40] Gentner to Wagner, 7.10.1963 and 13.12.1963, MPGB II/1A Az. I A 5/6 Wagner VIII. The tension goes back at least to the Hechingen high-voltage apparatus at issue in WGB, 133, and WHM, Hechingen 1946. The letters against DESY are part of the accelerator builders' mythology. They have not been documented, while letters for DESY have. See Carson, Heisenberg, ch. 9; also interview with Wolfgang Jentschke (6.1.2001, Pinneberg).

[41] Heisenberg, Zum Stand der experimentellen Kernphysik in der Bundesrepublik 1963, [17.12.1963], WGB, 92; Gentner to Wolf, 14.2.1964, WGB, 50. Gentner understood that the old MPG had been hostile to large institutes, but he could not be aware that Heisenberg had been the figure who had most pushed for change.

[42] Gentner to Wolf, 14.2.1964, WGB, 50.

[43] Ibid.; cf. note to Ringmann, 14.2.1964, Nachlaß Otto Hahn, MPGB, III/14A, 01529.

build the machine.[44] An optimistic fluctuation in the budgetary outlook encouraged plans for the K-meson factory earlier rejected. Then when the bad news soon came – the budget baseline was low and big cuts would be made – the physicists simply charged ahead. Regarding the budget, they "expected from the [research ministry]," so their minutes recorded, "that it leave no stone unturned to efficaciously counter these difficulties, which conflict with the federal government's public averrals of being supportive of research."[45] The high-energy committee took an expansionist line on the proton accelerator and a wealth of other projects. In fact, the Arbeitskreis Kernphysik went out on a limb to make the K-meson factory a precondition for German participation in the next-generation proton synchrotron at CERN.[46]

By this point, clearly, Heisenberg was out of sync with the Arbeitskreis. May 1965 would in fact be his last meeting. In typical conflict avoidant fashion, and also simply growing old and tired of the meetings, he had already handed in his resignation in order to give "the younger generation" their say.[47] Ultimately he agreed to be kicked upstairs to the Atomkommission presidium but resigned from all other posts. Gentner continued to speak for that younger generation – though his colleagues declined to make him Heisenberg's successor as chair of the Arbeitskreis Kernphysik.

ARGUING FOR AND AGAINST THE "MONSTER"

In 1965, ECFA's national component – the base of the pyramid – looked like it would be left to the physicists. However, the rest of the program – the international apex – promised complications. Politics had always been at stake in CERN. Now, however, for the first time in West Germany, openly political considerations might not work to the physicists' benefit. For the moment, let us look at the arguments the physicists developed, coming back later to the question of what difference they made.

Following ECFA strategy, the two international machines represented a "both/and" recommendation. With the first-decade estimate for West Germany's

---

[44] Schmelzer, report on 1st meeting of Studiengruppe Hochenergiebeschleuniger, 16.1.1965, WGB, 98; Brix et al., Memorandum zur Frage eines Hochenergie-Protonenbeschleunigers im Raum Karlsruhe-Heidelberg, 27.4.1965, WGB, 96. The memo was written by Schoch and sent out by Citron.

[45] Minutes of AKKP, 8.2.1965, WHM under BMAt AKKP II/3d.

[46] Minutes of AKKP, 24.5.1965, WGB, 94. Cf. Pestre, Gargamelle, pp. 43–49. For low-energy accelerators alone, the projected average annual expense ballooned in the space of two years (June 1963 to 5.1965) from 8 million to more than 35 million marks.

[47] Heisenberg to Lenz, 21.5.1965, WHM under BMAt Bundesminister – Schriftwechsel b); [Prior], Vermerk, Betr.: Frage des Ausscheidens von Prof. Heisenberg, n.d. [mid-1965]; Prior, Bericht über meine Dienstreise, 22.10.1965, BAK, B138/3923. The succession required two rounds of voting; Gentner and Wilhelm Walcher were the chief candidates, and Walcher came out ahead. Minutes of AKKP, 15.10.1965, WGB, 94.

contribution climbing above a billion marks, however, the proposal threatened to split back into an "either/or." Each machine had its reasons. The 300-GeV proton synchrotron, to start with, was meant to compete with a Soviet machine under construction in Serpukhov and a giant American accelerator ultimately to be built outside of Chicago. The European version was so big that the existing Geneva site was presumably ruled out. In principle, this was the 28-GeV PS, just ten times as large. Before the official ECFA proposal, the whole approach had actually been the subject of some European distaste. In mid-1961, for instance, weighty voices on the Arbeitskreis Kernphysik had curbed Gentner's advocacy by remarking that one simply did not know whether a tenfold energy increase would bring anything worthwhile.[48] Or could one find any clever new construction principle, instead of just more of the same? CERN's Scientific Policy Committee, its historian observes, was initially against a "monster," where

> the "monster" ... connoted at once a feeling that it was not urgent to increase available energies, a hope for a radically new idea (like the superconducting magnets of which people were speaking), a rejection of needless excess (be it in terms of ambition or cost), a criticism of a lack of imagination or of the urge to go ahead just for the sake of going ahead.[49]

Of course the Americans announced that their strategy would be to push the envelope: new phenomena would just as surely be manifest in the next energy regime, their hopeful argument went. Effectively they set the scale of the machine by the money they hoped to line up. In Europe, for the moment, the split opinion on CERN's choices remained.[50]

On the other hand, the intersecting storage rings (ISR) were easier to sell, to a certain set of audiences at least. The project would be inexpensive, relatively, and could be built in a comparatively short time. Starting from CERN's existing equipment (taking the 28-GeV PS as source for countercirculating beams), the ISR would deliver a center-of-mass energy comparable to a fixed-target machine reaching 1700 GeV. Who were the ISR's advocates? CERN's accelerator builders, for the novel construction principles, and the scientific statesmen on the CERN Council, for the timescale and the price. Of course, astronomical energies had their price: the low intensity associated with colliding two beams, and the head-on design that ruled out research with secondary beams (pions, neutrinos, etc.). At least the big proton synchrotron had intensity and flexibility; nearly all experimental particle physicists who actually wanted to use the accelerators thought the ISR a poor choice.[51] But in 1964–65, CERN leadership, whose decision it was, chose to deviate from the ECFA "both/and" scheme and put its weight behind the ISR. By December 1965 the political partners in the member states (including

---

[48] Minutes of AKKP, 15.6.1961, WHM under BMAt AKKP II/3c.

[49] Pestre, Generation, pp. 710–711. If the complications expressed here are replaced by a natural drive towards higher energies, the real dynamic is lost.

[50] Ibid., pp. 712–726.

[51] Ibid., pp. 712–715, 746–754.

Germany) had nearly unanimously signed on, and funding discussions for the 300-GeV option were postponed.[52]

Among other things, it was the prospect of getting only one machine, and that the ISR, that helped turn one-time scientific doubters to the 300-GeV cause. Since many European colleagues had previously been skeptical about American-style arguments, how did they now make the case? The trouble was that high-energy theory as a whole, through the whole decade of the 1960s, was truly in no position to make good predictions.[53] For the high-energy end of hadron physics, Regge theory gave smooth soft cross-sections. Here dispersion relations and the S-matrix bootstrap had little concretely useful to say. Certainly, quarks were proposed in 1964, but their success in organizing the resonance regime did not automatically help much above it. Physical evidence for proton substructure started coming in only at the end of the decade; and before a reasonable theory of quarks seemed secure, Feynman's parton explanation of deep inelastic scattering, even the neutrino experiments of the early 70s, still had to wait for asymptotic freedom and charm. Gell-Mann's current algebra had phenomenological offspring like sum rules, but it remained a recondite pursuit. For the weak interaction, models with intermediate vector bosons had certainly been a subject of interest since 1960 or so. However, hand-added masses did not settle the energy regime in which the new particles would appear, and the solution of Weinberg and Salam in 1967 (via the Higgs mechanism) attracted little attention before the electroweak theory's renormalizability was revealed a half-decade later.

But as Gentner had remarked, the theoreticians were at bat, and in a difficult situation they did what they could. The 300-GeV machine might find quarks or weak W bosons, though no one could say at what energy. Perhaps a super-heavy particle would set the scale against which mass differences among the known hadrons would be small. Neutrino experiments could explore the weak interactions, though concrete predictions to test were still few. Various particles favored by particular theorists might be identified, or maybe even the magnetic monopole. New data might show whether unknown leptons existed, whether quantum electrodynamics failed, … whether anything (for this was the gist of the argument) was different from the domain already known. To a skeptic the exercise was completely unpersuasive. It was surely equally frustrating to those who carried it out. "What we are anxiously awaiting," one West German report noted, "are exactly those discoveries that are not even framed as problems today."[54]

If this was the hour of the theorists, one of very few scientific skeptics was the most visible theorist in the West German camp.[55] The experimental and phenomenological side of particle physics was thriving, Heisenberg had suggested as early

[52] Ibid., pp. 755–758.

[53] See, e. g., Pickering, Constructing; Hoddeson et al., Rise. The course of thinking is well-registered in the biennial International Conferences on High Energy Accelerators.

[54] Citron, Lehmann, Jentschke, Paul, Rollnik, Soergel, and Prior, Über die Bedeutung des 300 GeV-Protonenbeschleunigers, 6.1968, BAK, B138/3303.

[55] Goldsmith and Shaw, Accelerator, p. 45, speak of "[t]he unanimity amongst the physics community … there was really only one dissenting centre of opinion," namely Heisenberg.

as 1962, with an abundance of new accelerator-generated data.[56] But contrary to
the approach picked up from the Americans, he balked at the easy extrapolation
that nature would prove bountiful all the way up. The dissent had two reasons,
physical and political, and he did not try to disentangle the one from the other.
[Fig. 2]

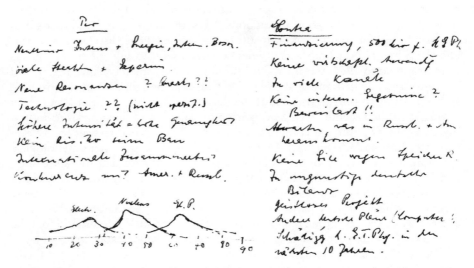

Fig. 2. From Heisenberg's notes on the 300-GeV project, ca. 1967–68. Source: WHM under
    BMAt unlabeled, probably for Beratenden Ausschuß für Forschungspolitik.

As Heisenberg saw it, the profusion of particles up to a few GeV indicated there
were no fundamental particles to be found. The high-energy spectrum was the
analogue of electronic energy levels in atoms. But above 10 or 20 GeV center-
of-mass, he asked, who could be sure that anything novel would be found, instead of something like an atom's continuous spectrum? Scientists had access to
cosmic rays of thousands of GeV; had they given any sign? Through the 1950s,
Heisenberg had been working on a unified theory of the elementary particles,
representing them all as excitations of a single underlying field. By the time of
the budget troubles, his theory had developed enough that he felt sure that the
general picture was sound. It was only towards the end of 1964, with a great
leap in his theory, that his conviction grew that his particular field equation was
definitively right. Thereafter he indeed started suggesting that the basic structure
of particle physics was known. However, it was the general picture, not the par-

[56] Heisenberg, Situation, pp. 2, 18.

ticular theory, that he used to question whether the highest energies would bring anything new.[57]

And exactly the scale of the project raised the stakes. Heisenberg, articulate in his political caution, put it this way:

> [W]ith this 300-GeV proton machine we would be able to do a lot of interesting and use-ful experiments. For the experimental physics of this period and for European cooperation, it would certainly be significant and fulfill important tasks. But when such an installation would require billions of marks, then showing that it will allow us to do many nice experi-ments is not enough. We need still more compelling grounds.[58]

What if the machine were authorized but, heaps of money and rhetoric later, nothing revolutionary came out? What if in the meantime it pulled resources from DESY? The question was about timing as well. In the 300-GeV case, with nine years to the payoff, the problem of putting results in front of the funders was compounded by the U.S. decision to push ahead with competing plans. And with the onrushing output of accelerators already operating, what was the hurry? If the reason was competition with the Americans, well, the Americans were bewitched by technical possibilities and size. A few years would allow European builders to explore other designs than an unimaginative super-PS.[59]

This was not an objection to accelerators in general, on the basis of physics alone. It was an objection to a specific accelerator in its political context. Heisen-berg supported the storage rings: he expected little physical novelty there, either, but the ISR was quick, clever, and cheap. He thought the heedless construction of low-energy accelerators in West German institutes at least as objectionable as high-energy ones. For reasons of physics and politics he had supported DESY when he was interested in the particle spectrum; he supported it even after his theory was fixed.[60] Finally, there were factors to weigh that had little to do with particle physics at all. It would be remarkable, even ministry staffers noted, if the project did not come at the cost of other branches of science.[61]

The most delicate issue was probably the site, because it was tied to ques-tions of international equity. And on this matter, Heisenberg was indeed better situated than his colleagues to make a political argument. The accelerator was supposed to be located somewhere other than Geneva. But where? A Munich option having fallen through early, Heisenberg made clear in the Atomkommis-sion that the project would have his undivided support if the German proposal

[57] That picture was not entirely idiosyncratic; compare Chew's S-matrix program. The con-ventional wisdom, moreover, was that 300 GeV might give little exciting information on protons, since they already appeared to behave like structureless clouds.

[58] Heisenberg, Europas Kernforschung.

[59] Heisenberg to Born, 21.1.1965; Heisenberg to Schramm, 18.3.1964, WHM.

[60] Minutes of MPG Verwaltungsrat, 11.3.1965, MPGB, II/1A; on DESY, Heisenberg to Jentschke, 27.10.1966, WHM; minutes of Beratende Ausschuß für Forschungspolitik, 20.11.1967, Nachlaß Ludwig Raiser (LRK), BAK, N127/28. Admittedly, Heisenberg's later enthusiasm had something to do with his interest in things like photon and electron propagators and the connec-tion between baryons and leptons.

[61] E.g., Prior in minutes of AKKP, 11.5.1964, 2.10.1964, WHM under BMAt AKKP II/3d.

of Drensteinfurt came out on top.[62] As he saw it, through the reconstruction period the Federal Republic had been dealt with inequitably in CERN personnel, contracts, and siting, and the time for putting up with that was passing. Other West German physicists might agree, but they chose other forums and means to press the case.[63]

ENDGAME

The Arbeitskreis Kernphysik went on record backing the 300-GeV proposal – though "in a circle of the interested parties," of course, another outcome would have been strange.[64] In 1967 the next German Atomprogramm would be up for discussion. As it was assembled from pieces drafted by its technical commissions, it looked like the Atomkommission would again rubber-stamp the requests. When that body assembled in the fall, it did not really prioritize, but it did put out a statement on the 300-GeV project. While praising CERN's work, the statement asked a number of pointed questions: Could new technologies be used? Could costs be reduced? Could the European program be coordinated with the U.S. and the U.S.S.R.? Could siting options be priced? CERN and ECFA experts, who wanted a quick statement of approval, took the inquiries as a major threat. But to call the project into question seems less the German intent than to get answers to real concerns about costs.[65] Indeed, when most of the issues were addressed, the Federal Republic seemed satisfied. In the next Atomkommission meeting half a year later, the body recommended Germany commit to the project, conditional on satisfaction of specifications for financial planning and an equitable allocation of contracts and other benefits.[66]

Colleagues were hardly in the dark about Heisenberg's critical stance in these meetings – though he also signed on to the resolution recommending the project. German physicists who had moved into leading roles in CERN, such as

---

[62] Minutes of DAtK, 25.6.1968, BAK, B138/3303, reformulated in Heisenberg to Heyne, 23.10.1968, WHM under BMAt Bundesministerium – Schriftwechsel div. c). On Drensteinfurt, see Rusinek, Europas 300-GeV-Maschine; on Munich, Deutinger, Europa in Bayern.

[63] Heisenberg to Hass, 21.1.1965, WHM; Klaus Gottstein, diary entry of conversation, 5.3.1970 (with thanks to Prof. Gottstein). This went together with a broader initiative to formalize standards for international equity: Heisenberg to Stoltenberg, 18.8.1969, WHM under BMAt Bundesminister – Schriftwechsel b).

[64] The phrase is Schmelzer's: Schmelzer to Heisenberg, 8.6.1967, WHM under BMAt Gutachterausschuß "Hochenergiephysik." Minutes of AKP, 1.6.1967, WGB, 95.

[65] Minutes of DAtK, 14.11.1967, BAK B138/3303, specifically Anlage 2; Pestre, Decision, pp. 75–77.

[66] Minutes of DAtK, 25.6.1968, specifically Empfehlung; minutes of Beratender Ausschuß für Forschungspolitik, 8.7.1968, LRK, 28.

Gentner and Wolfgang Paul, were under pressure to put an end to the trouble.[67] Colleagues began looking for ways to neutralize Heisenberg's influence, on the double assumption that he remained a power broker and that the issue was the theorist's conceit of preempting accelerators. For instance, a Heidelberg professor launched a signature campaign designed to show that nearly all theorists agreed that elementary particle physics was still open, and nearly all stood behind the accelerator.[68] Pulling few punches, Heisenberg's colleagues also drew up a report that contrasted his theory with a range of alternatives and set him up as the wishful spokesman of a physics come to its end. "Today voices are heard," the report commented,

> suggesting that we forego the further expansion of high energy physics because of the great material cost. But would this not be capitulation? We would be leaving behind a field of study in the knowledge that we had failed at its penetration. A deliberate braking of the drive with which these problems are taken up, by restricting them primarily to theoretical examination, would be against the essence of science, which can only flourish in the interplay of theory and experiment![69]

Heisenberg's colleagues assumed he was dangerous because of the science and because of his influence. When they imagined him so, they displayed their own belief that key scientists still drove the dynamic. But that misconceived the balance of forces, as they were only starting to learn. For early on, the research ministry had started raising its own concerns – and growing suspicious of the self-interest inherent in its scientists' recommendations.[70] The research minister through the critical phase (1965–1969) was none other than Gerhard Stoltenberg, one-time parliamentary inquisitor of the scientists' demands. As Stoltenberg framed the issue for the Bundestag, there were indeed many good reasons why the huge accelerator would be "a very useful scientific investment; still, this leaves open," he said, "whether other scientific investments of the same order of magnitude may have a much higher value for humanity."[71] The minister and his staffers were increasingly capable of forming a position internally; this is nowhere more visible

---

[67] Heisenberg, Braucht Europas Kernforschung, published in Die Welt on 29.6.1968 (in part a rectification of an earlier "interview," Si j'avais; Giese to Weber, 14.6.1968, WHM); interview with Ulrich Schmidt-Rohr (5.1.2001, Heidelberg). For other colleagues causing trouble besides Heisenberg (Jentschke, Maier-Leibnitz), see Citron to Jentschke, 5.4.1968 and 8.4.1968, WGB, 76.

[68] Asked to sign the petition, Heisenberg wrote up his own statement, alongside the science making his usual political points. The organizer thanked him but explained that the document was meant to stick to the physics. Heisenberg to Stech, 23.3.1970; Stech to Heisenberg, 8.4.1970, WHM; Stech to Bundesminister für Bildung und Wissenschaft, 21.4.1970, WHM.

[69] Citron, Lehmann, Jentschke, Paul, Rollnik, Soergel, and Prior, Über die Bedeutung des 300 GeV-Protonenbeschleunigers, 6.1968, BAK, B138/3303.

[70] E.g., Cartellieri, Bildungs- und Forschungspolitik, p. 398, arguing for a Gesamtplan because the 300-GeV project would damage other branches of science; or staffers' comments in minutes of HEP committee, 4.11.1966, BAK, B138/3328, and 1.6.1967, Nachlaß Heinz Maier-Leibnitz, Deutsches Museum, Munich, NL 111, 117. On suspicions see Stoltenberg, note, 30.3.1966, B138/3328; cf. Prior to Gentner, 4.2.1968, WGB, 50.

[71] Stoltenberg in Verhandlungen, 5. Wahlperiode, 12.10.1966, 3090 D –3091 A.

than in the transition from the first (1965) to the second (1967) Bundesbericht Forschung. So Heisenberg's views on budgets, priorities, and international ventures could resonate with ones present among government officials. However, in advancing a point against his colleagues' objections, staffers rarely felt a need to cite his support.[72] And Heisenberg's intuitions about the future of physics had, it seems clear, almost no influence on official decisions. What they did was cause problems for public relations and irritate other powerful physicists.[73]

What happened to the accelerators? Plans for a national 40-GeV proton machine were drawn up by a Karlsruhe study group. Toying with the idea of a joint French-German venture, discussions in the Arbeitskreis Kernphysik stretched out. Ultimately, resource conflicts, disagreements among builders, and the death of a proponent prepared its demise. The base of the pyramid was the one part of the ECFA scheme that did not survive.[74] By contrast, the CERN storage rings did get built. In 1971 Heisenberg came to Geneva to give the dedication address. However, the ISR would not be a great physics success, due partly to experimenters' slow accommodation to new types of detectors, partly to phenomena no one had expected when it was planned. The ISR's real legacies were its technical advances and novel construction principle as the possibilities of colliding-beam machines were explored.[75]

The big proton synchrotron was ultimately built, too. Along with the cost, the biggest problem was the site. On the basis of a revised proposal answering technical questions and giving solid cost estimates, the West German cabinet conditionally reaffirmed its participation in December 1969.[76] Unfortunately, it soon became obvious that Drensteinfurt's chances were slim, and that ran negotiations into the ground. What finally broke the deadlock was a new pan-European compromise, building the accelerator more quickly and cheaply by siting it in Geneva (after all) to use infrastructure there. The Federal Republic's research minister commented, significantly, that the delay meant that "the modified project is now visibly cheaper and more progressive ... The Federal Republic has made it clear

---

[72] On the DAtK resolution of 14.11.1967, see the draft version worked out in BAK, B138/3302; also Erwägungen zum Projekt eines europäischen 300 GeV-Protonen-Beschleunigers, 10.10.1967, and Rembser's rebuttal of 13.11.1967 to Heisenberg's letter of 8 November, BAK, B 138/3302.

[73] The conflict spilled over into the choice of a successor for Heisenberg as director of the MPI for Physics and Astrophysics. Gentner chaired the MPG's search committee; discussions grew acrimonious. See Carson, Heisenberg, ch. 10.

[74] Habfast, Großforschung, pp. 207, 240–243; interviews with Anselm Citron (4.1.2001, Karlsruhe), Ulrich Schmidt-Rohr (5.1.2001, Heidelberg), and Herwig Schopper (8.1.2001, Geneva).

[75] Heisenberg, Aufgabe; Russo, Rings.

[76] Rusinek, Europas; minutes of DAtK, 25.6.1968, BAK, B138/3303, and Heisenberg to Stoltenberg, 18.8.1969, WHM under BMAt Bundesminister – Schriftwechsel b). Bundesminister für Bildung und Wissenschaft, Mitteilungen für die Presse, 12.12.1969, BAK, B138/3303 (cabinet meeting of 11.12.1969); cf. Heisenberg to Leussink, 11.11.1969, WHM under BMAt Bundesminister – Schriftwechsel b).

that it intends to have a real voice in future joint European scientific projects."[77] Heisenberg greeted the final decision, calling it "in light of the many factors that have to be taken into account, a true optimum."[78]

The SPS (Super Proton Synchrotron) went on line in 1976, its operating energy upgraded to 400 GeV. Ultimately, any skepticism about high energies was shown to be erroneous, and there was more to be seen with secondary beams than anyone could have foreseen. But in fact, the interesting physics done at the SPS was only indirectly connected with its proponents' original visions. In this case, the advocates were well served by their sense of the future's openness. They were lucky, however, that the physics was not really of interest to the politicians, who made their decisions on other grounds.

CONCLUSION

What is new in the CERN second-generation story is not the interweaving of politics and science – for that was characteristic of CERN from the start. New is rather the changing relationship between physicists and their funders, once principally ministerial and parliamentary advocates, now increasingly ministerial and parliamentary overseers as well. In particular, the West German debates show the slow process of coming to terms with the fact that the two parties' interests were starting to diverge. Even if the drive towards ever bigger accelerators was viewed skeptically by some scientists, it was clearly a scientists' drive. As far as ministry staffers and politicians were concerned, however, other scientific projects might be equally worthy. So the postwar funding regime for nuclear and particle physics truly set patterns for massive expansion of state support for science in domains far afield. But when budgetary forces external to science put a cap on spending in the 1960s, the established routine of policymaking was pulled into the flux – even as ever huger sums were devoted to government funding for science.

In this way, particle physics and accelerator construction make manifest the shifting approaches to science policy and funding in the West German 1960s. To sum up the argument, we see here the immediate postwar period in transition to a new kind of post-postwar normalcy. It was a normalcy in which, first of all, scientific competition with the United States seemed conceivable (albeit on a European basis) – and in which, all the same, the postwar rhetoric of reconstruction, "catching up," still found powerful use. It was a normalcy characterized, simultaneously, by new fiscal circumstances, confronting for the first time since World War II the fact that economic cycles were real, that postwar prosperity was not indefinite, that expansion in government expenditures would eventually hit a

---

[77] Pestre, Decision, pp. 86–91, giving to a degree the CERN perception, as also Goldsmith and Shaw, Accelerator, pp. 56–58; Leussink in minutes of DAtK, 23.10.1970, BAK, B138/3304 (quotation). Leussink referred to the European Molecular Biology Laboratory, which ended up in Heidelberg.

[78] Heisenberg to Jentschke, 3.3.1971, WHM under CERN (file, not drawers).

limit. It was a normalcy, moreover, that forced physicists out of their completely privileged position into something more like that of many interest groups – displaying a robust sense of entitlement grounded in postwar experience, but now more often obliged to manage internal dissidence and defend their position vis-à-vis other interests. And it was a normalcy, finally, that developed new policy tools: the first of the plans for science, as voluntaristic and haphazard as they proved to be, and the maturation of expertise within the federal bureaucracy to challenge the primacy of elite scientists in their own organizations.

BIBLIOGRAPHY

Adami, Nikolaus: Die Haushaltspolitik des Bundes von 1955 bis 1965, Bonn 1970.

Bartz, Olaf: Der Wissenschaftsrat. Entwicklungslinien der Wissenschaftspolitik in der Bundesrepublik Deutschland 1957–2007, Stuttgart 2007.

Boenke, Susan: Entstehung und Entwicklung des Max-Planck-Instituts für Plasmaphysik 1955–1971, Frankfurt am Main 1991.

Braun, Dietmar: Die politische Steuerung der Wissenschaft. Ein Beitrag zum "kooperativen Staat", Frankfurt am Main 1997.

Bundesbericht Forschung II. Bericht der Bundesregierung über Stand und Zusammenhang aller Maßnahmen zur Förderung der wissenschaftlichen Forschung und Entwicklung in der Bundesrepublik Deutschland, Bonn 1967.

Carson, Cathryn: Heisenberg in West Germany. Science and the Public Sphere, Cambridge, forthcoming.

Carson, Cathryn: Heisenberg als Wissenschaftsorganisator, in: Kleint, Christian, Helmut Rechenberg, and Gerald Wiemers (eds.): Werner Heisenberg 1901–1976. Beiträge, Berichte, Briefe. Festschrift zu seinem 100. Geburtstag, Stuttgart 2005, pp. 214–222.

Carson, Cathryn: Nuclear Energy Development in Postwar West Germany. Struggles over Cooperation in the Federal Republic's First Reactor Station, in: History and Technology 18, 2002, pp. 233–270.

Deutinger, Stephan: Europa in Bayern? Der Freistaat und die Planungen von CERN zu einem Forschungszentrum im Ebensberger Forst bei München 1962–1967, in: Schneider, Ivo, Helmuth Trischler, and Ulrich Wengenroth (eds.): Oszillationen. Naturwissenschaftler und Ingenieure zwischen Forschung und Markt, München 2000, pp. 297–324.

Eckert, Michael: Primacy Doomed to Failure. Heisenberg's Role as Scientific Advisor for Nuclear Policy in the FRG, in: Historical Studies in the Physical and Biological Sciences 21 (1), 1990, pp. 29–58.

Eckert, Michael and Maria Osietzki: Wissenschaft für Macht und Markt. Kernforschung und Mikroelektronik in der Bundesrepublik Deutschland, München 1989.

Fischer, Peter: Atomenergie und staatliches Interesse. Die Anfänge der Atompolitik in der Bundesrepublik Deutschland 1949–1955, Baden-Baden 1994.

Frese, Matthias, Julia Paulus, and Karl Teppe (eds.): Demokratisierung und gesellschaftlicher Aufbruch. Die sechziger Jahre als Wendezeit der Bundesrepublik. Paderborn 2003.

Goldsmith, Maurice and Edwin Shaw: Europe's Giant Accelerator. The Story of the CERN 400 GeV Proton Synchrotron, London 1977.

Habfast, Claus: Großforschung mit kleinen Teilchen. Das Deutsche Elektronen-Synchrotron DESY 1956–1970, Berlin 1989.

Hars, Florian: Wenn Forschung zu groß wird. Internationalisierung als Strategie nationaler Forschungsplanung am Beispiel der Hochenergiephysik, in: Ritter, Gerhard A., Margit Szöllö-si-Janze, and Helmuth Trischler (eds.): Antworten auf die amerikanische Herausforderung.

Forschung in der Bundesrepublik und der DDR in den "langen" siebziger Jahren, Frankfurt am Main 1999, pp. 286–312.

Haushaltsreden Dr. Heinz Starke, Dr. Rolf Dahlgrün, Dr. h. c. Kurt Schmücker 1962 bis 1966. Dokumente – Hintergründe – Erläuterungen, Bonn 1995.

Heisenberg, Werner: Die Aufgabe der Speicherringe. Rede bei der Einweihung der Speicherringe in CERN am 16.10.1971, in: Physikalische Blätter 28, 1972, pp. 107–111, in: Gesammelte Werke, vol. C.V, München 1989, pp. 333–337.

Heisenberg, Werner: Braucht Europas Kernforschung das "Super-PS"?, in: Die Welt, 29.6.1968, in: Gesammelte Werke, vol. C.V, München 1989, pp. 313–319.

Heisenberg, Werner: The Present Situation in the Theory of Elementary Particles, in: Commemoration of the Fiftieth Anniversary of Niels Bohr's First Papers on Atomic Constitution, pt. 2, Session on Elementary Particles, Copenhagen 1963, pp. 1–13, in: Gesammelte Werke, vol. B, Berlin 1984, pp. 610–622.

Heisenberg, Werner: Si j'avais vingt ans …, in: Le nouvel observateur, 30.4.1968, pp. 22–25, in: Gesammelte Werke, vol. C.V, München 1989, pp. 608–614.

Hermann, Armin: Germany's Part in the Setting Up of CERN, in: Hermann, Armin, John Krige, Ulrike Mersits, and Dominique Pestre (eds.): History of CERN, vol. 1, Launching the European Organization for Nuclear Research, Amsterdam 1987, pp. 383–429.

Hoddeson, Lillian, Laurie Brown, Michael Riordan, and Max Dresden (eds.): The Rise of the Standard Model. Particle Physics in the 1960s and 1970s, Cambridge 1997.

Hohn, Hans-Willy and Uwe Schimank: Konflikte und Gleichgewichte im Forschungssystem. Akteurkonstellationen und Entwicklungspfade in der staatlich finanzierten außeruniversitären Forschung, Frankfurt am Main 1990.

Krige, John: Finance Policy. The Debates in the Finance Committee and the Council over the Level of the CERN Budget, in: Hermann, Armin, John Krige, Ulrike Mersits, and Dominique Pestre (eds.): History of CERN, vol. 2, Building and Running the Laboratory, Amsterdam 1990, pp. 571–635.

Lessmann, Sabine: Budgetary Politics and Elections. An Investigation of Public Expenditures in West Germany, Berlin 1987.

Metzler, Gabriele: Konzeptionen politischen Handelns von Adenauer bis Brandt. Politische Planung in der pluralistischen Gesellschaft, Paderborn 2005.

Müller, Wolfgang D.: Geschichte der Kernenergie in der Bundesrepublik Deutschland, vol. 1, Anfänge und Weichenstellungen, Stuttgart 1990.

Pestre, Dominique: The Difficult Decision, Taken in the 1960s, to Construct a 3–400 GeV Proton Synchrotron in Europe, in: Krige, John (ed.): History of CERN, vol. 3, Amsterdam 1996, pp. 65–96.

Pestre, Dominique: Gargamelle and BEBC. How Europe's Last Two Giant Bubble Chambers Were Chosen, in: Krige, John (ed.): History of CERN, vol. 3, Amsterdam 1996, pp. 39–64.

Pestre, Dominique: The Second Generation of Accelerators for CERN, 1956–1965. The Decision-Making Process, in: Hermann, Armin, John Krige, Ulrike Mersits, and Dominique Pestre (eds.): History of CERN, vol. 2, Building and Running the Laboratory, Amsterdam 1990, pp. 679–780.

Pickering, Andrew: Constructing Quarks. A Sociological History of Particle Physics, Chicago 1984.

Rechenberg, Helmut: Gentner und Heisenberg – Partner bei der Erneuerung der Kernphysik- und Elementarteilchenforschung im Nachkriegsdeutschland (1946–1958), in: Hoffmann, Dieter and Ulrich Schmidt-Rohr (eds.): Wolfgang Gentner (1906–1980). Festschrift zum 100. Geburtstag, Heidelberg 2006, pp. 63–94.

Ruck, Michael: Ein kurzer Sommer der konkreten Utopie – Zur westdeutschen Planungsgeschichte der langen 60er Jahre, in: Schildt, Axel, Detlef Siegfried, and Karl Christian Lammers (eds.): Dynamische Zeiten. Die 60er Jahre in den beiden deutschen Gesellschaften, Hamburg 2000, pp. 362–401.

Rusinek, Bernd-A.: Europas 300-GeV-Maschine. Der größte Teilchenbeschleuniger der Welt an einem westfälischen Standort?, in: Geschichte im Westen 11, 1996, pp. 135–153.

Russo, Arturo: The Intersecting Storage Rings. The Construction and Operation of CERN's Second Large Machine and a Survey of its Experimental Program, in: Krige, John (ed.): History of CERN, vol. 3, Amsterdam 1996, pp. 97–170.

Schmidt-Rohr, U.: Die Aufbaujahre des Max-Planck-Instituts für Kernphysik, Heidelberg 1998.

Schmidt-Rohr, U.: Die deutschen Teilchenbeschleuniger von den 30er Jahren bis zum Ende des Jahrhunderts, Heidelberg 2001.

Schmidt-Rohr, U.: Erinnerungen an die Vorgeschichte und Gründerjahre des Max-Planck-Instituts für Kernphysik, Heidelberg 1996.

Stamm, Thomas: Zwischen Staat und Selbstverwaltung. Die deutsche Forschung im Wiederaufbau 1945–1965, Köln 1981.

Stucke, Andreas: Institutionalisierung der Forschungspolitik. Entstehung, Entwicklung und Steuerungsprobleme des Bundesforschungsministeriums, Frankfurt am Main 1993.

Trischler, Helmuth: Planungseuphorie und Forschungssteuerung in den 1960er Jahren am Beispiel der Luft- und Raumfahrtforschung, in: Szöllösi-Janze, Margit and Helmuth Trischler (eds.): Großforschung in Deutschland, Frankfurt am Main 1990, pp. 117–139.

Trischler, Helmuth: Wolfgang Gentner und die Großforschung im bundesdeutschen und europäischen Raum, in: Hoffmann, Dieter and Ulrich Schmidt-Rohr (eds.): Wolfgang Gentner (1906–1980). Festschrift zum 100. Geburtstag, Heidelberg 2006, pp. 95–120.

Verhandlungen des Deutschen Bundestages, 4. Wahlperiode, München 1981, 5. Wahlperiode, München 1982.

Weingart, Peter and Niels C. Taubert (eds.): Das Wissensministerium. Ein halbes Jahrhundert Forschungs- und Bildungspolitik in Deutschland, Weilerswist 2006.

Weiss, Burghard: Harnack-Prinzip und Wissenschaftswandel. Die Einführung kernphysikalischer Großgeräte an den Instituten der Kaiser-Wilhelm-Gesellschaft, in: vom Brocke, Bernhard and Hubert Laitko (eds.): Die Kaiser-Wilhelm-/Max-Planck-Gesellschaft und ihre Institute. Studien zu ihrer Geschichte. Das Harnack-Prinzip, Berlin 1996, pp. 541–560.

SCIENTISTS AS INTELLECTUALS:
THE SOCIOPOLITICAL ROLE OF FRENCH AND WEST GERMAN NUCLEAR PHYSICISTS IN THE 1950s

Martin Strickmann

I. INTRODUCTION

Historians of science frequently ask how the relationship of scientists to society, politics and public should be understood, described and assessed, and which terms, concepts, methods and instruments are most appropriate to comprehend their social and political role, sociopolitical articulation and intervention.[1] The preferred approach has been to discuss social figures and types of scientists like the expert, the mediator, the policy advisor, and the "German mandarin" (Fritz Ringer), and generally to analyze scientists as professors or scholars in the context of their corresponding scientific community – both national and international.[2]

The question (of) which attitudes can be assessed as "apolitical" or "political" is of paramount importance to this undertaking. Of equal importance is where the boundaries of the disciplines and of professional and individual competence have to be drawn, and when these boundaries are probably transgressed, what happens in the moment when a scientist "transgresses" this boundary with political intervention. Since the approach and perspective mentioned above has run into obstacles and gradually lost explanatory power, this paper will attempt to find alternatives by introducing a concept that is new and untested in the history of science. Using the example of French and German nuclear physicists in the 1950s as a case study, the concept of the "intellectual" will be applied to scientists. The objects of investigation in my study are the leading French nuclear physicist intellectuals Frédéric and Irène Joliot-Curie and the leading West German quantum, nuclear and atomic scientists of the "Göttingen Eighteen", including Max Born, Carl-Friedrich von Weizsäcker, Walther Gerlach, Werner Heisenberg, Otto Hahn, and their antipode Pascual Jordan. Strictly speaking, not all of these are nuclear physicists, but they are all close to this discipline.

At first glance, this approach of conceptually linking nuclear physicists and intellectuals may seem surprising. After all, nuclear physicists are generally connected with the spheres of natural science and technology, whereas intellectuals are associated more with the antipodean spheres of culture, spirit, the social sciences, and humanities. In 1959, exactly fifty years ago, the physicist, writer

---

[1] Recent examples include the contributions to this book.
[2] Cf. Harwood, Forschertypen.

and politician Charles P. Snow correspondingly coined the concept of the "two cultures," describing the two spheres as separated by a deep, insurmountable gulf, which confronted each other diametrically in mutual ignorance, in silence, or vendettas: "Literary intellectuals at one pole – at the other scientists, and as the most representative, the physical scientists. Between the two a gulf of mutual incomprehension – sometimes … hostility and dislike, but most of all lack of understanding."[3]

Nuclear physicists and intellectuals do not necessarily have to confront each other from generally diametric and irreconcilable positions, as the social figure of the intellectuel scientifique (scientist-intellectual) or physicist-intellectual demonstrates (see below). This figure can also help bridge the gulf between the two cultures in historical research, or at least reduce it, by establishing a kind of "third culture" in Snow's sense, as will be demonstrated below.

An independent historiography of intellectuals emerged from France in the 1980s, which historicizes the phenomena and history of intellectuals in France and Germany.[4] Since this has been, and continues to be, an activity occupying primarily political scientists, literary scholars, sociologists and historians,[5] its works exhibit a heavy predominance of literary and humanistic intellectuals as opposed to the marginalized scientist-intellectuals.[6]

Correspondingly, this paper pursues a two-pronged approach: reinserting a narrative of scientist-intellectuals into the general historiography of intellectuals; and introducing the concept of the intellectual into the history of science. Thus this is to be understood historiographically as the nexus between the history of science and an individual-focused political and cultural history of intellectuals.

In order to do this, first of all a brief historical derivation and definition of the notion of the intellectual is required. The hour of birth and genesis of the (French) intellectual as a substantive term, concept and milieu was the Dreyfus Affair in 1898, when the writer Emile Zola published J'accuse to great political effect in order to advocate the rehabilitation of the Jewish Captain Alfred Dreyfus, unjustly convicted and banished for espionage.

Since then – and not only in France – creators of ideas like writers, philosophers, scientists and artists, who use the official reputation earned in their professional fields as "symbolic capital" (Pierre Bourdieu) to publicly articulate and actively participate in political and social debates extending far beyond their specialized fields, are considered to be intellectuals.[7] The "left-wing intellectuals" around Zola confronted the "right-wing intellectuals" Charles Maurras and Maurice Barrès. But because many "right-wing intellectuals," despite their original anti-German motivation, collaborated with the fascist Vichy regime from

---

[3] Snow, Two Cultures.

[4] Winock, Intellektuelle; Mommsen/Hübinger, Intellektuelle; Strickmann, Französische Intellektuelle; Marmetschke, Intellektuelle.

[5] Dominique Pestre refers to this as historiens-littéraires.

[6] "L'histoire et la sociologie des intellectuels ont privilégié les ‚littéraires' aux dépens des scientifiques …," Matonti, Joliot-Curie, pp. 107 ff.

[7] Cf. Winock/Juillard, Dictionnaire.

1940 on, this group was largely discredited in France after the Libération and Epuration. Among other things, the "myth" of the Résistance, which consisted of left-wing Catholics, socialists and a majority of Communists, led to the rise of intellectuals sympathizing with the Communists, with Jean-Paul Sartre and Simone de Beauvoir as their best-known representatives. They found their liberal-conservative counterpart in the intellectual Raymond Aron, whose tract L'opium des intellectuels (The Opium of the Intellectuals) railed in particular against the "infatuations" of the ubiquitous philo-Communist type.[8]

Because of the Franco-German "hereditary enmity" and rivalry, a historical fiction cemented in three wars, the term and concept of the intellectual did not become established in Germany with a more positive connotation until after World War II, as a consequence of Franco-German reconciliation and cultural transfer. Previously the historiography of intellectuals had been dominated by the following dictum: "France is considered to be the native land of intellectuals, Germany as the land where intellectuals are chastised."[9] The dichotomy of German culture versus French "civilization" and the image of the German academic mandarin (Fritz Ringer) as an introspective, apolitical, spiritual aristocrat and scholar above party lines, floating above the sociopolitical sphere, versus the French rational-politicized intellectual involved in public life, remained very influential for a long time.[10] Corresponding to the ideal, apolitical, "value-free" science was considered to be something noble.

Yet the "German mandarin" is only one side of the coin, for despite the historical and national-psychological stereotype of a supposed division between intellect and power, intellect and state, Germany also has a tradition that is nearly as long of political involvement by figures like writers and scientists. However, this tradition was often associated with quite national or even nationalist activity, as, for instance, in the "Appeal to the Cultural World" of 14 October 1914 and the "War of Intellects" at the beginning of World War I,[11] or under the guise of fascism during the Nazi period.

For this reason, a "normalization" of politically engaged intellectuals in the sense of, and following from "Westernization" (Anselm Doering-Manteuffel) – thus, among other things, not nationalistic – first took place in the Federal Republic of Germany. As Jürgen Habermas conceded in "Eine Art Schadensabwicklung" (A Sort of Claims Settlement) "Not until the Federal Republic of Germany did a class of intellectuals take shape that accepted itself as such. Now, years after France did so in the Dreyfus Affair, writers, and increasingly, scientists are catching up by taking the step toward the normalization of public involvement."[12]

According to Karl Mannheim and Alfred Weber's sociology of knowledge, the ideal type of the intellectual embodies unattached and critical "free-floating

---

[8] On the history of intellectuals in France, cf. Strickmann, Intellektuelle, pp. 40–69.
[9] Hübinger, Intellektuelle, p. 198.
[10] Ringer, Mandarin; Forman, Independence; Carson, New models; Bering, Intellektuelle.
[11] Cf. Wolff, Krieg der Geister; Flasch, Geistige Mobilmachung.
[12] Habermas, Intellektuelle, p. 46.

intelligence independent of society." This type stands up for universal values like truth, justice and freedom, based on moral integrity, whether as the moral conscience of the nation, or even of humanity, and functions as a sentry in state and society by warning, reminding, and enlightening. Historically speaking, however, the ideal type briefly outlined here is better regarded as a projection. Therefore this study refers not to the ideal type, but more pragmatically to the real type of the intellectual, so that the concept of the intellectual can be operationalized. The ideal-type definition does not do justice to the fact that many intellectuals in the 20th century succumbed to the ideological "infatuations" with and "promises" of fascism and/or (Soviet) Communism, which Julien Benda stigmatized in his pamphlet of the same name as the "intellectuals' betrayal" of universal values. For obvious reasons this definition cannot be applied just like that in this way to the historical reality of (physicist) intellectuals.[13]

Of particular interest for this study is the methodological set of instruments used by Jean-François Sirinelli, who coined three concepts for the history of intellectuals: sociabilité, génération and itinéraire.[14] He uses the term sociabilité to designate the formation of groups and milieus around focal points such as journals, publishing houses and cafés, which are determined by structures of communication. This term appears quite applicable for associations and organizations as well. Its main focus is on the internal factors that bind these groups and entities together, and mechanisms of cohesion in the formation of stable intellectual milieus. According to the concept of the génération intellectuelle, every generation is marked formatively by a collective événement fondateur, a key event or experience that shapes its identity and forms an "ideological system" in the sense that the members of the generation pose the same questions because of this event, even though they may end up reaching different conclusions. The concept of the itinéraire[15] can be used to trace individual or group biographies that allow shared stations and "crossroads" to be distinguished in the individual life paths.

In the following section, first the leading French nuclear physicist-intellectuals[16] Frédéric and Irène Joliot-Curie will be examined, and their socio-political involvement derived from and located in terms of their historical biographies. The subsequent section does the same for the leading West German nuclear physicist-

---

[13] Michel Winock differentiates between the party-intellectual connected to a political party, who is accused by Julien Benda of betrayal; the critical intellectual, who acts in the name of a "noble ethics" or "ethics of conscience" (Max Weber: Gewissensethik); and the "organic" intellectual like Barrès, who defends an established system through affirmation in the name of an "ethics of responsibility" (Verantwortungsethik). However, I consider these categories somewhat sketchy and hardly enlightening here.

[14] Sirinelli, Hasard.

[15] Path of development, resumé.

[16] To clarify terminology: The scientist-intellectual – abbreviated form for the natural scientist-intellectual – is the more general name, which contains the (nuclear) physicist-intellectual. The physicist-intellectual is the "leading" type of scientist-intellectual, also in the sense of physics as a leading science. Since the leading physicists of the 1950s were usually (also) nuclear physicists, I speak of the nuclear physicist-intellectual to designate the physicist-intellectual of this concrete epoch.

intellectuals, depicting and analyzing their development and thereby facilitating a comparative analysis of the two examples.

II. FRÉDÉRIC AND IRÈNE JOLIOT-CURIE AS FRENCH PHYSICIST-INTELLECTUALS

While Jean-Paul Sartre, as intellectuel littéraire, was regarded as the absolute embodiment of the intellectual in postwar France, Frédéric Joliot-Curie was considered the embodiment of the intellectuel scientifique. While Sartre's public standing was based on his philosophical and literary works on existentialism, the Joliot-Curies' influence was built on their pioneering research on "artificial radioactivity," for which they won the Nobel Prize for Chemistry in 1935, and Frédéric's subsequent professorship at the Collège de France in Paris from 1937 onwards.[17]

Born in 1900, Frédéric Joliot-Curie[18] belonged to the same generational cohort as his wife Irène Joliot-Curie, as well as Werner Heisenberg and Pascual Jordan. Common to these contemporaries is that they had already reached the pinnacle of their profession as leading scientists and members of the research avant-garde of atomic, nuclear and quantum physics during the period between the wars. The colleagues he worked closely with later in the French Atomic Energy Commission (Commissariat à l'energie atomique, CEA), Pierre Auger (1899–1993) and Francis Perrin (1901–1992), son of the Nobel laureate Jean Perrin, with whom Frédéric and Irène comprised the "quartet" of French postwar nuclear physicists, are also members of this generation.

It is significant that over generations the French nuclear physicists had constituted a family of Communist sympathizing "pacifist" scientists, practically a closed society when one considers that, by marrying Irène Curie, Frédéric Joliot, the pupil and intellectual heir of Paul Langevin (1872–1946), became a member of the famous family of Marie and Pierre Curie, whose family tradition continues today through the work of physicist Hélène Langevin-Joliot (* 1927), granddaughter of Marie Curie and Paul Langevin, at the National Center for Scientific Research (Centre national de la recherche scientifique, CNRS), and that Pierre Auger and Francis Perrin were also brothers-in-law. The focal points of their sociabilité included institutions of higher learning and research as well as (left-leaning) political associations.

The public political activities of Frédéric and Irène Joliot-Curie as intellectuals began in 1934: Just like Paul Langevin, Pierre Auger and Jean Perrin, they joined the Vigilance Committee of Antifascist Intellectuals (Comité de vigilance

---

[17] Archives du Collège de France, dossier Joliot-Curie, CDF 16–26.

[18] Authoritative archives on the Joliots-Curies in Paris: Archives Musée Curie, Fonds Irène et Frédéric Joliot-Curie, dossiers I et F; Archives du Collège de France, dossier Joliot-Curie, CDF 16–26; Archives départementales de la Seine-Saint-Denis, Paris-Bobigny; Archives du CEA; Archives du Mouvement de la Paix.

des intellectuels antifascists, CVIA) against the rise of fascism in Europe. This committee is thought to have contributed to the electoral victory of the Popular Front (Front Populaire) in 1936, in whose government Léon Blum appointed Irène Joliot-Curie as State Secretary for Science and Research – even before the introduction of women's suffrage in France![19] In 1937 she became a professor at the Sorbonne. In 1939, along with Langevin, she signed the Appeal of the Union of French Intellectuals (Appel de l'union des intellectuels français)[20] against the Hitler-Stalin pact and the Soviet "about face."

In contrast to other Communist intellectuals, who first became great and "influential" in and through the French Communist Party (PCF), Joliot-Curie approached Communism via a detour through the French Socialist party SFIO, which he joined in 1934. While intellectuals in postwar France like Jean-Paul Sartre supported the cause of the Communists as a Communist sympathizer – Sartre especially in the period from 1952 to 1956, Joliot-Curie abandoned the role of the fellow traveler (compagnon de route) and took the decisive step of joining the PCF, as will be related briefly below.

After France was defeated in 1940, Joliot remained for the most part in his laboratory at the Collège de France,[21] where at the time, according to the testimony of the nuclear physicist and laboratory director Wolfgang Gentner, German scientists were working with Joliot's cyclotron, the destruction of which he and the physicists Wolfgang Riezler and Walther Gerlach supposedly prevented at the end of German occupation in 1944.[22] However, from 1942 to 1944 Joliot was also active in the committee of the National University Front (Front national universitaire), founded in June 1941 at the initiative of the underground PCF, and also as the Front's president.[23]

Deeply impressed by the important role of Communists in the Résistance, in spring 1942 he joined the underground PCF, going underground himself in June 1944. He subsequently even succumbed to the PCF propaganda and martyr-like self-stylization of the PCF as the resistance "party of the (75,000) martyrs." For Joliot-Curie the experience of occupation and résistance by (and with) the PCF was key to his later political involvement. After the announcement that he had joined the PCF, the party organ L'Humanité wrote melodramatically: "Thus superior intelligence, spirits molded by the most rigorous analytical methods, now joins our party."[24]

From 1945 on Joliot-Curie was also active in the Communist National Union of Intellectuals (Union nationale des intellectuals, UNI), which was part of the National Resistance Council (Conseil National de la Résistance, CNR).[25] The

---

[19] Strickmann, Französische Intellektuelle, pp. 47–48.
[20] Appel de l'union des intellectuels français, in: L'Œuvre, 30 August 1939.
[21] Archives du Collège de France, dossier Frédéric Joliot-Curie, CDF 16–26.
[22] Gentner, Joliot-Curie; Pinault, Joliot-Curie, pp. 256 ff.
[23] AMCP, Fonds Joliot-Curie, F 124. Pinault, Joliot-Curie, pp. 246 ff.
[24] Cachin, Marcel: Bienvenue à Joliot-Curie!, in: L'Humanite, 1 September 1944, p. 1.
[25] Discours pour l'UNI, 1945, AMCP, Fonds Joliot-Curie, dossier F 31, Mappe 61; F 143, UNI 1945–1949.

PCF made intensive efforts to instrumentalize the geniuses of the age, among them intellectuals like Louis Aragon, Paul Eluard and Pablo Picasso, in order to cultivate its own image as the "party of the intelligentsia." Prominent scientists and Nobel laureates like Joliot-Curie were supposed to lend a "progressive" scientific veneer to the political doctrine of (Soviet) Communism.

There was a sort of unwritten pact between Joliot-Curie and the PCF: He lent the PCF his reputation and a "scientific appearance" in return, he was courted by the PCF and honored in its media, which provided him a national and international forum for his political activities.[26] Through joining and participating, Joliot-Curie, in keeping with the typological classification by Michel Winock, ultimately became a "partisan intellectual,"[27] who was pilloried in retrospect by many critics in Julien Benda's tradition during the age of totalitarianism for "betraying" universal values by belonging to the party. While he maintained a certain degree of independence up to 1939, he later became an increasingly loyal "party soldier" of the PCF and its global ideology and strategy.

From 1944 to 1946 Joliot-Curie was director of the CNRS.[28] In 1945 he was installed by President Charles de Gaulle as the High Commissioner of the CEA, founded by de Gaulle in 1945. Irène Joliot-Curie, Pierre Auger[29] and Francis Perrin were appointed as CEA commissioners.[30] He was enraptured as were his colleagues (and the majority of the nuclear physicists of his day) by the apparently unlimited possibilities of the "peaceful use of atomic energy." In 1948 his CEA team succeeded in setting up the first French nuclear reactor, Zoë.[31] At the same time, indeed as early as June 1946, he also formulated his refusal in principle personally to participate in building an atomic bomb.[32] Thus he answered the burning question faced by nearly all nuclear physicists of the day with an unequivocal "No."

Also in 1946, Joliot-Curie, together with the London physicist Desmond Bernal (1901–1970), founded the World Federation of Scientific Workers (WFSW), and became its presidents. Joliot also used this organization to fight against atomic weapons.[33] This federation was quite controversial in the Western world. For example, as Otto Hahn indicated in a letter to Joliot-Curie: "Thus many accuse the WFSW, whose president you are, of one-sidedly representing only the ide-

---

[26] On his relationship with the PCF cf. AMCP, Fonds Joliot-Curie, dossier F 141.

[27] Winock, Intellektuelle, p. 54.

[28] AMCP, Fonds Joliot-Curie, dossier F 70, F 97–100.

[29] From 1937 on, professor at the Sorbonne in Paris.

[30] Archives du CEA, fonds du haut commissaire, there also Archives Francis Perrin.

[31] Cf. Hecht, Nuclear power; Hecht, Nuclear Reactors; AMCP, Fonds Joliot-Curie, dossier F 71–87, F 96.

[32] Pinault, Joliot-Curie, p. 565; A propos de la bombe atomique, 10 August 1945, AMCP, Fonds Joliot-Curie, dossier F 31, Mappe 46.

[33] Pour la cessation immédiate d'armes nucléaires, AMCP, Fonds Joliot-Curie, dossier F 121, on the FMTS cf. esp. dossier F 119–122.

ologies of the Eastern-oriented peoples."[34] In August 1948 the Communist World Congress of Intellectuals for Peace (Congrès mondial des intellectuels pour la paix/Weltkongress der Intellektuellen für den Frieden) was held in Wroclaw (formerly Breslau), Poland. Joliot did not attend because of illness, despite pressure from the PCF, but Irène Joliot-Curie did participate. After verbal tirades against Sartre, who was denounced as a "desk hyena," it took some convincing to prevent Irène from returning back to Paris immediately.

Also at this congress, the International Liaison Bureau of Intellectuals for Peace (Bureau international de liaison des intellectuels pour la paix) was set up, which, in turn, founded the World Congress of Fighters for Peace (Congrès mondial des partisans de la paix/Weltkongress der Kämpfer für den Frieden).[35] Joliot-Curie became its president and gave the inaugural speech of the first Congress in the Salle Pleyel of Paris on 20 April 1949, before an audience of delegates reported to be 2,000 strong from 75 countries. At the margins of the congress, Joliot participated in a National Conference of Intellectuals for Peace (Conférence nationale des Intellectuels pour la paix) reserved for French delegates.[36] The presidency of the World Congress of Fighters for Peace had a decisive influence on his final years: On 19 March 1950 he launched as the first signatory the "Stockholm Appeal": "We demand an absolute ban on atomic weapons, terrible weapons for the mass extermination of populations ... We consider that the government which is the first to utilize atomic weapons against any other country commits a crime against humanity and should be treated as a war criminal."[37]

Symptomatic of the later Communist exegesis of the Appeal was the interpretation of the word "war criminal," which was always a potential charge against the United States. According to the PCF propaganda press and Cominform, this appeal, for which Pablo Picasso drew a dove as a symbol of peace, was supposedly signed by 400 to 600 million people – a very huge exaggeration – fourteen million of whom were supposedly French. However, the actual signatories included such illustrious intellectuals and artists as Marc Chagall, Pierre Renoir, Marcel Carné, Gérard Philip, Jacques Prévert and Maurice Chevalier, but not Sartre and de Beauvoir.[38] Joliot-Curie was able to convince even the quantum physicist Pascual Jordan to sign the appeal. Yet when Jordan noticed that the appeal emerged not from a scientific elite, but from a worldwide Communist signature campaign controlled by Moscow, he quickly rescinded his signature and wished nothing more than to "... accelerate the growth of the proverbial grass over it."[39]

---

[34] Letter from Hahn to Joliot-Curie of 9 February 1955, AMCP, Fonds Joliot-Curie, dossier F 121.

[35] AMCP, Fonds Joliot-Curie, dossier F 125–140; Chebel d'Appollonia, Intellectuels, p. 174.

[36] Pinault, Joliot-Curie, pp. 441 ff.

[37] AMCP, Fonds Joliot-Curie, dossier F 125.

[38] Winock, Intellektuelle, pp. 615–616. However, a CMP brochure does list Sartre as a member: AMCP, Fonds Joliot-Curie, dossier F 127, 1950.

[39] Letter from Pascual Jordan to Ernst Brüche of 28 July 1950, Landesmuseum für Technik und Arbeit, Mannheim, DE Ernst Brüche.

The Stockholm Appeal was followed in April 1950 by a quite remarkable speech at the XIIth Convention of the PCF in Gennevilliers near Paris, in which Joliot-Curie announced: "Yes, the Soviet atomic scientists work with enthusiasm! ... They know the blessings of the peaceful applications of atomic power and have already tried them out, but they also know that their science and technology, if the just and acceptable proposals for banning atomic weapons continue to be rejected again and again, and if war criminals were to decide to drop atomic bombs on their country, they would certainly be able to carry out decisive retaliatory strikes ... Never will the progressive scientists, the Communist scientists ever hand over even an iota of their science in order to go to war against the Soviet Union."[40]

After international critique of his one-sided Soviet Communist involvement, this unambiguous speech provided the government of Georges Bidault with the final external reason for recalling him from his office as High Commissioner of the CEA in April 1950. His successor was Francis Perrin, who later was regarded the father of the French atomic bomb, detonated in 1960.[41] The preliminary pinnacle of his anti-nuclear activities as a public, political "party" intellectual collided head-on with his function as an official in a French authority that also was in charge of sensitive defense research, and, moreover, was subordinated to the United Nations Atomic Energy Commission (UNAEC).[42] Thus 1950 was a turning point for Joliot, after which his activities and statements took place nearly exclusively under the auspices of the PCF. His work for the communist World Peace Movement (Mouvement de la Paix/Weltfriedensbewegung) presumably took up a considerable portion of the energy he still had.[43]

In 1951 he was awarded the "Stalin Peace Prize," the Soviet counterpart to the Nobel Peace Prize, for his involvement.[44] In 1956 he finally became a member of the Central Committee of the PCF. As president of the World Peace Movement he conducted an extensive correspondence, also internationally, in order to win over reputed personalities covering the entire spectrum of society for the work of the Movement and the communist World Council for Peace (Conseil mondial de la Paix, CMP), in particular for the CMP events in East Berlin[45] from 1 to 6 July 1952 and from 24 to 28 May 1954, in Vienna[46] in December 1952, in Helsinki in June 1955, and for the Vienna Appeal against the Preparations for Atomic War[47]

---

[40] Joliot-Curie, Wissenschaft, p. 121.

[41] Estate archives on Francis Perrin and on atomic policy in: Archives de l'Institut Pierre Mendès France. At the same time Perrin actively advocated scientific-industrial collaboration between France and West Germany in the framework of EURATOM. Cf. Perrin, Atomenergie, p. 26.

[42] Cf. Hecht, Nuclear Power.

[43] On Joliot-Curie and the Mouvement de la Paix: AMCP, Fonds Joliot-Curie, dossier F 125–140; Archives départementales de la Seine-Saint-Denis, Paris-Bobigny.

[44] Bibliothèque littéraire Jacques Doucet, Archives de la modernité, Papiers de Vercors.

[45] AMCP, Fonds Joliot-Curie, dossier F 40, 210 b, F 41, 243.

[46] At the Congress Meeting in Vienna, this time Sartre was fêted enthusiastically.

[47] Présentation de l'Appel contre la préparation de la guerre atomique, Bureau du CMP à Vienne, 19 January 1955, AMCP, Fonds Joliot-Curie, dossier F 42, Mappe 253.

of 19 January 1955. He wrote to church representatives, politicians of every stripe, journalists, writers, and, of course, scientists. His letters to Pope Pius XII, Thomas Mann, ex-NSDAP politician Hermann Rauschning and Ludwig Zimmerer of the journal Ende und Anfang apparently remained unanswered. However, Joliot did receive diplomatically formulated refusals from Paul Distelbarth, a publicist of France, from Gustav Heinemann, Minister of the Interior of the Federal Republic of Germany from 1949–1950, and from Martin Niemöller, president of the Protestant Church.[48]

Additional correspondence was directed to the (pacifist-neutralist) milieu of Germans sympathizing with the Communists, among them Josef Wirth, Chancellor of the Reich from 1921/22, and now member of the national-neutralist Federation of the Germans (Bund der Deutschen, BdD) and the 1955 winner of the Stalin Peace Prize; Wilhelm Elfes, also a member of the BdD and president of the CMP from 1964 on; the Peace Committee members Walter Diehl[49] from the German Communist Party (Kommunistische Partei Deutschlands, KPD) and Walter Friedrich[50] (Council member), Siegmund Schulze of the "International Federation for Conciliation" (Internationalen Versöhnungsbund) and the writer Arnold Zweig, deputy to the East German parliament and speaker at the World Council for Peace Congresses in Paris and Warsaw.[51] The writer Anna Seghers (The Seventh Cross) became president of the German Committee of the CMP.[52] Apparently Martin Niemöller and Bishop Otto Dibelius of the "Confessing Church" impressed Joliot-Curie, yet Niemöller diplomatically rejected all of Joliot's congress invitations and increasingly doubted the sincerity of Joliot's commitment to peace.[53]

The year 1952 presented another turning point: In 1951 the PCF launched a campaign against a supposed "bacteria war" by the Americans in Korea, which later was exposed as a propaganda lie. Despite his potential expertise as a scientist, his allegiance to the PCF was so strong that Joliot-Curie had no qualms about lending his name and his scientific reputation to a public protest against the U.S. "bacteria war" on 8 March 1952.[54] In so doing he once again damaged his already fragile reputation in the Western world, entering into a conflict that dragged on until 1954. In direct response, the American U.N. Ambassador Warren R. Aus-

---

[48] Correspondance avec l'Allemagne 1951, 1953, 1955, AMCP, Fonds Joliot-Curie, dossier F 126.

[49] As a student, Diehl had been Président des jeunes partisans de la paix de l'Allemagne de l'ouest, AMCP, dossier F 125, 1950.

[50] Friedrich, physics professor at the Humboldt University in Berlin, Council Member and Président du conseil allemand de la paix in 1950, AMCP, dossier F 125, 1950.

[51] Correspondance avec Sigmund Schulze, 1952–1955, AMCP, Fonds Joliot-Curie, dossier F 126.

[52] Bertold Brecht, Helene Weigel, and Johannes R. Becher were also members of the comité allemand of the CMP, AMCP, Fonds Joliot-Curie, dossier F 125, 1950.

[53] Correspondance avec Martin Niemöller, 1951–1956, AMCP, Fonds Joliot-Curie, dossier F 126.

[54] AMCP, Fonds Joliot-Curie, dossier F 129–135.

tin published a sharp open rejoinder to the accusations and corresponded most intensively with Joliot-Curie.[55] Now Joliot increasingly faced the accusation of "prostituting" the sciences.[56] In 1952 the American Committee of the anti-Communist Congress for Cultural Freedom – whose international presidency was held by such figures as Raymond Aron and Eugen Kogon – published a sharp "open letter"[57] from American Nobel laureates. West German scientists, too, wrote in an urgent open letter: "The undersigned direct to you, Professor Joliot-Curie, the request to join your voice with ours in the demand that the assertion that bacterial weapons have been used in the Far East ... will be verified with all methods of modern science ... by experts from the International Red Cross."[58] Its signatories included Otto Hahn, Werner Heisenberg, Max von Laue and Adolf Butenandt.

In the years from 1955 to 1958, an extremely intensive correspondence took shape between Joliot-Curie and the British mathematician and philosopher Lord Bertrand Russell, winner of the 1950 Nobel Prize for literature.[59] Some of the highlights of this correspondence were Russell's incessant and critical efforts to "build a bridge" over ideological-political boundaries, in particular for Joliot-Curie to add his signature to the Russell-Einstein Manifesto of 9 July 1955, initiated by Russell and Max Born.[60] After lengthy "negotiations" and a struggle about the wording, in May 1955 Russell saw only two obstacles to Joliot-Curie's participation: He was irritated by the new Communist party line on the hydrogen bomb, which had been introduced by Khrushchev and adopted by the PCF Chief Maurice Thorez, and he found the new changes to the text proposed by Joliot-Curie far too extensive.[61] Since Einstein had already signed the previous version of the Manifesto shortly before his death, Russell adamantly refused to implement the additional changes demanded by Joliot-Curie. Using pressure and diplomatic skill Russell ultimately succeeded in convincing Joliot-Curie, the Communist representative of the international scientific community, to sign the Manifest unconditionally after all, more or less at the last minute.[62]

Irène Joliot-Curie's political activity as a nuclear physicist-intellectual in postwar France was increasingly restricted by her extremely fragile state of health due to radiation damage incurred many years previously.[63] When her husband went underground in 1944, she had to flee to Switzerland for tuberculosis treatment.

[55] AMCP, Fonds Joliot-Curie, dossier F 129.

[56] Letter from Saint Louis to Joliot-Curie of 11 May 1952, AMCP, Fonds Joliot-Curie, dossier F 129.

[57] Letter of 2 May 1952, ibid.

[58] Open letter by German Nobel laureates ... to Prof. Joliot-Curie, in: Hahn, Otto Hahn, p. 225.

[59] Correspondance, dossiers 'Russell', mouvement Pugwash, AMCP, Fonds Joliot-Curie, dossier F 121.

[60] Press Conference by Earl Russell, 9 July 1955, AMCP, Fonds Joliot-Curie, dossier F 121.

[61] Letter from Russell to Joliot-Curie of 12 May and 17 June 1955, AMCP, Fonds Joliot-Curie, dossier F 121.

[62] Letter from Joliot-Curie to Russell of 13 May and 4 July 1955, ibid.

[63] On Irène as an intellectual, cf. esp., AMCP, Fonds Irène et Frédéric Joliot-Curie, dossiers I, 14, 15, 31.

Already a professor at the Faculté des Sciences, in 1946 she also became director of the Institut du Radium.

In addition to supporting her husband's efforts to ban and condemn atomic weapons[64] – she was awarded the Prize of Honor (Prix d'honneur) of the CMP in 1950 – her main focus was on the struggle for strengthening women's rights and feminism, so that it is tempting to describe her as a Simone de Beauvoir in the field of science. She belonged to the National Committee of the Union of French Women (Union Des Femmes Françaises, UFF),[65] and was invited to participate in national and international (that is, Communist-leaning) women's conventions and women's associations, such as the National French Federation of Womens' Clubs for Liberal and Commercial Careers (Fédération Nationale Française des Clubs de Femmes de carrière libérales et commerciale), the Congress of German Women for Peace of the Democratic Women's League (Kongress deutscher Frauen für den Frieden des Demokratischen Frauenbundes Deutschlands)[66] in East Berlin in May 1948, the International Assembly of Women and the Congress of American Women in New York, International Women's Day in London, and the International Democratic Federation of Women (Fédération démocratique internationale des Femmes).[67] Yet she was seldom able to attend due to poor health.[68] However, she did give a speech against atomic weapons, for peace, and also for the USSR at the National Assembly of Women for Disarmament (Assemblée nationale des Femmes pour le désarmement) of the UFF on 11 March 1951 in Gennevilliers near Paris.[69]

Like her husband, she was also active in Communist sympathizing associations for (Soviet-oriented) international understanding, like the French-Polish Friendship (Amitié franco–polonaise), in part due to family tradition,[70] but also the Association France–Soviet Union (Association France–l'URSS)[71] and the French-Yugoslavian Association (Association franco–yougoslave).[72] In contrast to her husband, she hardly cultivated any German or German-French contacts and correspondence.

---

[64] "La puissance toujours accrue des moyens de destructions fournis par la science a rendu la guerre toujours plus dévastatrice. Aujourd'hui, avec les armes atomiques ... la guerre menace l'existence même de l'humanité ... c'est d'abord l'interdiction de toutes les armes de destruction massive ...", AMCP, Fonds Joliot-Curie, I 15, dossier 35, texte sur la bombe atomique d'octobre 1950.

[65] Ibid., I 14. Letter from the UFF to Irène of 16 June 1949, I 14, Mappe 25. Allocution au congrès de l'UFF, mai 1947.

[66] Letter from the DFD to Irène of 3 May 1948; Irène's refusal of 7 June 1948, AMCP, Fonds Joliot-Curie, dossier I 14.

[67] AMCP, Fonds Joliot-Curie, dossier I 14, Mappe 29, 17 May 1949.

[68] AMCP, Fonds Joliot-Curie, dossier I 33.

[69] AMCP, Fonds Joliot-Curie, dossier I 15, Mappe 37. Speech of 8 March 1951. An article about the bomb in: ibid., I 15, 35, October 1950.

[70] AMCP, Fonds Joliot-Curie, dossier F 118.

[71] AMCP, Fonds Joliot-Curie, dossier F 123.

[72] AMCP, Fonds Joliot-Curie, dossier I 14, I 33.

In June 1949 she attempted to convince Albert Einstein to participation in the communist Movement for Peace, by requesting that he chair its Committee for the Arts, Sciences, and Professions (ASP).[73] However, Einstein amicably rejected this request in a letter that mentioned the dissent about the Movement for Peace as follows: "Dear Irene Joliot-Curie … I am also convinced that the cause which is so near to our heart cannot be saved effectively in the proposed way … they always act strictly in conformity with the 'party line'."[74] Her term as Commissioner of the CEA, reaching an end after six years, was not extended as a result of her involvement with Communist sympathizing organizations. In 1956 she died of leukemia.

III. REPRESENTATIVES OF THE GÖTTINGEN EIGHTEEN AS GERMAN PHYSICIST-INTELLECTUALS

Analogous to the Joliot-Curies and the "Stockholm Appeal," the leading West German scientists of the "Göttingen Eighteen," spoke out publicly in their "Göttingen Manifesto" of 12 April 1957 against nuclear arms and armament, albeit under a different political banner, with different motives, and against a completely different historical-biographical background. Unlike France, Germany was divided and at the center of the global East-West confrontation of the Cold War, including the menace from atomic weapons.[75] During the Nazi era leading German nuclear physicists had worked in the "Uranium Club" (Uranverein) on a "uranium machine" as a nuclear source of energy or power plant, and had also investigated the possibilities of building a German atomic bomb.[76]

The Göttingen Manifesto[77] cannot be understood without this collective experience of the Nazi Uranium Club, which constitutes the événement fondateur (Sirinelli), the key experience that served as the foundation for an identity shared by the participating scientists, including Walther Gerlach, Werner Heisenberg, Otto Hahn, and Carl-Friedrich von Weizsäcker. Like the Manhattan Project, the Uranium Club served as a kind of bridge from the classical "humanist" scientist that had existed since Galilei Galileo, to modern scientists capable of bringing to the world an inconceivable destructive potential: "Gerlach wanted the bomb, but

---

[73] Letter from Irène to Albert Einstein of 3 June 1949, AMCP, Fonds Joliot-Curie, dossier I 14.

[74] Letter from Albert Einstein to Irène of 24 June 1949, ibid.

[75] Cf. among others, Soutou, Guerre.

[76] To date, this chapter of the history of science has led to numerous research controversies, and its assessment remains a highly controversial minefield, which cannot be discussed in greater detail here, cf. among others, Walker, Uranmaschine; Karlsch, Hitlers Bombe.

[77] Unlike Joliot-Curie and the Mouvement de la Paix, the history of the "Göttingen Manifesto" and the physicists of Göttingen has already been subject of intensive research and discussion; for this reason the framework of this paper will not include the known biographical data and life events and will note trace the history of the "Göttingen Manifesto," cf. most recently: Kraus, Göttinger Erklärung; Stölken-Fitschen, Atombombe; Rese, Wirkung.

not as a means to wage war; instead as a potential bargaining chip in negotiations with the Allies."[78]

This investigation of physicist-intellectuals will include all of those West German physicists (as well as some scientists from related disciplines) during the 1950s, the so-called atomic age, who applied their reputations and their names – even beyond the boundaries of their specialized fields – to the purpose of getting involved politically and socially, especially concerning the consequences of their (atomic) research. For reasons of limited space, this paper focuses on the example of the initiatives of several leading West German scientist-intellectuals, namely Carl-Friedrich von Weizsäcker, Max Born, Walther Gerlach, Otto Hahn, Werner Heisenberg, and their antipode Pascual Jordan.

These scientists were members of four different generational cohorts, each separated by an age difference of around ten years. The first and oldest cohort, born around 1880, included the "elders" Max von Laue, Otto Hahn and Max Born. This study cannot and should not elaborate any further on Max von Laue, however, because although he regarded the initiatives of his colleagues benevolently, for reasons of health and age he was barely able or willing to take part in them.[79]

Walther Gerlach represents the second cohort, born around 1890. The third cohort, born around 1900, includes Werner Heisenberg and Pascual Jordan, along with their French colleagues Frédéric and Irène Joliot-Curie. Von Weizsäcker, born in 1912, belongs to the fourth and youngest cohort. What members of these cohorts have in common, despite the often considerable age differences, is that they experienced the period of National Socialism in Germany as adults, consciously and actively, and generally at a rather high level of their careers. Only Max Born, in part because of his German-Jewish pedigree, was forced to emigrate to Britain and thus played a special role. Because of the political regime changes of 1933 and 1945/49, these physicists possess "broken biographies" in contrast to the subsequent stable career paths in the final third of the 20th century.

The "breaks" in their biographies are manifested in the leading scientific role they played during the Nazi period, their imprisonment and detention at Farm Hall near Cambridge in 1945, and their subsequent successful second scientific career in the Federal Republic. These broken biographies reflect the transition from the predominant tradition, which extended well into the Nazi period, of the scientist as a supposedly apolitical "German mandarin" (Ringer), whose every action serves only science and the scientific community, through aspiring to the role of an expert political consultant, to the physicist-intellectual type in the Federal Republic.

---

[78]　Karlsch, Hitlers Bombe, p. 270.

[79]　"I … find it quite gratifying that you want to act … in order to propagate the Pugwash idea … I ask only to renounce on my personal contribution. I am 79 years old, and my productivity has already dropped sharply", letter from Max von Laue to Gerlach of 23 April 1959, DMA, DE 80 Gerlach, 143, 1.

Along the long path that led to the articulation of a protest petition, also excellent scientists like Werner Heisenberg had to learn lessons in democracy. In a painful process of accommodation, Heisenberg experienced the failure of his efforts to establish himself as Adenauer's personal scientific advisor. These efforts, which began in 1949, were influenced by Heisenberg's elitist, meritocratic, and authoritative understanding of how to advise politicians, and carried out by means of the "German Research Council" (Deutscher Forschungsrat) he himself founded and presided over. His Research Council, which was also suspiciously similar in name to the National Socialist "Reich Research Council" (Reichsforschungsrat), ultimately failed to overcome the federal West German states' responsibility for science policy.[80]

Once his research council had merged with the "Emergency Foundation for German Science" (Notgemeinschaft der Deutschen Wissenschaft) to form the "German Research Foundation" (DFG), governed by the principle of self-administration, Heisenberg's elitist conception of advising was rather out of place and not appropriate for the times. The Commission for Nuclear Physics, officially appointed by the DFG and composed of former members of the Uranium Club, played a role in the consultations about the European Defense Community (Europäische Verteidigungsgemeinschaft, EVG) and the European Council for Nuclear Research (Conseil européen pour la recherche nucléaire, CERN). Yet Heisenberg's expert opinion was no longer consulted with regard to planning Adenauer's atomic policy, including the stationing of nuclear weapons in the Federal Republic.

In a precarious and confusing situation in the context of the Cold War, a variety of motives ultimately caused leading West German scientists, after a process of adaptation and "learning," to intervene in politics repeatedly and collectively. These included: the boundaries and setbacks experienced over and over in the course of science advising and the direct influence of science policy on the politics of the young Federal Republic in the form of Adenauer's authoritative "Chancellor Democracy"; the personally virulent involvement as nuclear scientists in the debate over nuclear weapons (and of course, previously as scientists under Hitler); Adenauer's public comments downplaying of "tactical nuclear weapons" as a mere "further development of the artillery"; moreover, ensuring a successful career for themselves in an area of the so-called "peaceful use of atomic energy" that appeared very promising.

The Nuclear Physics Working Group of the German Atomic Commission reacted to the planned "nuclearization" of NATO, and the attendant atomic armament of the newly founded Federal German Army, with a coordinated letter-writing campaign to the ministers responsible, Franz Josef Strauß and Siegfried Balke. In response, Strauß invited them to a personal meeting in January 1957, which ultimately remained inconclusive.

Adenauer's unfortunate words, the "further development of the artillery", functioned as the catalyst for the "Göttingen Manifesto" of 12 April 1957, com-

---

[80] Cf. Eckert, Primacy; Schirrmacher, Dreier Männer; Carson, New models.

posed by the "Göttingen Eighteen," a group that included four Nobel laureates and took its name in a clear reference to the "Göttingen Seven," a group of professors, including the brothers Grimm, who had protested against authoritarianism at the university in 1837. The Eighteen sought to educate the as yet almost completely uninformed West German public that: "Tactical nuclear weapons have the destructive effect of normal nuclear bombs."[81] Furthermore it advanced a political recommendation for foreign and defense policy: "We believe that a small country like the Federal Republic can protect itself most effectively and best advance the cause of world peace by expressly and voluntarily renouncing the possession of nuclear weapons of any kind."[82] At this juncture at the latest, they transgressed the limits of their discipline and intervened politically, legitimated by their status as intellectuals.

Broad sectors of the public were impressed by the scientists' sense of responsibility, and especially by the personal refusal of the signatories "to participate in the production, testing, or deployment of nuclear weapons in any way," in part because the public initially did not know about the connection with the Nazi Uranium Club, and thereby the scientists' involvement in nuclear weapons research for the Nazis, or knew only distorted facts presented in a flattering light.[83]

As intellectuals, they used a public protest petition to warn and enlighten. This intervention, which can be interpreted as an important step in the transition toward the type of the intellectuel scientifique/physicist intellectual, enjoyed a resounding success in the public's initial perception. Jürgen Habermas even declared the scientists as a model of critical reflection.[84] In form and content the Manifesto was primarily composed by the "atomic writer" Carl-Friedrich von Weizsäcker, who took on the leading role in shaping a phalanx of scientist-intellectuals and in the formulation of public campaigns.

Of all their colleagues, the physicist Pascual Jordan,[85] also a CDU backbencher in the Bundestag from 1957 to 1961,[86] was practically the only one who tried with his polemics to cut a swath through the phalanx of the nuclear physicists in Göttingen: "In Germany there is a tendency to regard the atomic physicist as a kind of higher being, endowed with omniscience … For the assessment of serious political issues of the day, the average atomic physicist is doubtlessly less competent and capable than the average democratic citizen."[87] Moreover, he

---

[81] Göttinger Manifest, in: Kraus, Göttinger Erklärung, pp. 202–203.

[82] Ibid., p. 202.

[83] Cf., for instance, the glorifying, heroic historical patchworks by Robert Jungk.

[84] Habermas, Hochschulreform, here pp. 282–283.

[85] Cf., among other sources, Staatsbibliothek Berlin, DE Pascual Jordan; Schirrmacher, Dreier Männer; Beyler, Demon.

[86] Letter from Jordan to Brüche of 18 September 1957, Landesmuseum für Technik und Arbeit, Mannheim, DE Ernst Brüche; letter from Weizsäcker to Gerlach of 31 May 1957, pp. 1–3, DMA, DE 80 Gerlach, 347, 1.

[87] Jordan, Frieden, pp. 14–15; letter from Gerlach to von Weizsäcker of 7 September 1957, ibid.

continued, the nuclear physicist had "absolutely no … professional responsibility [for politics]!"[88] These nearly classic (German) counterarguments, which were advanced by the theologian Helmut Thielicke and the philosopher Karl Jaspers among others,[89] thus can be juxtaposed to the social and political legitimation of the (physicist-) intellectual. What is remarkable is that it was precisely Jordan who presented himself to the public as a physicist-intellectual by commenting on social-political themes in countless books and articles, and thereby naturally claimed competence and "responsibility" for himself in this regard.

The conflict between the Göttingen eighteen and their antipode Jordan continued to smolder, finally escalating with the comments by Walther Gerlach,[90] vice president of the DFG from 1951 to 1961, on "the Jordan affair, which is gradually becoming a tragicomedy."[91] After Gerlach publicly criticized Jordan, Jordan's attorney threatened him with "legal action,"[92] unless he retracted his criticism and provided the "explanations demanded." The defamation case went before the District Court of Munich in 1959 was thrown out that same year.

Jordan's relationship to his former teacher Max Born and Born's wife Hedwig[93] was also destroyed, especially because of his anti-Göttingen activities. This damage is expressed particularly strongly in a passionate letter from Born to Minister of Defense Franz Josef Strauß, in which Born even alludes to Jordan's previous membership in the NSDAP and SA[94]: "If it comes to war and H bombs fall, nothing else remains … In this we, who understand something about physical forces, are all in agreement (apart from half-pathological phantoms like Pascual Jordan, who is always to be found where the power is, in Hitler's Reich as today …)."[95] Obviously depressed, Born summed up Jordan's political role: "Adenauer brought Pascual Jordan into the Bundestag as 'his physicist' and in so doing paralyzed the Eighteen."[96]

In fact the "Göttingen Manifesto" was the preliminary culmination of a series of public initiatives and interventions, which began in West Germany as early as 1955. The events can be divided into periods as follows: The years from 1955 to 1957 constituted the incubation phase for the emergence of the West German physicist-intellectual type: Max Born and Bertrand Russell initiated the Russell-Einstein Manifesto of 9 July 1955 against the development of the hydrogen bomb, which was signed by a total of eleven scientists, including nine Nobel

---

[88] "gar keine … fachliche Zuständigkeit", Jordan, Verantwortung, pp. 50 ff.

[89] Cf. Jaspers, Atombombe, pp. 268–277.

[90] A Walther-Gerlach biography by Bernd A. Rusinek is apparently about to be written.

[91] Letter from Walther Gerlach to Max Born of 7 July 1958, p. 2, DMA, DE 80 Gerlach, 347, 3.

[92] Ibid.

[93] Letters from Hedwig Born to Walther Gerlach of 4 February 1958, 2 March 1958, 14 April 1958, DMA, DE 80 Gerlach, 347, 2.

[94] Letter from Jordan to Born of 23 July 1948, Staatsbibliothek Berlin, DE Born, 353, 7/8.

[95] Letter from Max Born to Franz Josef Strauß of 7 March 1959, p. 3, DMA, DE 80 Gerlach, 347, 3.

[96] DMA, DE 80 Gerlach, 347, 2. Max Born.

laureates, among them Joliot-Curie (see above).[97] While Werner Heisenberg and Otto Hahn did not want to sign this appeal, shortly thereafter they did participate, along with Max Born and Carl-Friedrich von Weizsäcker, in the Mainau Conference of eighteen Nobel laureates in Lindau at Lake Constance on 15 July 1955, where they warned about the dangers of an atomic war. German physicists, through their main professional organization, the German Physical Society (DPG), finally joined both initiatives en bloc in September 1955.

A visible consequence of the Russell-Einstein Manifesto was the first "Pugwash Conference," which took place in Canada in July 1957, with international scientists discussing such topics as "the dangers of a nuclear arms race." While leading scientists all over the world declined to participate in the first conference,[98] among them Joliot-Curie, Werner Heisenberg, Otto Hahn, Adolf Butenandt and Gustav Hertz (from the Communist German Democratic Republic), and Heisenberg apparently did not even nominate the requested representative of the "Göttingen 18," the invitation to the second Pugwash Conference in Lac Beauport, Canada, in April 1958 was accepted by von Weizsäcker and the Frenchman Bernard Grégory.[99] During this trip von Weizsäcker also met Edward Teller, the father of the hydrogen bomb.

Weizsäcker's intensive discussions with Teller appear to have been a turning point concerning the issue of nuclear weapons, bringing him into conflict with several of his colleagues in Göttingen for a time. As Born remarked: "He reported about his visit to Canada (participation in the Pugwash Conference) and a trip to California, where he discussed with Teller four days long. This apparently had a strong effect on him. I do not want to say that he sold out, but he is now primarily interested in the technical-political questions rather than the clear and unambiguous ethical aspects …".[100] Weizsäcker discussed and analyzed these questions in a series of articles for the weekly journal Die Zeit entitled "Living with the Bomb. The current prospects of limiting the danger of an atomic war" (Mit der Bombe leben. Die gegenwärtigen Aussichten einer Begrenzung der Gefahr eines Atomkriegs).[101] Gerlach remarked in this regard: "Hahn was extremely dubious at the time. He thought that, 'this good cause has probably gone up in smoke'. I myself was appalled , but I believed it could be explained by von Weizsäcker's manner of discussing all such problems so long that one can no longer see his standpoint."[102]

Born, too, complained to Ernst Brüche about "von Weizsäcker's somewhat capricious stance."[103] This was an expression of von Weizsäcker's behavior, dip-

---

[97] AMCP, Fonds Joliot-Curie, dossier F 121, Correspondance, dossiers 'Russell', 1955.

[98] Participating from France was the biomedical scientist Antoine Lacassagne, AMCP, Fonds Joliot-Curie, dossier F 121, dossier Cyrus Eaton et Pugwash, 1957.

[99] Conférences Pugwash, 1958, ibid.

[100] Letter from Born to Ernst Brüche of 4 July 1958, DMA, DE 80 Gerlach, 347, 3.

[101] Linus Pauling responded with "Mit der Bombe kann man nicht leben" ("One cannot live with the bomb"), DMA, DE 80 Gerlach, 143, 2.

[102] Letter from Gerlach to Born of 7 July 1958, DMA, DE 80 Gerlach, 347, 3.

[103] Letter from Born to Ernst Brüche of 4 July 1958, ibid.

lomatically strategizing and maneuvering, always intent on the external effect on his political image, and somehow unable to take an enduring clear position. In 1957, in Weizsäcker's own words, he had "outwardly appeared to have left physics ... and switched over to philosophy."[104] He accepted a chair for philosophy in Hamburg. Thus he moved back and forth, interceding between the roles of nuclear physicist, philosopher, intellectual, "politician"[105] and "diplomat." Weizsäcker, who once called himself a "conformist,"[106] always aspired to a course of action "coordinated"[107] among all eighteen as a bloc.

In April 1958 he presented the draft of the "Neuenahr Declaration"[108] as a possible reissue of a declaration of the Eighteen to his now "former" colleagues in Göttingen. This posed many political and strategic questions, demanded information, public education, and civil protection, but not that nuclear weapons should be renounced. In response Born and Hahn "expressed reservations."[109] Born even called it a "feeble" and "limp, meaningless declaration that costs us our reputation."[110] Gerlach criticized, "Responsibility, morals, and ethics founder on considerations of utility and difficulty."[111]

Von Weizsäcker then admitted that he had not succeeded, "in bringing the opinions to a common denominator and therefore the declaration cannot be published."[112] According to Born, moreover, "the indecision that this brought to light [was] exploited by several gentlemen from the Atomic Energy Ministry in order to thwart the cause."[113] Born then demanded: "We should draw up a new declaration of the 18 and present it to the rest ... in the form of an ultimatum. Then we will see whether our group still holds together or falls apart."[114] However, this demand had no effect. The "Göttingen Declaration" as such thus remained de facto unique.

At the third international Pugwash Conference in September 1958 in Kitzbühel in Tyrol, 70 scientists, including Max Born as well as the physicists Gerd Burkhardt and Werner Kliefoth, approved the differentiated "Vienna Declaration"[115] against the (nuclear) arms race and for more "responsibility of science." The West German press hardly took notice of this, "until the Göttingen Eighteen joined

---

[104] Letter from Weizsäcker to Gerlach of 8 May 1957, DMA, DE 80 Gerlach, 347, 1.

[105] He also sought a mandate as a CDU delegate to the Bundestag. Negotiations on this failed however.

[106] Letter from Weizsäcker to Gerlach of 16 December 1957, DMA, DE 80 Gerlach, 347, 1.

[107] Ibid.

[108] Letter from Weizsäcker to Göttingen 18 of 14 April 1958, ibid.

[109] Letter from Weizsäcker to Göttingen 18 of 25 April 1958, ibid.

[110] Letter from Born to Gerlach of 28 April 1958, ibid.

[111] Walther Gerlach: Bemerkungen zu Weizsäckers Abhandlungen, DMA, DE 80 Gerlach, 143, 2.

[112] Letter from Weizsäcker to Göttingen 18 of 25 April 1958, DMA, DE 80 Gerlach, 347, 1.

[113] Letter from Born to Ernst Brüche of 4 July 1958, DMA, DE 80 Gerlach, 347, 3.

[114] Letter from Born to Gerlach of 4 July 1958, DMA, DE 80 Gerlach, 347, 2.

[115] "Verantwortung der Wissenschaft", in: Physikalische Blätter 11/12, 1958, pp. 482–488. The French participants were Bernard Grégory and Antoine Lacassagne.

in and facilitated the publication of the full text"[116] by attaching an introductory remark to the declaration in order to use their prestige to attract more attention to it.[117]

The Party Chairman of the Social Democratic Party (Sozialdemokratische Partei Deutschlands, SPD) Erich Ollenhauer wanted to win over the Göttingen Eighteen for the "Fight against Nuclear Death" (Kampf dem Atomtod) movement, which was created in the spring of 1958 .[118] However, von Weizsäcker rejected this in the name of the Eighteen, offering the following explanation, "The majority appears to wish to remain neutral in party politics" and "not to participate in ... political demonstrations."[119] Only Max Born, who was also already involved in the "Battle League against Atomic Damage" (Kampfbund gegen Atomschäden) and its journal Das Gewissen (The Conscience), was involved in the Fight against Nuclear Death movement, [120] side by side with politicians like Ollenhauer and Gustav Heinemann, church leaders like Martin Niemöller and intellectuals like Heinrich Böll and Alfred Weber.[121]

Born became much more influential after he won the Nobel Prize for physics in 1954. After emigrating back to West Germany from Great Britain to spend retirement in Bad Pyrmont, Born and his wife Hedwig fought tirelessly for the cause of eliminating nuclear weapons. Two elements were particularly characteristic of his commitment: As a devout Quaker he had a "Testimony against War;"[122] as a pacifist he always used this "Peace Testimony"[123] to appeal for "conscientious objection to war."[124] In contrast to colleagues like von Weizsäcker, Gerlach, Hahn and Heisenberg, his exile in Britain had also given him long years of experience with democracy as a responsible citizen, as he emphatically reminded Franz Josef Strauß with regard to the Federal Republic: "After hearing your lecture in Frankfurt you will perhaps contest my right to participate in the discussion of such matters ... After living over twenty years in England and Scotland, I have learned something about the nature of democracy. Every citizen feels himself partially responsible for the state and justified to have a say in it ... and we scientists also have the right to speak out ... Concern about the future of humanity was my sole reason for turning to the public."[125] With this plea Born also struck a blow for

---

[116] Radio lecture by Max Born in Süddeutscher Rundfunk of 17 January 1959, pp. 1–9, here p. 8, DMA, DE 80 Gerlach, 347, 3.

[117] Letter from Weizsäcker to Göttingen 18 of 20 November 1958, DMA, DE 80 Gerlach, 347, 1.

[118] Letter from Erich Ollenhauer to von Weizsäcker of 15 February 1958, ibid.

[119] Letter from von Weizsäcker to Erich Ollenhauer of 18 February 1958, ibid.

[120] On Max Born, cf., for instance, Staatsbibliothek Berlin, DE Max Born; Schirrmacher, Dreier Männer; Greenspan, Born; Born, Mein Leben; Born, Verantwortung.

[121] Cf., for instance, Rupp, Opposition.

[122] Letter from Hedwig Born to Gerlach of 28 April 1958, DMA, DE 80 Gerlach, 347, 2.

[123] Letter from Born to von Weizsäcker of 15 July 1958, ibid.

[124] Kriegsdienstverweigerung. Letter from Hedwig Born to Gerlach of 28 April 1958, ibid.

[125] Letter from Born to Strauß of 7 March 1959, p. 4, DMA, DE 80 Gerlach, 347, 3.

the type of the socially and politically committed physicist-intellectual, and for a more sustainable democratization of the public in the Federal Republic.

In keeping with Bertrand Russell's "Appeal to European Intellectuals" of 1958, Born debated at the "European Congress Against Nuclear Armament" on 5 and 6 July 1958 in Basel with illustrious intellectuals like the former resistance fighter (ancien résistant) of the journal Combat Claude Bourdet, Jean-Marie Domenach of Esprit, Hans-Werner Richter of Gruppe 47, Robert Jungk, Erich Kästner, Karl Barth, John B. Priestley and Benjamin Britten.[126] At the "Congresses for Cultural Freedom," which, as we know today, were financed with funds from the CIA, Born and Hahn met with leading anti-Communist intellectuals such as Arthur Koestler and Ignazio Silone. In collaboration with Gertrud von Le Fort, Hedwig Born published the book "The Christian in the Nuclear Age" (Der Christ im Atomzeitalter) in the name of Hans-Werner Richter's "Committee for Disarmament" (Komitee gegen Abrüstung).[127]

Born regarded it as his "main task in retirement to enlighten the general public about the ... consequences of scientific research and to arouse scientists' sense of responsibility with regard to these ... consequences of their work."[128] In a letter to Joliot-Curie, Born described his committed agenda as follows: "I am deeply interested in the problem of the application of nuclear energy to the struggle for political power, and I do all I can to work for reason and peace, in close connection with Otto Hahn. For this end it is absolutely necessary to remain 'neutral' between the main contesting ideologies."[129] Joliot-Curie's response thereupon complimented Born's commitment.[130] Insisting on his "neutrality," however, Born was concerned about the fact that an East German radio station had broadcasted a lecture Born had given about responsibility in science as if he had read it personally.[131]

In early 1961 the "Göttingen Eighteen" were asked whether they would still sign the "Göttingen Manifesto" in its original form. Somewhat implausibly, all respondents answered yes. This inquiry was preceded by Born's thesis "that the silence of the 'Göttingen Eighteen' on the progression of nuclear armament of the Bundeswehr was not understood and often condemned."[132] The silence of

---

[126] DMA, DE 80 Gerlach, 143, 3. Bertrand Russell: Aufruf an die europäischen Intellektuellen, Europäischer Kongress gegen Atomrüstung, DMA, DE 80 Gerlach, 143, 4.

[127] DMA, DE 80 Gerlach, 143, 6.

[128] Friedenskomitee (ed.), Blaubuch. Reprint from: Das Gewissen 10 (1957).

[129] Letter from Born to Joliot-Curie of 4 April 1955, AMCP, Fonds Joliot-Curie, dossier F 121, Correspondance, dossiers 'Russell'.

[130] "Je crois que des progrès importants ont été accomplis vers la compréhension des dangers atomiques, en particulier, grâce à des actions comme la vôtre", letter from Joliot-Curie to Max Born of 24 October 1955, AMCP, Fonds Joliot-Curie, dossier F 121.

[131] Ibid.

[132] Göttinger Erklärung heute. Bericht über die Einstellung der Göttinger Physiker Anfang 1961, in: Physikalische Blätter 6, 1961, pp. 263–269, also in: DMA, DE 111 Heinz Maier-Leibnitz, 094.

the Eighteen as a group, however, did not mean the premature end of the still
new type of physicist-intellectual in the Federal Republic.

The center of committed physicist-intellectuals had instead shifted toward
a different, "loosely organized group,"[133] the Federation of German Scientists
(Vereinigung deutscher Wissenschaftler, VDW). Proposed in Kitzbühel and
founded as a German Pugwash group in 1959 along the model of the Federation
of American Scientists (FAS), this organization failed to achieve a public effect
comparable to that of the Göttingen Eighteen. The political agenda shifted as well.
While representatives of the Eighteen were initially reluctant to get involved, and
especially to take on positions of leadership,[134] in the end its founding members
included Born, von Weizsäcker, Hahn, Heisenberg, and Gerlach along with Gerd
Burkhardt and Werner Kliefoth. Burkhardt's proposal to win over "people from
industry"[135] and Atomic Energy Minister Balke for the VDW, "mainly to obtain
financial assistance"[136] seemed to many "very misguided" and was rejected on in
order to maintain a certain degree of independence – in particular with regard to
politics: "The very strength of our Pugwash movement is that people who are in
politics for reasons of ambition are not included."[137]

The goal pursued by all was supposed to be the acceptance of responsibility for the effects and consequences of science, and not only for nuclear weapons.[138] Heisenberg described the two possible paths of VDW work as follows:
"To be publicly effective with clear demands, and to work in silence by exerting
personal influence on authoritative personalities in politics and business. The
mission of the group must ... lie between the extremes, whereby the 'hidden
path' is quite essential."[139] Weizsäcker described the difficulty faced by a German Pugwash group: "Steering correctly between a passionate and ineffective sect
on the other hand, and a bourgeois association on the one hand, is apparently
very difficult."[140] Even during its founding phase, this undertaking was attacked
sharply from officials in Bonn.[141] Yet to a considerable extent Gerlach was apparently able to calm the waters.

---

[133] Werner Kliefoth to Gerlach of 6 July 1959 (circular), DMA, DE 80 Gerlach, 143,1.

[134] "... steht es bei den meisten so, dass es ihnen unangenehm ist nein zu sagen, aber ebenso
unangenehm, ja zu sagen und es ihnen am liebsten gewesen wäre, man hätte sie gar nicht ge-
fragt.", letter from Burkhardt to Gerlach of 26 June 1959, Gerlach to Otto Hahn of 24 June 1959,
Burkhardt to Gerlach of 22 May 1959, DMA, DE 80 Gerlach, 143, 1.

[135] Letter from Born to Gerlach of 24 April 1959, DMA, DE 80 Gerlach, 143,1.

[136] Ibid.

[137] Ibid.

[138] Ibid. Vorbereitende Besprechung zur Gründung einer 'Pugwash-Gruppe' in der Bundesrepublik. Sitzungsprotokoll, pp. 1–2, in: VDW-Rundbrief 6/7, 1960, pp. 1–12, in: DMA, DE 111
Heinz Maier-Leibnitz, 094 pugwash.

[139] Ibid., p. 3.

[140] Letter from Weizsäcker to Gerlach of 8 February 1958, DMA, DE 80 Gerlach, 347, 1.

[141] Gerlach wrote, he had "... received a rather sharp letter from an official office because
of the whole Pugwash undertaking ..., with quite grave attacks against a gentleman", letter from
Gerlach to Burkhardt of 8 September 1959, DMA, DE 80 Gerlach, 143, 1. Unfortunately, the letter mentioned is not included in the Gerlach estate.

In the end von Weizsäcker again took on the role of "speaker" of the VDW. Besides scientists' responsibility for the effects and consequences of science, he placed two other items at the top of the agenda: civil defense and air-raid protection. A small VDW Research Office was set up in Hamburg, where he taught philosophy. In fall 1961 a study group was founded to work out a civil defense program, which was presented in the form of a VDW memo in 1962. He urged that these projects always be conducted with a high degree of academic and scientific thoroughness. For this purpose he also founded the "Max Planck Institute for the Study of Living Conditions in the World of Science and Technology" (Max-Planck-Institut zur Erforschung der Lebensbedingungen der wissenschaftlich-technischen Welt) in Starnberg in 1970, with Jürgen Habermas and himself as directors. Although the VDW served as the German Pugwash group from 1959 on, the French branch of the international Pugwash movement, the Association Française pour le Mouvement Pugwash, was not founded until 1964.

At this juncture it should be noted that the members of the wartime Uranium Club, the Göttingen Eighteen group, the Nuclear Physics Working Group of the German Atomic Energy Commission, and the VDW as German Pugwash group were largely identical and were the central "focal points"[142] of the sociabilité of West German nuclear physicist-intellectuals. The factors holding this sociabilité together were an elitist class consciousness as outstanding scientists, the external insignia of awards, such as the Nobel Prize for Physics or Chemistry, or the Max Planck medal, as "symbolic capital" (Bourdieu), and a kind of esprit de corps. The nuclear physicist-intellectuals not only acted en bloc as a closed group, rather especially since 1955/1957 also as individuals, turning to the media on their own, admonishing and informing, and seeking contact with international colleagues like Joliot-Curie. The important example of Max Born has been discussed in detail above.

For example Otto Hahn, President of the Max Planck Society from 1948 to 1960, informed the public about the risks and benefits of nuclear energy in a radio speech broadcast throughout Europe on 13 February 1955, entitled "Cobalt 60: Danger or Blessing for Humanity?" (Cobalt 60: Gefahr oder Segen für die Menschheit), which produced a reaction in the international press. Four days before, Hahn had written to his colleague Joliot-Curie: "… I see how despondent you are about the current situation in the world and how you, too, attempt, with all means at your disposal, to prevent a world conflagration"[143] and endeavored to work with Joliot-Curie in "bringing a number of internationally known nuclear scientists together to the table" in a neutral country like Switzerland.[144] In 1957 Hahn gave a lecture in Vienna entitled "Atomic Science for Peace

---

[142] Hahn, VDW, in: VDW-Rundbrief DM 6/7, 1960, p. 10, in: DMA, DE 111 Heinz Maier-Leibnitz, 094 pugwash.

[143] Letter from Hahn to Joliot-Curie of 9 February 1955; letter from Joliot-Curie to Hahn of 1 February 1955, AMCP, Fonds Joliot-Curie, dossier F 121, Correspondance, dossiers 'Russell'. Their contact was resumed after the war through a letter of 12 August 1946, with Hahn's words: "Je n'avais aucun occasion jusqu'ici de renouveller nos relations agréables …", ibid.

[144] Ibid.

and War."[145] In July 1958, shortly before Joliot-Curie's death, Hahn met him one last time at the Nobel laureate convention in Lindau, about which he later noted, "In Lindau the friendly contact was resumed immediately."[146] In the face of Joliot's death, he was able to find much that was positive in his colleague's political and scientific ambivalence: "It is a sign of a truly statesmanlike disposition that the French government honored the memory of this great man, whose political activity had caused it some trouble, with a state funeral."[147]

Werner Heisenberg, too, no longer exclusively advocated the "silent diplomacy" of the "invisible path" (see VDW, above) of the "advisor to the prince" (conseiller du prince,) which had failed to achieve the hoped-for success, but also took the path of mobilizing the public (remaining opposed to the nuclear armament of smaller countries like the Federal Republic). In a letter to Gerlach, Heisenberg wrote: "Many thanks for sending me your lecture about the 'Responsibility of the Physicist,' which I very much enjoyed. It seems that these issues will have to be pointed out to the public repeatedly for quite some time, for there is the risk that the public will lose interest. The circles of the federal government that believe in defense with nuclear weapons will simply wait until the public has grown tired of talking about the problem, and then carry out their armament ... Perhaps I, too, will speak in radio once more myself, and attempt to make clear that a defense with nuclear weapons is for the federal government simply not a defense, but rather precisely the opposite ...".[148]

His brother-in-law Erich Kuby, a controversial intellectual, and as of 1947 Editor-in-Chief of the Gruppe 47's journal Der Ruf (The Call), repeatedly urged Heisenberg to engage more frequently in public as an intellectual. In 1957, von Weizsäcker, also lectured on "The Responsibility of Science in the Atomic Age." Walther Gerlach spoke about the "responsibility of the physicist" in 1957, warning, "But the greatest calamity of all for humanity lies in the use of the energy of the atomic nucleus as an explosive, in 'atomic weapons'."[149] The catchphrases from the mouths or pens of highly regarded atomic physicist-intellectuals, "responsibility" and "atomic age," often combined as "responsibility in the atomic age," "responsibility of the scientist in the atomic age," or "peaceful use of atomic energy," thus experienced a boom as the expression of the specific Zeitgeist of the epoch.[150]

---

[145] "Atomwissenschaft für den Frieden und den Krieg", in: Der Mittag, 15 November 1957.

[146] Cited in: Hahn, Otto Hahn, p. 295.

[147] Ibid.

[148] Letter from Heisenberg to Gerlach of 12 February 1958, DMA, DE 80 Gerlach, 347, 11.

[149] Ibid., pp. 11–12.

[150] Elisabeth Kraus also argues with the responsibility of the scientist. Yet this study does not follow her typology, according to which the acceptance of responsibility in the Manifesto was humanitarian for Hahn (as well as for Born and von Laue), scientific for Heisenberg, and political for von Weizsäcker, because in my view initiatives like the Göttingen Manifesto are themselves a (social-) political act, so that at the most the motifs of their politicial action might have been formed corresponding to the typology of Kraus. Cf. Kraus, Göttinger Erklärung.

IV. CONCLUSION

The objective of this paper was to introduce to the history of science the concept of the intellectual, complete with an operational set of methodological tools, to study certain scientists, in particular nuclear physicists, not merely as they have been regarded to date – namely isolated as experts and parts of their national and international scientific community – but rather to open up a more fertile, broader perspective by regarding them as politically involved intellectuals interacting with other public intellectuals, and thus reintegrate them into the history of intellectuals so dominated by literati. This case study shows that, by transgressing previous boundaries, without ignoring or relativizing existing differences, there need not be any contradiction or diametric difference between nuclear scientists and intellectuals, between a history of scientists and of intellectuals; instead they can exhibit considerable intersections and commonalities.[151]

A brief German-French comparison results in the following rough sketch: While it initially appears more obvious for French nuclear physicists like Frédéric and Irène Joliot-Curie to be regarded and studied as intellectuals, this is no less the case for nuclear physicists in the early Federal Republic, such as representatives of the "Göttingen Eighteen." Their exertion of influence on the public and their social and political involvement was not only based on and justified by their scientific expertise as experts, but rather after a sometimes painful "learning curve," they also asserted the public "responsibility" of the scientist (even if this explanation, according to Ulrich Herbert, may be interpreted as a kind of "cover story"). For the Joliot-Curies, the German Occupation of France and (philo-)communist experience of the French Resistance against Nazism were the événements fondateurs, the key experiences and events shaping their identity, whereas for the "Göttingen Eighteen," some of whom belonged to similar generation cohorts, the événement fondateur was the personal experience in the Nazi uranium club and its confrontation with the possibilities of building an atomic bomb. While the French nuclear physicist-intellectuals were politically affiliated with Communism in the age of the (emerging) Cold War, and their West German colleagues were for the most part anti-Communist (yet generally without any party political affiliation), the two groups nevertheless arrived at the same fundamental conclusion despite completely different premises in this "ideological system" (Sirinelli) of past dictatorship and the issue of nuclear weapons: the refusal to participate personally in the building of a nuclear bomb, and the decision to get involved as politically articulating and intervening intellectuals interacting with other intellectuals and social groups against rearmament and/or nuclear armament.

---

[151] Although nuclear physicists, because of their sometimes controversial scientific and technical know-how, often doubtlessly possess much greater involvement, proximity and importance for the given political powers than do, for instance, writers or philosophers, and tenured professors have resources of research and a different obligation than free-lance literati and artists.

BIBLIOGRAPHY

Arbeitsgemeinschaft sozialdemokratischer Akademiker (ed.): Weltmacht Atom, Frankfurt am Main 1955.

Aron, Raymond: Opium für Intellektuelle, Paris 1957.

Aron, Raymond and Daniel Lerner (eds.): La querelle de la C.E.D., IEP, Paris 1956.

Bald, Detlef: Die Atombewaffnung der Bundeswehr, Bremen 1994.

Benda, Julien: Der Verrat der Intellektuellen, Paris 1927, Repr.: München 1978.

Bering, Dietz: Die Intellektuellen. Geschichte eines Schimpfwortes, Stuttgart 1978, Repr.: Frankfurt am Main 1982.

Beyler, Richard H.: The demon of technology, mass society and atomic physics in West Germany 1945–1957, in: History and Technology 19, 2003, pp. 227–239.

Bordry, Monique and Pierre Radvanyi (eds.): Œuvre et engagement de Frédéric Joliot-Curie, actes du colloque d'octobre 2000 au Collège de France, Paris 2001.

Born, Hedwig and Gertrud von Le Fort: Der Christ im Atomzeitalter, München 1958.

Born, Max and Hedwig Born: Der Luxus des Gewissens. Erlebnisse und Einsichten im Atomzeitalter, München 1969.

Born, Max: Physik und Politik, Göttingen 1960.

Born, Max: Von der Verantwortung des Naturwissenschaftlers. Gesammelte Vorträge, München 1965.

Born, Max: Mein Leben. Die Erinnerungen des Nobelpreisträgers, München 1975.

Carson, Cathryn: New models for science in politics. Heisenberg in West Germany, in: Historical Studies in the Physical and Biological Sciences 30, 1999, pp. 115–172.

Carson, Cathryn: Nuclear Energy Development in Postwar West Germany: Struggles over Cooperation in the Federal Republic's First Reactor Station, in: History and Technology 18 (3), 2002, pp. 233–270.

Carson, Cathryn: Nuklearpolitik im Zeichen des Kalten Krieges: Deutsche Wissenschaft zwischen Kernenergie und Protestbewegung, in: Aurora. Magazin für Kultur, Wissen und Gesellschaft (online magazine), 2003, http://www.aurora-magazin.at/gesellschaft/atom_carson_frm.htm (accessed on 24 June 2009).

Cassidy, David: Uncertainty: The Life and Science of Werner Heisenberg, New York 1992.

Chebel, d'Appollonia, Ariane: Histoire politique des intellectuels en France 1944–1954, Vol. I and II, Paris 1991.

Eckert, Michael: Primacy doomed to failure: Heisenberg's role as scienctific advisor for nuclear policy in the FRG, in: Historical Studies in the Physical and Biological Sciences 21, 1990, pp. 115–172.

Eckert, Michael: Die Anfänge der Atompolitik in der Bundesrepublik Deutschland, in: VfZ 1, 1989, pp. 113–142.

Eckert, Michael: Werner Heisenberg: controversial scientist, in: Physics World, December 2001, pp. 35–40.

Fischer, Ernst Peter: Werner Heisenberg. Das selbstvergessene Genie, München 2001.

Flasch, Kurt: Die geistige Mobilmachung. Die deutschen Intellektuellen und der Erste Weltkrieg. Ein Versuch, Berlin 2000.

Fölsing, Albrecht: Albert Einstein. Eine Biografie, Frankfurt am Main 1993.

Forman, Paul: Scientific Internationalism and the Weimar Physicists: The Ideology and its Manipulation in Germany after World War I, in: Isis 64, 1973, pp. 151–180.

Forman, Paul: Independence, not Transcendence, for the Historian of Science, in: Isis 82, 1991, pp. 71–86.

Friedenskomitee der Bundesrepublik Deutschland (ed.): Blaubuch über den Widerstand gegen die atomare Aufrüstung der Bundesrepublik, Düsseldorf n.yr. (1958).

Füßl, Wilhelm (ed.): Der wissenschaftliche Nachlass von Walther Gerlach, 2 vol., München 1998.

Gentner, Wolfgang: Gespräche mit Frédéric Joliot-Curie im besetzten Paris 1940–1942/Entretiens avec Frédéric Joliot-Curie à Paris occupé 1940–1942, Max-Planck-Institut für Kernphysik, Heidelberg 1980.

Gollwitzer, Helmut: Die Christen und die Atomwaffen, München 1957.

Greening, Richard H. and Jack M. Holl: Atoms for Peace and War 1953–1961. Eisenhower and the Atomic Energy Commission, Berkeley 1989.

Greenspan, Nancy T.: Max Born, Baumeister der Quantenwelt. Eine Biografie, Heidelberg, Berlin 2005.

Habermas, Jürgen: Heinrich Heine und die Rolle des Intellektuellen in Deutschland, in: Habermas, Jürgen: Eine Art Schadensabwicklung, Frankfurt am Main 1987, pp. 27–54.

Habermas, Jürgen: Das chronische Leiden der Hochschulreform, in: Merkur 11, 1957, pp. 265–284.

Hahn, Dietrich (ed.): Otto Hahn. Begründer des Atomzeitalters. Eine Biografie in Bildern und Dokumenten, München 1979.

Hahn, Otto: Cobalt 60– Gefahr oder Segen für die Menschheit? (Radio lecture on 13 February 1955, published by MPS, Dokumentationsstelle).

Harwood, Jonathan: Forschertypen im Wandel 1880–1930, in: Bruch, Rüdiger vom (ed.): Wissenschaften und Wissenschaftspolitik, Stuttgart 2002, pp. 162–168.

Hecht, Gabrielle: Enacting cultural identity. Risk and ritual in the French nuclear workplace, in: Journal of Contemporary History 32, 1997, pp. 483–507.

Hecht, Gabrielle: Political Designs. Nuclear Reactors and National Policy in Postwar France, in: Technology and Culture 35 (4), 1994, pp. 657–685.

Hecht, Gabrielle: The Radiance of France. Nuclear Power and National Identity after World War II, Cambridge 1998.

Heipp, Gunther (ed.): Es geht ums Überleben!, Hamburg 1965.

Heisenberg, Werner: Der Teil und das Ganze. Gespräche im Umkreis der Atomphysik, München 1969.

Hentschel, Klaus and Gerhard Rammer: Kein Neuanfang. Physiker an der Universität Göttingen 1945–1955, in: Zeitschrift für Geschichtswissenschaft 8, 2000, pp. 718–741.

Hentschel, Klaus and Gerhard Rammer: Physicists at the University of Göttingen 1945–1955, in: Physics in Perspective 3, 2001, pp. 189–209.

Hentschel, Klaus: Die Mentalität deutscher Physiker in der frühen Nachkriegszeit 1945–1949, Heidelberg 2005.

Hermann, Armin: Werner Heisenberg, Reinbek 1976.

Hoffmann, Klaus: Otto Hahn. Stationen aus dem Leben eines Atomforschers, Berlin 1978.

Hudemann, Rainer and Hélène Miard-Delacroix (eds.): Wandel und Integration. Deutsch-französische Annäherungen der fünfziger Jahre, München 2005.

Jaspers, Karl: Die Atombombe und die Zukunft des Menschen. Politisches Bewusstsein in unserer Zeit, München 1958.

Joliot-Curie, Frédéric: Cinq années de lutte pour la paix, articles, discours et documents, 1949–1954, Paris 1954.

Joliot-Curie, Frédéric: Wissenschaft und Verantwortung. Ausgewählte Schriften, Berlin 1962.

Jordan, Pascual: Der gescheiterte Aufstand. Betrachtungen zur Gegenwart, Frankfurt am Main 1956.

Jordan, Pascual: Wir müssen den Frieden retten! (brochure), Köln 1957.

Jordan, Pascual: Die Verantwortung des Wissenschaftlers, in: Internationales Jahrbuch der Politik 4, 1957, pp. 50–55.

Judt, Tony: Past Imperfect. French Intellectuals, 1944–1956, Berkeley 1992.

Jungk, Robert: Heller als tausend Sonnen – Das Schicksal der Atomforscher, Stuttgart 1956.

Jurt, Joseph: Der deutsch-französische Dialog der Intellektuellen, in: Kühn, Thomas et al. (eds.): Dialogische Strukturen, Tübingen 1996, pp. 232–259.

Jurt, Joseph: Status und Funktion der Intellektuellen in Frankreich im Vergleich zu Deutschland, in: Krauß, Henning (ed.): Offene Gefüge, Tübingen 1994, pp. 329–345.

Karlsch, Rainer: Hitlers Bombe, München 2005.

Koeppen, Wolfgang: Das Treibhaus, Stuttgart 1953.

Kraus, Elisabeth: Von der Uranspaltung zur Göttinger Erklärung. Otto Hahn, Werner Heisenberg, Carl Friedrich von Weizsäcker und die Verantwortung des Wissenschaftlers, Würzburg 2001.

Latour, Bruno: Joliot: Geschichte und Physik im Gemenge, in: Serres, Michel (ed.): Elemente einer Geschichte der Wissenschaften, Frankfurt am Main 1994, pp. 869–904.

Marmetschke, Katja et al. (eds.): Der Intellektuelle und der Mandarin, Kassel 2005.

Martens, Stefan and Maurice Vaïsse (eds.): Frankreich und Deutschland im Krieg. Okkupation, Kollaboration, Résistance, Bonn 2000.

Mommsen, Wolfgang and Gangolf Hübinger (eds.): Intellektuelle im Deutschen Kaiserreich, Frankfurt am Main 1993.

Perrin, Francis: Die Verwendung der Atomenergie für industrielle Zwecke, in: AG für Forschung des Landes NRW, Köln 1958, pp. 5–27.

Pinault, Michel: Frédéric Joliot-Curie. Le savant et la politique, Paris 2000.

Pinault, Michel: L'intellectuel scientifique: du savant à l'expert, in: Sirinelli, Jean-François and Michel Leymarie (eds.): Où en est l'histoire des intellectuels? Paris 2003, pp. 229–254.

Raulff, Helga (ed.): Strahlungen. Atom und Literatur, Marbacher Magazin (123/124), Marbach 2008.

Rese, Alexandra: Wirkung politischer Stellungnahmen von Wissenschaftlern am Beispiel der Göttinger Erklärung zur atomaren Bewaffnung, Frankfurt am Main 1999.

Ringer, Ritz: The Decline of the German Mandarins. The German Academic Community, 1890–1933, Cambridge/Mass. 1969, Repr.: 1990.

Rupp, Hans Karl: Außerparlamentarische Opposition in der Ära Adenauer. Der Kampf gegen die Atombewaffnung in den fünfziger Jahren, Köln 1970.

Schirrmacher, Arne: Dreier Männer Arbeit in der frühen Bundesrepublik: Max Born, Werner Heisenberg und Pascual Jordan als politische Grenzgänger, MPI für Wissenschaftsgeschichte, Preprint 296, 2005.

Sime, Ruth Lewin: Otto Hahn und die Max-Planck-Gesellschaft. Zwischen Vergangenheit und Erinnerung, MPG-Präsidentenkommission "Geschichte der KWG im NS", Ergebnisse 14, 2004.

Sirinelli, Jean-François: Le hasard ou la nécessité? Une histoire en chantier: l'histoire des intellectuels, in: Vingtième siècle 9, 1986, pp. 97–108.

Snow, Charles P.: The Two Cultures, Cambridge 1959.

Soutou, Georges-Henri: L'alliance incertaine. Les rapports politico-stratégiques franco-allemands 1954–1996, Paris 1996.

Stölken-Fitschen, Ilona: Atombombe und Geistesgeschichte. Eine Studie der fünfziger Jahre aus deutscher Sicht, Baden-Baden 1995.

Strickmann, Martin: L'Allemagne nouvelle contre l'Allemagne éternelle. Die französischen Intellektuellen und die deutsch-französische Verständigung 1944–1950. Diskurse, Initiativen, Biografien, Frankfurt am Main, New York et al. 2004.

Strickmann, Martin: Französische Intellektuelle als deutsch-französische Mittlerfiguren 1944–1950, in: Oster, Patricia and Hans-Jürgen Lüsebrink (eds.): Am Wendepunkt. Deutschland und Frankreich um 1945 – zur Dynamik eines 'transnationalen' kulturellen Feldes, Bielefeld 2008, pp. 17–48.

Strickmann, Martin: Die französischen Atomphysiker Frédéric und Irène Joliot-Curie als politische Intellektuelle am Ende der 1930er Jahre, in: Boyer-Weinmann, Martine and Olaf Müller (eds.): Das Münchener Abkommen und die Intellektuellen. Literatur und Exil in Frankreich zwischen Krise und Krieg, Tübingen 2008, pp. 257–266.

Strickmann, Martin: 'L'Allemagne nouvelle' oder 'l'Allemagne éternelle'? Die französischen Intellektuellen und die deutsch-französische Verständigung 1944–1950, in: Francia 32 (3), 2005, pp. 139–160.

Thielicke, Helmut: Christliche Verantwortung im Atomzeitalter, Stuttgart 1957.

Trischler, Helmuth: Geschichtswissenschaft – Wissenschaftsgeschichte: Koexistenz oder Konvergenz, in: Berichte zur Wissenschaftsgeschichte 22 (1999), pp. 239–256.

Trischler, Helmuth: Beyond 'Weimar Culture' – die Bedeutung der Forman-These für eine Wissenschaftsgeschichte in kulturhistorischer Perspektive, in: Berichte zur Wissenschaftsgeschichte 31, 2008, pp. 305–310.

Vom Bruch, Rüdiger (ed.): Wissenschaften und Wissenschaftspolitik. Bestandsaufnahmen zu Formationen, Brüchen und Kontinuitäten im Deutschland des 20. Jahrhunderts, Stuttgart 2002.

Walker, Mark: Otto Hahn. Verantwortung und Verdrängung, MPG-Präsidentenkommission "Geschichte der KWG im Nationalsozialismus", Ergebnisse 10, 2004.

Walker, Mark: Die Uranmaschine. Mythos und Wirklichkeit der deutschen Atombombe, Berlin 1990.

Walker, Mark and Dieter Hoffmann (eds.): Physiker zwischen Autonomie und Anpassung. Die DPG im Dritten Reich, Stuttgart 2006.

Weizsäcker, Carl Friedrich von: Die Verantwortung der Wissenschaft im Atomzeitalter, Göttingen 1957.

Weizsäcker, Carl Friedrich von: Mit der Bombe leben. Die gegenwärtigen Aussichten einer Begrenzung der Gefahr eines Atomkriegs, Hamburg 1958.

Winock, Michel: Das Jahrhundert der Intellektuellen, Konstanz 2003.

Winock, Michel and Jacques Juillard (eds.): Dictionnaire des intellectuels français, Paris 1996.

Wolff, Stefan L.: Physiker im "Krieg der Geister," working paper MZWTG 2001.

THE GERMAN RESEARCH FOUNDATION AND THE EARLY DAYS OF LASER RESEARCH AT WEST GERMAN UNIVERSITIES DURING THE 1960s[1]

Helmuth Albrecht[2]

INTRODUCTION

Much has been said and written about the impact the Emergency Foundation for German Science (Notgemeinschaft der deutschen Wissenschaft, NGW), or the German Research Foundation (Deutsche Forschungsgemeinschaft, DFG) as it was later renamed, has had on the development of German science since the 1920s. Recently the DFG itself has initiated and funded a program for confronting the history of the DFG from the 1920s to the 1970s, which has led to many new findings regarding the structure, methods, focused programs for support, and significance of the DFG for German academia, in particular for the post-1945 era. Numerous scientific fields that the DFG was instrumental in advancing in Germany, especially during the second half of the 20th century, remain poorly studied or unexamined. These include high-frequency physics and technology, which began to establish itself in the 1930s, and in particular led to the development of radar technology during World War II, and the Microwave Amplification by Stimulated Emission of Radiation (= Maser), and the Light Amplification by Stimulated Emission of Radiation (= Laser) as new applications during the 1950s and 1960s. Above all since 1960 the laser has been turned into an appliance that was in equal measure fascinating for scientific basic research, as well as for technical applications, always opening new perspectives on cognition and applications; its part as a key technology of the future was undisputed from the start.

The following will scrutinize the early stages of foundational research in this new key technology in the Federal Republic of Germany (Bundesrepublik Deutschland) during the 1960s, using as examples the roles played by the West German universities and DFG, as well as their connection, in this process of establishment.

In other words, the subject of analysis will be academic research. It should be pointed out, however, that virtually at the same time laser research also began to take hold in West German non-university and industrial research, but initially

---

[1] The following article is based on the author's habilitation thesis, Albrecht, Laserforschung in Deutschland (Laser Research in Germany) from 1997, as well as upon new research within the framework of the VolkswagenStiftung-funded project "Emergence and Evolution of a Spatial-Sectoral System of Innovation: Laser Technology in Germany, 1960 to Present".

[2] TU Bergakademie Freiberg.

without major overlaps.[3] Though it was not until the 1970s and 1980s that laser technology in the Federal Republic led to practical applications or commercially successful products worth mentioning, even so the 1960s represented something like an important stage of trial and error, during which above all at universities the basis was established in several areas of laser research, as well as especially in the education of the academic experts needed.

THE INITIAL SITUATION: WEST GERMAN HIGH-FREQUENCY RESEARCH AROUND 1960

The news, long-awaited in expert circles, that during the summer of 1960 in the USA a laser had been achieved for the first time spread rather reluctantly at first. The difficulties Theodore H. Maiman faced when he published his test results bespeak the skepticism with which the first reports were received among experts.[4] Samuel Goudsmit, editor-in-chief of the leading physics journal Physical Review Letters, to whom Maiman had first submitted his paper, rejected publication, saying in his opinion it hardly offered anything new. Maiman finally had to make do with the – in professional circles – less distinguished British journal Nature, which accepted and published his article "Stimulated Optical Radiation in Ruby" in its edition from the 6th of August 1960.[5]

Magazines, both for popular science as well as more technical expert publications, initially responded only hesitantly to the announcement of the laser discovery.[6] The reaction of physicists was notably cautious, since Maiman's short article in Nature was devoid of details regarding the structure and results of his tests, and given that in spite of the first symptoms of laser activity, he obviously failed to prove the expected abrupt appearance of laser radiation (laser threshold), probably due to the bad quality of pink ruby crystals he used.[7] Despite these obvious defects, however, Maiman's work made an impact on the other US working groups dealing with the laser issue at that time. In the very same summer of 1960 most of the approximately half a dozen groups at least partially readjusted their research approaches based on Maiman's design, and as a result also met with suc-

---

[3] Cf. Albrecht, Laserforschung in Deutschland.

[4] Cf. Bromberg, Laser, pp. 91 ff.

[5] Maiman, Radiation.

[6] See e.g. Laser – Coherent Light Source, in: British Communications and Electronics 7 (10), 1960, p. 766; Light-Beam Amplifier, in: Sky & Telescope 20 (4), 1960, p. 203; Optical Maser, in: Perspective 2 (4), 1960, pp. 390–392; Optical Maser, in: Scientific American 203 (6), 1960, pp. 80 and 82.

[7] It was not until 1961 that Maiman together with his colleague Irnee J. D'Haenens were able to prove without doubt a laser activity in his test series in a more detailed publication. See Maiman/D'Haenens, Emission.

cess.[8] By the end of the year no less than four different kinds of laser were put into operation: Maiman's pink ruby laser (Hughes Aircraft Company), Arthur Schawlow's dark ruby laser (Bell Laboratories), Peter Sorokin's uranium calcium fluoride laser (IBM, Thomas J. Watson Research Center), and Ali Javan's helium-neon gas laser (Bell Laboratories).[9] But it was only after these successes were gradually recognized, and the October 1960 paper from the Schawlow-Townes group demonstrated for the first time laser action in ruby beyond doubt, that laser made its breakthrough in the world of science and outside the USA. At that time, many physicists believed that the Bell group had won the great laser race.[10]

A year after the Schawlow-Townes' article, the Karlsruhe high-frequency engineer and university professor Horst Rothe concluded his plenary address in October 1961 at the 26th German Physicists' Conference (Deutscher Physikertag) "State and New Developments in Maser Research", which dedicated a passage to the emergent laser development, with a dramatic statement:

> It is disconcerting, however, that German research is not involved in this development initiated by Heinrich Hertz in my hometown Karlsruhe. We have to face the fact, that without supreme mastery of those experimental tools and aids research in those frequency ranges will be doomed to stagnation, if not failure.[11]

Given that working groups in high-frequency physics and high-frequency techniques had already existed for a couple years, concerned among other things with the development of the maser and aware of the fundamental Schawlow-Townes article regarding the optical maser in the Physical Review of December 1958,[12] Rothe's statement seems startling at first. Several institutions were active in the field of maser research in the Federal Republic. These included, for example, the Institute for High-Frequency Technology and Physics of the Technical University Karlsruhe,[13] headed since 1956 by Rothe himself. The Federal Physical-Technical Institute (Physikalisch-Technische Bundesanstalt, PTB) in Braunschweig[14] had been involved since 1958, initially concerned with the construction of ammonia masers, and soon afterwards with hydrogen masers in connection with building

---

[8] That concerned above all the laser media used, the application of a Fabry-Perot structure to create the reaction coupling, as well as the pumping sources to maintain the required population inversion in the laser media.

[9] Cf. Bromberg, Laser, pp. 92 ff.

[10] Ibid.

[11] Cf. Physikalische Verhandlungen 12, 1961, p. 182. The lecture was published in: Brüche, Physikertagung Wien, pp. 68–81. The laser is subject on the pp. 76–81.

[12] Several German laser pioneers have brought the author's attention to that in interviews. The author holds the audiotape recordings, as well as transcripts of the interviews (= Laser Archive Albrecht, henceforth cited as LAA).

[13] Rothe (1899–1974), a graduate and doctorate engineer (Dipl.-Ing. and Dr.-Ing.), had worked from 1927 to 1956 with Telefunken, lastly as director of the Röhrenwerk Ulm, before he was appointed professor at the TH Karlsruhe. His work mainly focused on electron tubes and semiconductors, as well as on the maser. See also: Bedeutung.

[14] The national metrology institute providing scientific and technical services.

atomic clocks.[15] Industrial circles, including the company Siemens, were also interested. At Siemens the solid-state physicist and subsequent head of the conglomerate's central research and development, Walter Heywang, had already applied in 1958 for a – at that time admittedly rather speculative – patent with the title "Arrangement to amplify and excite very high-frequency radiation according to the maser principle", providing the use of a semiconductor diode in the radiation field.[16] Soon afterwards the research lab of Siemens & Halske AG in Munich, founded in 1959, successfully developed a ruby maser, which would be put to use from 1962 onwards in satellite communication.[17]

The founding dates of the groups and departments make clear that, in the Federal Republic, research in the field of high-frequency physics and high-frequency technology did not begin until the mid-1950s. Research restrictions imposed by the Allies after the lost war, which were not lifted until the mid-1950s, caused this late entry into a research field in which Germany had been among the leading nations during the 1930s and early 1940s in connection with radar development.[18] The Control Council Law No. 25 issued on April 29th 1946 prohibited any kind of basic or applied research in fields of military importance. This restriction included research concerning "electro-magnetic, infrared and acoustic radiation" as well as "tubes or other electron-emitting appliances". The prohibition, which prevailed for a decade, disconnected German high-frequency research from the general trend, for example, in the USA or Great Britain, and beginning in the mid-1950s the Federal Republic had to struggle to catch up. For this reason, and acting on a suggestion made by Horst Rothe in 1958, the DFG set up a special priority program "High-Frequency Physics" with the aim "[to promote] theoretical and experimental studies in connection with the development of molecular amplifiers".[19] Until this program expired, West German high-frequency research received funds of more than 10 million DM (see chart 1); a wide array of high-frequency technological and physical work, as well as developments ranging from resonance spectroscopy over relaxation measurements, spectroscopic and calorimetric work through to quantum theoretical studies about the matter's energy state in all aggregate states, were supported. On the first DFG Focus Program Colloquia, which were held annually in the fall, methodical and theoretical basic issues were at the centre of attention. Finally, the symposium in October 1960 included a report "about the first amplifying effect of a maser arrangement."[20]

---

[15] Cf. Becker/Fischer, Beitrag.

[16] Yet the laying-out document of the patent was not published until 1967, when the semiconductor laser diodes had long since appeared on the market. Cf. Lemmerich, Geschichte, p. 52.

[17] The research lab concentrated on both, gas, as well as solid-state masers. The ruby maser developed at Siemens found use in the satellite radio station Raisting established in 1963/64. Cf. interview with Röß from 6.12.9, LAA; Kleen, Arbeit; Kinder/Stöhr, Funkstelle.

[18] Cf. Kern, Entstehung, esp. pp. 106 ff.

[19] Cf. Mitteilungen der Deutschen Forschungsgemeinschaft 2/3, 1959, pp. 16 ff.

[20] Cf. Mitteilungen der Deutschen Forschungsgemeinschaft 2/3, 1960, p. 37.

Chart 1: Funds within the framework of the DFG priority program „High-Frequency physics" 1958–1967[21]					
Year	Number of applications	Funds granted (in DM)	Subjects granted		
			Maser	Laser	Others
1958	16	1.436.440	3	–	12
1959	20	589.703	3	–	16
1960	39	1.462.482	6	–	20
1961	22	984.776	3	–	?
1962	22	1.322.965	4	5	?
1963	28	1.203.398	?	4	?
1964	28	1.234.224	?	4	?
1965	33	1.341.445	?	7	?
1966	33	1.253.853	?	7	?
1967	4	35.700	?	–	?

Prior to the fall of 1960 no German working group concerned itself with testing or even the theoretical aspects of an optical maser or laser. More than 30 years later Hermann Haken, whose theoretical work about the laser was funded in the early 1960s within the scope of the DFG High-Frequency Physics priority program, cited as reasons for this late start communication problems and the unwillingness of the German scientists involved to take risks:

> I believe one reason for this is that researchers in America communicate better with each other, that they talk a lot more about new developments, and they are much less risk-averse in taking up something basically new. Building this laser had been an immense challenge, in which one was aware of the fact that maybe it might not work, but one was stretching the limits of feasibility and at least, had given it a try. Whereas, I think, in Germany we kind of let it slip.[22]

The close cooperation between physicists and engineers, that is, representatives of different faculties and research fields, which was characteristic for the American maser and laser research obviously posed a problem in the Federal Republic of the late 1950s. Thus it had been the professed aim of the DFG High-Frequency Physics priority program to advance cooperation between high-frequency technologists and physicists and to overcome the existing communication problems

[21] Source: Berichte der Deutschen Forschungsgemeinschaft über ihre Tätigkeit 1958/59. In this context should be also mentioned the scientists at the Heinrich Hertz Institute of the TU Berlin, who also conducted microwave maser development.

[22] Albrecht, Hermann Haken im Gespräch, here in particular p. 258.

among the two groups of researchers. This becomes apparent in a DFG evaluation from 1963:

> Thus the separate working groups differed from the outset in their academic training and their technical terminology. For that reason one decided immediately in the first year to establish an annual one-week colloquium. It was supposed to serve the following purposes:
> 1. Mutual introduction into the experimental and theoretic methods,
> 2. Exercising a language that should become familiar for everyone involved,
> 3. Exchange of individual experiences among working groups,
> 4. Reports about the individual work, promoting friendly competition in doing so.[23]

Thus the DFG priority program was first of all supposed to provide the basis regarding methods, models and language for said interdisciplinary communication, cooperation, and constructive competition among theoreticians, experimental physicists and engineers. In so doing, the program responded in 1958 to the maser's history of development in the USA four years before. The maser – and soon afterwards the laser, too – combined physical concepts, like the concept of stimulated emission or the one of the Fabry-Perot interferometers, following the principle of a feedback oscillator familiar to every high-frequency engineer. The extraordinary achievement of Charles H. Townes had been his capability as a physicist and as an engineer to go beyond disciplinary boundaries in realizing the maser and to combine concepts, which at first sight did not belong together, as Joan Bromberg emphasizes in her book Laser in America.[24]

Yet cooperation comparable in quantity and quality to the USA between scientists and technicians of different disciplines, or between university and industry research, between basic and applied research as well as between university and industry on side and state and military research institutions and donors on the other did not exist in the Federal Republic. Until the mid-1960s the lion's share of the existing research potential focused on higher education with its traditional concepts[25] of "freedom and unity of science and research." Above all the natural scientists frequently remained quite attached to the neo-humanistic ideal of not being goal-driven (Zweckfreiheit), which caused substantial reservations regarding cooperation in particular with engineers. A DFG report about the colloquia held within the framework of the High-Frequency Physics priority program in November 1963 conveys indirectly the kind of difficulties German high-frequency research had to overcome in doing so:

> After the fifth colloquium it became obvious that notable progress had been made. In specific, more physicists have dealt with the technical problems of molecular amplifiers, and an increasing number of electrical engineers have pondered basic issues of the material used. Furthermore, a theoretical approach has favorably become a permanent feature of the work conducted by experimentalists. And finally a standardized language can be observed in debates and lectures, thus turning communication difficulties into a thing of the past.[26]

---

[23] Mitteilungen der Deutschen Forschungsgemeinschaft 4, 1963, pp. 13 ff.
[24] Bromberg, Laser, p. 19.
[25] See generally: Weingart, Wissensproduktion, pp. 93 ff.; as well as Hirsch, Fortschritt, p. 128.
[26] Mitteilungen der Deutschen Forschungsgemeinschaft 4, 1963, p. 14.

In retrospective, however, Hermann Haken, himself a participant of the DFG program's colloquia, judged their result clearly more negatively:

> We gave lectures every so often, but it did not lead to direct contacts, like people saying, for instance, in Karlsruhe, where there were a lot of high-frequency people, and in Stuttgart, let us for once join forces in this new field. Unfortunately that never occurred, though I do not want blame anybody for not taking the initiative. Well, neither did I.[27]

Even though since the mid-1950s the development of nuclear energy and the establishment of large-scale research facilities[28] in the Federal Republic demonstrated a gradual change in how science was managed and conducted, German high-frequency research in 1960 was far behind such revolutionary transformations. It remained, at least until the discovery of the laser, more of a physics-technical borderland, which, on top of everything, was totally overshadowed by the actual big research topics of the time, including nuclear power, space travel, and semi-conductor development. Nuclear physicist Peter Brix implied with the benefit of hindsight that high-frequency research suffered from this lack of attractiveness during the 1950s. When he gave a series of lectures at the University of Heidelberg in November 1991 under the title "Emeriti remember", he answered his own question, "Why didn't we invent the laser at that time in Göttingen (or later in Heidelberg)?":

> We dare say, because the Americans were smarter. But maybe there is a bit more to it than meets the eye. In Germany one started neglecting nuclear physics, you see; it was widely regarded as identical with the spectroscopy of free atoms, and that was considered as quite worked out. To put it bluntly: atomic physicists had the image of outdated scientists. In the early 1950s the nuclear physicists were regarded as young and hot (later it would be the high energy physicists). … We probably all ogled too much at nuclear physics.[29]

But there were others reasons too, why this research field did not appeal to many physicists, conveyed in a remark by Hermann Haken:

> In fact high-frequency is more engineerish. Hardly any physicists are doing high-frequency technology. The laser as such, however, has become a purely physical instrument.[30]

---

[27] Albrecht, Hermann Haken im Gespräch, p. 258.

[28] In the DDR this development was above all reflected in 1957 in the foundation of the Forschungsrat der DDR chaired by Peter-Adolf Thießen, as well as the establishment of the For-schungsgemeinschaft der naturwissenschaftlichen, technischen und medizinischen Institute der Deutschen Akademie der Wissenschaften zu Berlin, which not least aimed at a closer interlocking of science and economy. See Rexin, Entwicklung, here pp. 97 ff.; Landrock, Akademie, pp. 61 ff. In the case of the FRG should be adverted to the foundation of the Wissenschaftsrat (1957), as well as a whole series of large-scale research institutions, which included until 1960 the Kernfor-schungszentrum Karlsruhe (KfK, 1956), the Kernforschungsanlage Jülich (KFA, 1956), the Ge-sellschaft für Kernenergieverwertung in Schiffbau und Schifffahrt in Geesthacht (GKSS, 1956), the Deutsche Elektronen-Synchrotron in Hamburg (DESY, 1959), the Hahn-Meitner-Institut in Berlin (HMI, 1959), the Gesellschaft für Strahlen- und Umweltforschung in Munich (GSF, 1960), as well as the Max Planck Institute for Plasma Physics in Garching (IPP, 1960). Cf. Szöllösi-Janze/Trischler, Großforschung (1990).

[29] Brix, Erinnerungen, here pp. 17 ff.

[30] Albrecht, Hermann Haken im Gespräch, p. 258.

Words like that confirm what the English physicist and novelist Charles Percy Snow had called the breakdown of communication between the "two cultures" (at just the same time that the theoretical foundations for the laser were developed) with regard to the gulf between the sciences and the humanities.[31] A similar rift in language, methods and "culture" can be detected in the early stages of high-frequency research among physicists and engineers.

Even a priority program like the one for high-frequency physics offered no short-term solution for such problems. Because of the general development gap, the comparatively low funds, the rather poor occupational outlook, and the lack of promise of rapid scientific validation in this research field, up-and-coming junior researchers did not exactly push into high-frequency physics and high-frequency technology, at least not in Germany. Only with the laser could a trend reversal be achieved, and likewise more young physicists became fascinated by the now "new" high-frequency physics.

But given the available resources and mentalities, it was hardly possible to make up the American advantage or even take the lead within a couple of years – regardless of this "trend reversal". It was not until late 1966 that the DFG came to the conclusion that the program helped at least "some papers to catch up on international level" and that now such a "great number of younger and promising employees [dealt] with issues of this specific field" that the program could be concluded.[32]

In 1960, that is, at the time the first laser was realized in the USA, high-frequency research in the Federal Republic was on the level of a follow-on country, which had not yet managed to catch up. Nonetheless, the laser research that now commenced in Germany by no means started at zero, rather instead could build upon the initial achievements in laser and maser research. It soon became apparent that the know-how required for a rapid development of laser research was available in small but nonetheless innovative research groups, and that the DFG priority program was responsible in no small part for the existence of these groups.

FROM HIGH-FREQUENCY TO LASER RESEARCH:
THE LASER EUPHORIA OF THE 1960s

Like in the USA and Great Britain, in 1960 brief comments on the realization of the laser in the German-speaking region were mainly found in technical and popular science magazines.[33] Hermann Haken was probably one of the first to

---

[31] Snow, Kulturen.

[32] Mitteilungen der Deutschen Forschungsgemeinschaft 2, 1966, p. 13.

[33] For instance in the articles: Laser – neuartiger Verstärker für monochromatisches Licht, in: Elektronik 9 (11), 1960, p. 345; Nach dem „Maser" auch der Laser, in: Technica 9 (17), 1960, p. 1016; Der „Laser" – ein neuer Lichtverstärker, in: Elektron 9, 1960, pp. 245 and 254; Laser-Rubin – ein neuer Lichtverstärker, in: Orion 15 (12), 1960, p. 958; Kosmos 56 (10), 1960, p. 374. The first to speculate in public about laser applications in telecommunication engineering was Dietrich in 1960/61; Dietrich, Licht.

bring detailed information to Germany, when on his return from the USA he took up the chair in theoretical physics at the Technical University Stuttgart in early October 1960– on the eve of the publication of the Schawlow group's article.[34] Since the fall of 1959 Haken had been working as a solid-state theoretician in an advisory role at the Bell Telephone Laboratories in Murray Hill, New Jersey and had thus come directly in touch with the laser development there. The German physicist Wolfgang Kaiser, then on the staff of Schawlow's laser group at Bell and a friend of Haken, had alerted the latter to the secret laser research of the Americans. Shortly afterwards Haken himself became involved in the theoretical debates at Bell on solving the problem of the solid-state laser. Haken brought the ideas and concepts developed in the process to Stuttgart in the fall of 1960, where, together with his staff, he developed them further and successfully into a quantum laser theory.

The better part of German physicists[35] and their colleagues in other countries, however, did not learn details about the laser until the publication of the article of the Schawlow group in the Physical Review Letters or in discussions and talks on the relevant symposia in fall and winter 1960/61. The Alsatian physicist and later Nobel laureate (1966) Alfred Kastler[36], whose technique of "optical pumping" developed in the 1950s was essential for the stimulation of masers and lasers, gave a plenary address under the title "Orientation of Atomic Nuclei by Optical Pumping" on the 25th Physicists' Conference held from October 17th to 21st 1960 in Wiesbaden.[37] Even though neither this nor any of the other lectures of the Wiesbaden meeting mentioned optical maser or laser with as much as a word, it is safe to assume that this topic had been the subject of many unofficial discussions and talks among the participating physicists, especially since a number of participants in Wiesbaden came from the USA.[38] Some weeks later German physicists could inform themselves in more detail about the laser development in the USA through a professional article written by Wolfgang Kaiser in the Physikalische Blätter, that is, in a manner of speaking get insider information.[39] The first lectures on the subject of "Lasers", given during a larger gathering of German physicists, were Horst Rothe's (Karlsruhe) survey lecture on the "State of Maser Research" and a single lecture by Hermann Haken (Stuttgart) titled "Quan-

---

[34] See the interview with Hermann Haken conducted by the author together with a group of students at the department for the History of Science and Technology in Stuttgart on the 13.2.1992 (LAA: Interview 1 with Haken), as well as a conversation the author had with Haken on the 28.6.1994 (LAA: Interview 2 Haken), see Albrecht, Hermann Haken im Gespräch.

[35] For the simultaneous development in the DDR, see Albrecht, Laser Sozialismus; Albrecht, Laserforschung Jena.

[36] Albrecht, Kastler.

[37] Published in: Brüche, Physikertagung Wiesbaden, pp. 62–73.

[38] The meeting in Wiesbaden counted 1.700 participants, including 110 physicists from the DDR, as well as a number of physicists from the USA. See the short report of the meeting in: Physikalische Blätter 16 (11), 1960, pp. 588–589.

[39] Kaiser, Optische Maser.

tum Theory of the Optical Maser" at the 26th Physicists' Conference in Vienna
from October 15th to 21st 1961.[40]

The increased interest in lasers within science and technology in the 1960s is
clearly illustrated by the history of publications and patent numbers in this sector
(see chart 2). Though numerical data of single laser bibliographies varies considerably[41] – depending on the kind of publication – they altogether show the same
trend internationally as well as nationally.

Chart 2: History of Laser Publications and Laser Patent Granting during the 1960s in the FRG			
Year	Publications	FRG Patents	
	international	total	in FRG
1960	17	–	–
1961	136	–	–
1962	304	3	1
1963	752	15	4
1964	957	42	8
1965	1.290	51	14
1966		44	14
1967		26	9
1968		94	46
1969		256	88
1970		293	119

Source: Edward V. Ashburn (ed.): Laser Literature Vol. 1. A permuted Bibliography
1958–1966, North Hollywood 1967, S. VI, Zusammenstellung in- und ausländischer Patente
(1966, 1967, 1972).

As one can see from the registers of such relevant organs like Physics Abstracts or
Physikalische Berichte, following a rather hesitant start in 1960/61 the number of
publications literally exploded from 1962 to 1965 and then leveled out at about
1,000 laser publications a year in the late 1960s. According to the Jena literature

---

[40] See Physikalische Verhandlungen 12, 1961, pp. 182 and 220.
[41] More than anything else that depends on the individual understanding of a "laser publication." Hence e. g. the literature compilation in Jena includes in addition to early maser studies
also popular (science) articles about the laser.

compilations, in 1963/64 merely eight German-language magazine articles appeared in 1960 on the subject of lasers.[42] Yet in 1961 there were already twice as many (17) and in the next two years (1962: 59; 1963: 203) the number of articles on lasers in German more than tripled each time. The total number of international magazine articles about lasers increased from approximately five to about eleven percent.

A similar development can be seen in the number of laser patents: the "Compilation of Domestic and Foreign Patents in the Area of Lasers" issued in Jena from 1966 to 1972 counted until mid-1971 alone 5,137 patents, including 1,100 patents applied for the Federal Republic and the DDR.[43] The number of patents granted, for instance, for the Federal Republic (see chart 2) escalated between 1962 and 1970 from less than ten to almost 300. Only in 1966/67 it suffered a temporary slump, probably triggered by the same general disillusionment that was almost bound to follow the exaggerated euphoria of the first years, the same that Joan Bromberg registered in the USA for 1964:

> Around 1964, laser euphoria began to show signs of wearing thin. Some of the projected applications were proving noticeably difficult to achieve, for a variety of reasons. In some cases, the lasers themselves were recalcitrant. ... In other cases, it was the auxiliary elements for laser systems that posed the problems. ... Sometimes unexpected side effects appeared. ... At other times, more detailed analyses revealed that the traditional technologies simply had imposing physical and economic advantages.[44]

Not only in the USA did the joke circulate that the laser was "a solution looking for a problem."[45] It was, however, primarily directed at the application of the new technology, whereas basic research was hardy affected at all, as the virtually undaunted boom in laser publications already indicated.

An essential reason for the constant boom in laser research during the 1960s was the discovery of ever-new laser types from 1963 to 1970.[46] Above all the USA realized about a hundred new types in the field of solid-state, semiconductor, gas and liquid lasers, including about 50 lasers with neutral or ionized gas laser media as well as more than twenty lasers with different semi-conductors, about ten with crystals and five with different kinds of glass as laser media. Yet in each of those groups only one or two laser types were significant for research and application.

Thus in the late 1960s the foundations for the physical understanding of lasers were laid in experimental as well as in theoretical terms. That was reflected,

---

[42] Compilation of laser literature (1963 und 1964).

[43] Compilation of National and International Patent Specifications in the Field of Laser (1966, 1967, 1972). The figure mentioned also includes the pending patens on holography, with a total number of 389 patents. It should be regarded that the numbers mentioned in chart 2 refer to the patents actually granted until the end of 1970 (minus the holography patents), which explains the considerably lower number of altogether 824 patents.

[44] Bromberg, Laser, p. 158.

[45] Cf. ibid., p. 159, as well as on the subject of Germany the remark by Fritz Peter Schäfers, quoted in: Anwendungen des Lasers, p. 7.

[46] See for that and the following: Bromberg, Laser, pp. 163 ff.

for instance, in the publication of comprehensive reference works on laser physics and laser theory, which included in Germany above all the handbooks by Dieter Röss quoted on every occasion "Laser. Lichtverstärker und -oszillatoren" published in 1966, Werner Kleen's and Rudolf Müller's "Laser. Verstärkung durch induzierte Emission. Sender optischer Strahlung, hoher Kohärenz und Leistungs-dichte" (1969) and Hermann Haken's "Laser Theory" from 1970.[47] It stands out, however, that in even books written by physicists active in industrial practice, like Röss and Kleen/Müller, practical laser applications take up relatively little room compared to theoretical and und experimental basics.[48]

The reason that laser technology fell noticeably behind laser research in the 1960s was the fact mentioned above that, against all expectations, in the early 1960s the realization of technical laser applications proved to be extremely difficult. On the laboratory scale it was comparatively easy to make a laser operate – provided one possessed the required know-how and the corresponding materials. It was, however, a whole different story to make such a bench-scale laser ready for the market, i. e. to devise it for operations outside the lab in an acceptable way regarding durability, manageability, operational reliability, and costs. In late 1963 about 20 to 30 US companies succeeded in this; altogether they could sell some hundred lasers for a total of about one million dollars.[49] According to the laser market magazine Laser Focus in late 1965 about 115 companies offered already lasers for sale. By the late 1960s the turnover amounted to about 100 to 120 million dollars, in which with about 49 million dollars the share of civil applications ranged between a third part and half of the total volume (1969).[50]

Purchase prices for the first commercial lasers were considerable. Hence the first commercial ruby laser of the Technical Research Group (TRG), for instance, cost in spring 1961 26,000 dollars and in July 1962 the first Helium-neon laser (HeNe laser) of Perkin-Elmer and Spectra-Physics cost 7,500 dollars.[51] The durability of these first lasers available on the open market still left a lot to be desired. Mostly the quality of the ruby crystals used was poor, and in many cases the high-energy laser beam quickly destroyed the optical coating of their end faces. The glass tubes at the resonators of the early gas lasers often proved to be leaky, their cathodes short-lived and the adjustment of their mirrors vulnerable to mechanical commotions. Thus the early HeNe lasers only possessed an operation period of some hundred hours. Even though by the end of the 1960s substantial progress could be made here by improved manufacturing and stimulation technologies, new materials for laser media, electrodes, laser tubes, and laser mirrors (thus, for example, the average life expectancy of the HeNe laser was increased to

[47] Röss, Laser; Kleen/Müller, Laser; Haken, Laser Theory.
[48] Röss dedicates only 34 of 722 pages to the laser and Kleen/Müller only 48 of 567 pages.
[49] See ibid., pp. 113 ff.
[50] See Laser Focus, January 1970, p. 27.
[51] Quotation of prices according to Meinel/Neubert/Wiederhold, Wirkungsweise, here p. 493.

several thousand hours), the heads of the Siemens research lab Werner Kleen and Rudolf Müller came in 1969 to the conclusion:

> Predictions regarding the future importance of laser in the different fields of technology are afflicted with considerable uncertainty. We are only in the early stages of laser applications, with the laser competing with 'classical' devices. For many suggested applications of lasers, it is still not clear whether lasers, or other appliances and procedures will be successful. That is not least due to the fact that applications of technology are essentially determined by economic factors, which in many cases are currently not clear.[52]

In 1969 Kleen's and Müller's description of possible fields of laser application included: optical communication, locating, optical data processing, holography, strobe photography, non-linear optics, plasma generation and plasma diagnostics, material machining, medical and biological application, as well precision measurement of geometrical and mechanical values.[53] But apart from numerous promising approaches they could only refer to initial successes in this connection in the fields of distance measurement, plasma diagnosis, material machining, medical applications (retina coagulation) and precision measurement.

THE BEGINNINGS OF LASER RESEARCH AT WEST GERMAN UNIVERSITIES

In the spring of 1961 there were a total of 147 physics professorships at the universities in the Federal Republic. These included, for physics and experimental physics, 38 full professors (ordinarius)[54] and 16 associate professors (extraordinarius),[55] for theoretical physics 36 and 12, as well as for applied and technical physics 36 and 9.[56] When the untenured (außerplanmäßige) professors, private lecturers and lectureships are included, the number of physicist positions in the three mentioned areas amounts to 281. At the same time there were altogether 36 positions in the field of (tele-)communications and high-frequency technology with 18 professorships, 17 and 1.[57]

Assuming that, because of previous knowledge and expertise, the entire faculty in these fields of physics, communication, and high-frequency technology was capable of overcoming the comparatively low thresholds to enter laser research, there was a potential of more than 300 scientists for laser research at West German universities in the early 1960s, at least theoretically speaking. If one included the research assistants, academic staff and graduate students assigned to said fac-

---

[52] Kleen/Müller, Laser, p. 510.

[53] See ibid., pp. 510 ff.

[54] This position was equivalent to a full professor, often director of an institute.

[55] This was a permanent, tenured position, but subordinate to the Ordinarius.

[56] See Die Lehrstühle an den Wissenschaftlichen Hochschulen, pp. 82 ff. and 141 ff.

[57] In 1961 there were altogether 160 chairs in physics (5.1 % of all chairs), as well as 60 chairs in electrical engineering (1.9 %) at West German universities. See Empfehlungen des Wissenschaftsrates; Albert/Oehler, Materialien, p. 330.

ulty, the academic potential for West German laser research would have reached a dimension of about thousand scientists, philosophically speaking.

But the actual number of university professors and scientists who entered laser research during the (early) 1960s was certainly considerably lower, despite the assumption made by the Munich laser physicist Wolfgang Kaiser in 1967: "All over the world several thousand scientists are working on or with the laser, and even in Germany there is hardly any university that does not work with the laser in some form or other."[58] It is not easy to obtain reliable data regarding the actual number of West German scientists in this field, due to the fact that the title "laser physicist" or "laser technician" was neither established nor wise to use, given the then still entirely unsettled field of research and the variety of disciplinary entry options.

The West German high-frequency researchers who received grants in 1962 within the framework of the DFG "High-Frequency Physics" priority program gives a first indication of the staff size in the early stages of university laser research and its subjects, given that the high-frequency community was particularly predestined for taking up laser research for the reasons already mentioned (chart 3). Only two physicists of the altogether 17 scientists with grants worked on typical laser subjects: Hans Boersch with a chair for experimental physics and director of the First Physical Institute at the Technical University (Technische Hochschule, TH) Berlin and Hermann Haken with a chair for theoretical physics at the Technical University (TH) Stuttgart.

Chart 3: Researchers/subjects who received grants within the framework of the DFG priority program "High-Frequency Physics" in 1962[59]		
Location	Researcher [name]	Research subject
TH Berlin	Prof. Dr. Hans Boersch	– Structure and analysis of different laser types – Laser applications
	Prof. Dr.-Ing. Friedrich W. Gundlach	– Utilizing the paramagnetic resonance of crystals to generate high-frequency signals – Radiation analysis of exited ammonia molecules – Properties of induced emission of atomic hydrogen – Manufacturing of backward wave tubes
TH Darm-stadt	Prof. Dr. Karl-Heinz Hellwege	– Term electron paramagnetic ions in crystals
Freiburg Uni	Prof. Dr. Wilhelm Maier	– Measurement and analysis of the microwave spectrum of dimethyl sulfide

[58] Kaiser, Laser, here p. 25.
[59] Source: Bericht der Deutschen Forschungsgemeinschaft über ihre Tätigkeit, Bad Godesberg 1962, pp. 262 ff.

Chart 3: Researchers/subjects who received grants within the framework of the DFG priority program "High-Frequency Physics" in 1962		
Location	Researcher [name]	Research subject
Göttingen Uni	Prof. Dr. Rudolf Hilsch	– Examining mono-crystals and the influence of disorder
Heidelberg Uni	Doz. Dr. Karl H. Hausser	– Hyperfine structure of free radicals and complex compounds and electron resonance of impurities
	Prof. Dr. Hans Kopfermann	– High-frequency spectroscopy of atoms – Atomic beam resonance studies of the hyperfine structure of the ground state of the stable isotopes – Double resonance studies of the hyperfine structure of excited atoms
TH Karlsruhe	Prof. Dr. Helmut Friedburg	– Term distances and relaxation times in doped rutile
	Prof. Dr. Günther Laukien	– Measurement of electron resonance spectra and paramagnetic relaxation times – Studying the magnetic interaction between nuclei and unpaired electrons
	Prof. Dr.-Ing. Horst Rothe	– HF properties of a solid-state amplifier – Influence of the crystal orientation and negative two terminal network properties
	Dr. Siegfried Wilking	– Application options of multiple quantum procedures for maser issues
Mainz Uni	Prof. Dr. Gerhard Klages	– Electron spin resonance for free radicals from gas discharges and in liquids
Munich Uni	PD Dr. Eugen Fick	– Thermodynamics and statistics of spin systems
TH Stuttgart	Prof. Dr.-Ing. Joachim Dosse	– Molecular beam clocks as frequency standard
	Prof. Dr. Hermann Haken	– Optical maser properties of the ruby – Double-quantum procedures – Theoretical analysis of light propagation in the optical maser
	Prof. Dr. Heinz Pick	– Optical maser properties of the ruby – Double-quantum procedures – Theoretical analysis of light propagation in the optical maser
Tübingen Uni	Prof. Dr. Hubert Krüger	– High-frequency transfers in excited atoms – Nuclear resonance in solid states and liquids – Lamb shift in the simply isolated atomic helium

Chart 4: Laser lectures within the framework of meetings of West German Physical Societies in 1961 and 1962[60]		
Contributor	**Institution**	**Lecture Subject (Meeting)**
H. Friedburg	TH Karlsruhe	Quantum mechanical amplifiers; survey lecture at the spring meeting of the Physical Society of Hessen-Mittelrhein-Saar in Bad Nauheim from April 19th to 22nd 1961
H. Rothe	TH Karlsruhe	State of and new developments in maser research; survey lecture at the 26th German Physicists' Conference in Vienna from October 15th to 21st 1961
H. Haken	TH Stuttgart	Quantum theory of the optical maser, single lecture at the 26th German Physicists' Conference in Vienna from October 15th to 21st 1961
H. Rothe	TH Karlsruhe	Optical maser, survey lecture at the spring meeting of the Physical Society of Hessen-Mittelrhein-Saar in Bad Nauheim from April 19th to 22nd 1961
H.-G. Häfele	Osram	Measurement of power distribution of different ruby lasers and analyses of non-linear optical effects, single lecture at the spring meeting of the Bavarian Physical Society from April 26th to 28th 1962 in Munich
H. Boersch u. a.	TU Berlin	Development of various simple construction laser types, single lecture at the spring meeting of the Berlin Physical Society on April 27th and 28th 1962
W. Kaiser	TH Stuttgart	The optical maser, survey lecture at the 27th German Physicists' Conference in Stuttgart from September 24th to 28th 1962
G. Koppelmann	TU Berlin	A microwave Perot-Fabry interferometer as a model for the laser resonator, lecture within the "Maser" section at the 27th German Physicists' Conference in Stuttgart on September 25th 1962
D. Röss	Siemens	Amplification and saturation of pulsed optical ruby masers, lecture within the "Maser" section at the 27th German Physicists' Conference in Stuttgart on September 25th 1962
K. Gürs	Siemens	Internal modulation of optical masers, lecture within the "Maser" section at the 27th German Physicists' Conference in Stuttgart on September 25th 1962
A. Lohmann	TH Braunschweig	Electron acceleration by light waves, lecture within the "Maser" section at the 27th German Physicists' Conference in Stuttgart on September 25th 1962

[60] Compiled according to: Physikalische Verhandlungen 12, 1961, pp. 74, 182 and 220; 13, 1962, pp. 53, 85, 103, 168 and 193 ff.

Looking at the meetings of the Physical Societies in Germany between 1961 and 1962 (chart 4) provides a further indication for the personnel and geographic foci of the early stages of West German laser research. While in 1961 only three lectures were given on laser subjects (and two of them were survey lectures), a year later there were eight lectures, with six of them dealing with specialist subjects. Early West German academic laser research was again concentrated at the Technical Universities in Stuttgart and Berlin, as well as in Karlsruhe and – at least to some degree – in Braunschweig. Again the names of Hermann Haken and Hans Boersch pop up. This double concentration – both, within the framework of the DFG priority program as well as at the DFG meetings – in Stuttgart (Haken) and Berlin (Boersch) suggests taking a closer look at these two universities.

In the late 1950s the engineer Joachim Dosse[61] and the physicist Heinz Pick[62] conducted joint research at the TH Stuttgart within the framework of the DFG "High-Frequency Physics" priority program, which received continuous funding from 1958 onwards. Since 1960 Dosse worked on the subject "Molecular beam clocks as frequency standard", a sub-field of maser research[63], without, however, expanding his research focus to the area of the optical maser. Likewise Pick initially refrained from doing laser research. This situation finally changed with Hermann Haken's[64] appointment to Stuttgart in the winter term 1960/61. As mentioned above, Haken brought specific laser knowledge back to Germany straight from the Bell Telephone Laboratories in the United States, where he now abandoned his previous research domain, solid-state physics, for the new research field of lasers. Haken, who by his own admission had prior to 1960 never been concerned with the maser, took up without delay theoretical studies about the optical maser that drew upon ideas that he himself had already developed during his advisory role in the spring and summer of 1960 in the research group for solid-state laser at the Bell lab.[65] Haken himself characterized this early Stuttgart work, which he conducted and published together with his colleague Herwig Sauermann, as:

[61] Dosse, born in 1911, since 1942 associate professor at the Technical University Berlin, had accepted a chair of high-frequency technology at the TH Stuttgart in 1958. Within the framework of the DFG priority program he conducted absorption measurements, resonance measurements and the determination of relaxation times.

[62] Pick, born in 1912, had habilitated in Göttingen in 1948 and became associate professor there in 1952. In 1954 he accepted a chair for experimental physics at the TH Stuttgart. For many years his work within the framework of the DFG priority program focused on electron spin resonances in alkali halide crystals.

[63] Cf. Bericht der Deutschen Forschungsgemeinschaft über ihre Tätigkeit, Bad Godesberg 1960, p. 190.

[64] Haken, born in 1927, had studied mathematics and physics (as a minor) from 1946 to 1950 at the universities of Halle and Erlangen. From 1952 to 1960 Haken was research assistant at the Institute for Theoretical Physics in Erlangen, where he habilitated in 1956. In spring 1960 he was offered the chair in Stuttgart. See also LAA: interview I with Haken from 13.2.1992.

[65] Albrecht, Hermann Haken im Gespräch, pp. 251 and 258.

First the idea was to develop a semi-classical theory. But actually, we always had a quantum theory in mind. I used the formalism of the so-called 'second quantization', in which light field and matter are quantized too, i.e. are governed by quantum law. Then I derived a consistent theory from that. This theory contained, however, an averaging. And then this averaging about the so-called quantum fluctuations meant, as it turned out afterwards, that we had a semi-classical theory...[66]

Finally in the mid-1960s a purely quantum mechanical laser theory arose from this "semi-classical" theory, which Hermann Haken and his staff in Stuttgart developed about the same time as the American physicist Willis E. Lamb jr.

Initially there was no experimental laser research in Stuttgart. When Haken invited experimental physicist Wolfgang Kaiser[67] as a visiting professor to Stuttgart in spring 1962 this marked the beginning of laser research there. "In surprisingly short time" Kaiser set up in Stuttgart "a small, but rather active group"[68] for experimental laser research, which probably put its first laser into operation in the spring or summer of 1962. However, these promising beginnings suffered a setback when Kaiser returned to the USA in the fall of 1962. Yet the laser group set up by Kaiser was able to continue its work at the Second Physical Institute under Professor Pick. In 1965 this group achieved substantial success with the construction of the first dye laser worldwide, which had already been stimulated in discussions with Wolfgang Kaiser.[69]

Shortly after the American laser experiments became known in late 1960, other West German universities aside from Stuttgart also took up experimental laser research. In no time at all the First Physical Institute at TU Berlin headed by director Hans Boersch[70] became one of the leading West German laser research centers. Boersch, who in the 1930s had been instrumental in developing electron microscopy at the AEG Laboratories under Ernst Brüche,[71] had been Carl Ramsauer's successor at the Berlin Institute in 1954, which he had turned within a couple of years into one of the biggest and best equipped institutes in West German academia. Until Boersch developed laser research as an additional domain of his institute in 1961, his main areas of work had been electron optics

---

[66] Cf. ibid., pp. 251 ff., as well as Haken/Sauermann, Frequency Shifts; Haken/Sauermann, Nonlinear Interaction.

[67] Kaiser, born in 1925, had studied physics until 1952. He then went to the USA, where he worked from 1957 to 1964 in the Bell Telephone Laboratories in Murray Hill/New Jersey. In 1964 he accepted the chair for experimental physics at the Technical University Munich. See also LAA: Interview with Kaiser from 19.1.1994.

[68] Thus Hermann Haken in: Albrecht, Hermann Haken im Gespräch, p. 252.

[69] The persons who realized this laser were the physicists Bernd Fritz and E. Menke. See LAA: Interview I with Haken from 13.2.1992.

[70] Hans Boersch, born in 1909, studied physics, took his doctoral degree at the University of Vienna in 1935, from 1935 to 1940 staff member of the AEG Research Institute in Berlin, from 1941 to 1946 lab director of the First Chem[ical] Laboratory of the University Vienna, 1942 habilitation, 1942 to 1946 associate professor for experimental physics, at last at the University of Innsbruck, 1946–1948 lab director in Tettnang/Württemberg at the Institute Recherches Scientific, 1948–1954 lab director at the PTB in Braunschweig. Boersch died on 9.6.1986.

[71] See Qing, Frühgeschichte.

and low temperature physics. During the following years Boersch established the biggest laser group in the Federal Republic at the First Physical Institute in Berlin[72], which achieved a leading position in experimental basic laser research in Germany during the 1960s.

According to the entries of Boersch's doctoral students Gerd Herziger and Horst Weber into their lab journals, their laser studies at the Berlin Institute started in June and November 1961 respectively, with the first attempts to realize a HeNe laser as well as a ruby solid-state laser.[73] Before the end of the year, on December 18th, Herziger made a first "oriented attempt to operate a laser" with the gas laser, and on the 29th of December 1961 Weber accomplished for the first time the generation of laser flashes with a ruby crystal.[74] Four months later, in late April 1962, Boersch and his two staff members reported for the first time publicly about their work with a lecture titled "Development of various simple construction laser types" at the spring meeting of the Berlin Physical Society.[75] The first Diplomarbeit (analogous to a Master's thesis) at the institute on lasers was also finished[76] in 1962, which soon would be followed by numerous more Diplomarbeiten and doctoral theses as well as Habilitationsschriften (analogous to a second Ph.D. dissertation) in this new research field.

Charts 3 and 4 have already indicated that, along with Stuttgart and Berlin, early laser research centers also existed at the universities of Karlsruhe and Braunschweig. Unlike the physical institutes in Stuttgart and Berlin, however, the institutes in Karlsruhe and Braunschweig were technical institutes, whose directors were not physicists but engineers. At the Technical University Karlsruhe it was professor Horst Rothe, who initiated the first laser researches at the Institute for High-Frequency Technology and High-Frequency Physics. Together with his staff Rothe was able to present experimentally at the DFG High-Frequency Physics colloquium in October 1961, that is, earlier than the Berlin physicists around Boersch, an "optical molecular amplifier (LASER)", which probably was a ruby laser.[77] Rothe's colleague in Karlsruhe, at that time associate professor for high

[72] See also the article: the ATU Institute possesses a "'Juwelen-Sammlung'. Unter Professor Boersch arbeitet die größte Laser-Gruppe Deutschlands", in: Die Welt, No. 73, dated 18.3.1967, p. 12.

[73] Physikalisches Beobachtungsbuch I (lab journal) by Gerd Herziger (19.4.1961–9.4.1962), undated entry on the testing of a laser (gas) tube (previous entry dated 10.6.61), p. 52, as well as Physikalisches Beobachtungsbuch I (lab journal) by Horst Weber (1.6.1961–19.4.1962), entry dated 6.11.1961 with the title "Laser": checking the condensers, pp. 53 ff.; Archive of the Optical Institute of the TU Berlin (OIB).

[74] Cf. Beobachtungsbuch Herziger I, p. 56, as well as Beobachtungsbuch Weber I., p. 92, OIB.

[75] Cf. Physikalische Verhandlungen 13, 1962, p. 103. See also Head, German Research.

[76] In 1962/63 Stefan Maslowski produced the first diploma thesis on the "Herstellung und Untersuchung von Wellenformen des Festkörper-Rubin-Lasers", see OIB: Liste der Diplomarbeiten, p. 7.

[77] See Mitteilungen der Deutschen Forschungsgemeinschaft 4, 1963, p. 15. It was probably a ruby solid-state laser.

frequency technology and electronics, Helmut Friedburg[78], had already given a
survey lecture in April 1961 at the spring meeting of the Physical Society of Hes-
sen-Mittelrhein-Saar in Bad Nauheim about "Quantum mechanical amplifiers",
which discussed the structure and operation mode of the ruby laser as well as the
HeNe laser, as documented in the notes of professor Harald Volkmann, at that
time head of the Zeiss physics laboratory in Oberkochen.[79] Shortly afterwards
close contact can be detected between Rothe's institute in Karlsruhe and Zeiss
in the field of laser research. Thus in the second half of 1961 Rothe made rubies
from his institute available to the company for its laser tests and likewise tested
in Karlsruhe rubies processed at Zeiss for their optical properties, that is regarding
their adequacy for the construction of lasers.[80] In late January of 1962 Volkmann
informed the Zeiss management in this regard: "All of the rubies manufactured
at our [company] worked at the Prof. Rothe's first go."[81]

Another example for early West German academic laser research is provided
by the Institute for High-Frequency Physics at the Technical University Braunschweig, headed by Professor Hans-Georg Unger[82]. Unger had just been appointed to the newly established chair of High Frequency Technology in Braunschweig
in the winter term 1960/61. Like Hermann Haken he had previously worked in
the Bell Telephone Laboratories, eventually as Department Head for Research
in Communication Techniques, where he concentrated on circular waveguides
for the realization of broadband long-distance traffic communication systems as
well as with theoretical studies about optical communication. Unger continued
to work on those research subjects in Braunschweig.[83] Based on the experiences
he made at Bell, "that given the few existing Electrical Engineering institutes it
was above all important to consolidate the field of communication technology"
for which high frequency technology would be "an important basic technology",
Unger came to the conclusion that research at his blossoming institute had to
range over ever higher frequencies. "Therefore optical communication belonged

---

[78] H. Friedburg, born in 1913, with a degree and a doctorate in physics, in 1958 habilitation
at the University of Heidelberg, in 1959 extraordinary professor and in 1964 appointed chair at
the TH Karlsruhe for High-Frequency Technology and Electronics.

[79] Cf. Physikalische Verhandlungen 12, 1961, p. 74.

[80] Interview with Volkmann on 8.12.1993, LAA, as well as Bericht über die Tätigkeit in
Phys-Lab für die Zeit vom 1.7.1961 bis 31.12.1961, dated 31.1.1962, Zeiss Archive Oberkochen
(ZAO), Phys-Lab/VN/8/62.

[81] Ibid.

[82] Unger had studied electrical engineering from 1946 to 1951 at the TH Braunschweig with
the focus on high frequency technology. He worked from 1951 to 1955 as designing engineer for
Siemens in Munich, at last as the director of the lab for microwave research. From 1956 to 1960
he worked at the Bell Telephone Laboratories in Murray Hill/USA. From 1960 until he became
professor emeritus in 1993 he was director of the Institute for High Frequency Technology at the
TU Braunschweig. Cf. Material Unger, LAA.

[83] See 25 Jahre Institut für Hochfrequenztechnologie, Braunschweig 1985 (ms.), as well as
Kertz, Forschung, here pp. 688 ff.

right from the start to the research field of this institute, and the physico-technical basics for that were taught here as well."[84]

According to the goal of developing optical communication systems laser research in Braunschweig, he focused on the continuously operating HeNe laser, which the institute used in 1963 for the first time to test communication via pipe systems either empty or filled with inert gas.[85] The institute's laser research had started, however, in 1961 with ruby lasers, even though their light pulse was admittedly

> not by far as constant and regular as desired for the signal transmission. Here the gas lasers provided much better options for transmission attempts. At that time we also quite quickly generated beams of highly coherent light in time and position with such a self-made gas laser.[86]

According to Professor Unger's memoires, the HeNe laser of the Braunschweig Institute was one of the first lasers of that type built in Europe, going into operation at the end of 1961 and beginning of 1962.[87] After having achieved the goal of a sufficiently powerful and mono-frequency gas laser, research at the institute focused on the application of this laser in optical telecommunication engineering.

What stands out in early laser research in West German academia is that its "strongholds" were almost exclusively located at technical universities. All lectures on laser research at the meetings of the German Physical Societies 1961/62 were given either by physicists or engineers employed at technical universities. Albeit not an astonishing fact where laser researchers in the field of high-frequency technology were concerned (given that this subject was not offered at the classical universities), it is surprising that physicists working at the classical universities were so visibly absent in the early West German laser research. Possible reasons for that could be the above mentioned reservations physicists had against the "engineerish" high-frequency physics as well as the greater appeal of seemingly more attractive (scientifically as well as professionally) disciplines like nuclear physics. Admittedly, the same reservations held true for the physicists employed at technical universities, even though the latter ones were often more open-minded towards "technical" problems due to their immediate vicinity to their engineer colleagues.

Then again, the same physicists occasionally regarded the "utility"-orientation or rather the need for application of the engineers – an almost inevitable result of their work at a technical university – as detrimental to their "true" vocation for physical basic research. Here the laser offered, especially for the physicists at technical universities, an elegant way out of this dilemma, given that for one thing it was of particular interest for the engineers due to its potential applications; but on the other hand it likewise opened a completely new field of experimental and

---

[84] Quotations taken from Professor Unger's salutary on the occasion of the 25th anniversary of his institute, in: 25 Jahre Institut für Hochfrequenztechnik, Braunschweig 1985, p. 7.

[85] See Sporleder, 25 Jahre Institut für Hochfrequenztechnologie.

[86] Unger, 25 Jahre optische Nachrichtentechnik, here pp. 25 ff.

[87] Material Unger, letter dated 18.5.1995, LAA. See also Gloge, Gesichtspunkte; Runge, He-Ne-Laser.

theoretical basic research in physics. Hence the fact, that the laser proved to be a "meeting place" (Hermann Haken) for so many different disciplines, may have been part of the appeal this new field of research held especially for physicists at technical universities.

The 27th Physicists' Conference from September 24th to 28th 1962 in Stuttgart provided impressive proof of the wide interest and attention that the subject "laser" soon began to attract in the Federal Republic and its scientific community. On the opening day of this major event, the plenary address delivered by Wolfgang Kaiser, who had arrived from the USA, on the "Optical Maser" emphasized the particular significance of the new research subject.[88] Unlike at the meeting in Vienna the year before, there was now a specific "maser" section[89], which for the first time provided a framework for an increasing number of specialist lectures on lasers, which were held on the 25th of September. On this occasion not only university but also industry representatives introduced the first results of their laser research. The supporting program of the Physicists' Conference included a visit to the Carl Zeiss Company in the nearby Oberkochen, where a demonstration of the first commercial Zeiss laser proved that, in addition to academic research, industrial research too had discovered the laser now as a new domain.[90]

INSTITUTIONALIZING LASER RESEARCH AT WEST GERMAN UNIVERSITIES

Taking a look at the Handbook of German Teaching and Research Institutions (Vademekum deutscher Lehr- und Forschungsstätten, VDLF) from 1964 to 1973 reveals how quickly laser research gained ground in West German academic research.[91] Whereas as late as 1964 none of the universities mentioned the laser as a specific research domain, in 1968 there were already five and in 1973 nine universities that explicitly listed laser, laser technology, laser spectroscopy, laser physics, optical coherence, laser diagnostics or similar subjects as their research foci.

The doctoral and professorial dissertations produced on lasers at German universities during the 1960s convey even more clearly the actual dimension of laser research at that time (see chart 5). Even if the exact number cannot be established[92], it still shows that active laser research was conducted at far more

---

[88] Kaiser, Der optische Maser, 1963.

[89] See Physikalische Verhandlungen 13, 1962, pp. 193/94, as well as chart 2.3.

[90] The 70 physicists, who participated in this excursion, were shown at Zeiss the company's first salable ruby laser, among other things. See Zeiss – aktueller dienst 8, 1962, p. 5.

[91] Vademecum deutscher Lehr- und Forschungsstätten (VDLF), issued by the Stifterverband für die Deutsche Wissenschaft, 4th edition Essen 1964, 5th edition Essen 1968, and 6th edition Düsseldorf 1973.

[92] Inspecting the number of doctoral dissertations at the TH Stuttgart and the TU Berlin in the respective archives/institutes revealed that the data used in the annual lists of Deutsche Hochschulschriften is obviously incomplete. Likewise the year specifications generally don't refer to the year the doctoral degree was taken but to the date of publication.

universities than mentioned in the VDLF. Of the 36 (technical) universities that existed until the late 1960s in the Federal Republic, 15 (41,7%) are represented with doctoral dissertations and/or Habilitationsschriften. If those universities are included that were represented by laser lectures at the different symposia of the German Physical Society (Deutsche Physikalische Gesellschaft, DPG), then this number increases by six to a total of 21 (58,3%).[93]

Chart 5: Doctoral dissertations and Habilitationsschriften on the subject of laser at West German universities 1961–1970[94] [Number of Habilitationsschriften in ( )]										
University	1961	1962	1963	1964	1965	1966	1967	1968	1969	1970	total
TH Aachen	–	–	–	–	–	–	–	2	2	–	4
TU Berlin	–	–	–	–	2(1)	–	3	1	(1)	4	12
TH Braunschweig	–	–	–	–	1	(1)	1	–	–	1	4
TH Hannover	–	–	–	–	–	–	(1)	–	–	–	1
TH Karlsruhe	–	–	–	–	–	–	1	–	1	–	2
TH Munich	–	–	–	1	–	2	2	2	–	3	10
TH Stuttgart	–	–	–	–	–	1	–	1(1)	1	2(1)	7
U Erlangen	–	–	–	–	–	–	–	–	1	–	1
U Frankfurt	–	–	–	–	1	–	–	–	–	–	1
U Gießen	–	–	–	–	–	–	–	–	1	–	1
U Heidelberg	–	–	–	–	–	1	1	(1)	2	–	5
U Munich	–	–	–	–	–	–	–	–	1	1	2
U Münster	–	–	–	–	–	–	–	1	1	2	4
U Tübingen	–	–	–	–	–	–	–	–	–	(1)	1
FRG total	–	–	–	1	5	5	9	9	10	16	54

Analyzing the DPG symposia, doctoral dissertations, and Habilitationsschriften shows significant differences between West German universities (see chart 6) regarding the intensity of research activities in the laser field. As had been the case

[93] These were the technical universities in Aachen, Berlin, Braunschweig, Darmstadt, Hannover, Karlsruhe, Munich and Stuttgart, as well as the universities in Berlin, Erlangen-Nürnberg, Frankfurt, Freiburg, Gießen, Göttingen, Heidelberg, Köln, Marburg, Munich, Münster, Saarbrücken and Tübingen.
[94] Compiled according to: Jahresverzeichnis Deutscher Hochschulschriften, ed. by the Deutsche Bücherei Leipzig, vol. 77 (1961) to vol. 86 (1970), Leipzig 1964–1979.

in the DFG High-Frequency Physics priority program or with the laser articles in professional journals, it also becomes apparent in this context that the hub of West German laser research was located above all at the technical universities, with 93 (72,7 %) of 128 DPG lectures and 40 (71,4 %) of 56 dissertations. By far the most outstanding was the TU Berlin, followed at a considerable distance by the TH Stuttgart and, in the middle of the field, the technical universities in Munich and Karlsruhe, and in Heidelberg and Frankfurt respectively. An analysis of meetings of the German Branch of the European Optical Society (Deutsche Gesellschaft für angewandte Optik, DGaO) during the 1960s confirms this picture regarding the position of technical universities as well as concerning the dominance of the TU Berlin.[95]

Chart 6: Ranking of West German universities regarding their number of laser lectures (DPG meetings) and laser dissertations (doctoral dissertations and Habilitationsschriften) for the 1960s

DPG Lectures		Dissertations		Overall Ranking		Rank
University	Number	University	Number	University	Number	
TU Berlin	49	TU Berlin	12	TU Berlin	61	1.
TH Stuttgart	14	TH Munich	10	TH Stuttgart	21	2.
U Heidelberg	11	TH Stuttgart	7	TH Munich	16	3.
TH Karlsruhe	10	U Heidelberg	5	U Heidelberg	16	4.
U Frankfurt	10	TH Aachen	4	TH Karlsruhe	12	5.
TH Darm-stadt	7	TH Braun-schweig	4	U Frankfurt	10	6.
TH Munich	6	U Münster	4	TH Braun-schweig	7	7.
U Munich	5	TH Karlsruhe	2	TH Darm-stadt	7	8.
TH Braun-schweig	3	U Munich	2	U Munich	7	9.
TH Hannover	2	TH Hannover	1	TH Aachen	4	10.

Looking at the institutional implementation and focal points of laser research at the leading universities shows that, in particular, Hans Boersch conducted

---

[95] Of altogether 40 laser lectures held between 1963 and 1970, 22 were given by representatives of technical universities, three by university representatives, ten by representatives of the industry and five by representatives of non-university research institutions. No less than eleven of the lectures of technical universities came from the TU Berlin.

basic experimental laser research at his First Physical Institute at the TU Berlin.[96] At the TH Stuttgart, however, laser research had clearly focused on laser theory since the early 1960s, conducted by Hermann Haken at the Institute for Theoretical and Applied Physics, which rapidly gained an international reputation. The appointment of laser pioneer Wolfgang Kaiser in 1964 to the chair in experimental physics marked the beginning of experimental laser physics studies at the TH Munich, which, due to the close cooperation between the newly established Physics Department and the Max Planck Institute for Plasma Physics (Max-Planck-Institut für Plasmaphysik, IPP) in Garching, focused mainly on laser applications in this field.[97] The laser spectroscopy research that Peter Toschek began to conduct at the Institute for Applied Physics at Heidelberg University in the mid-1960s was also application-oriented.[98] In contrast, the following technical universities focused on physical-technical basic research as well as on possible technical laser applications in communications engineering: Horst Rothe's Institute for High-Frequency Technology and High-Frequency Physics, as well as Wolfgang Ruppel's Institute for Applied Physics, both TH Karlsruhe;[99] the Institute for High-Frequency Technology directed by Hans-Georg Unger, TH Braunschweig; Karl-Heinz Hellwege's Institute for Technical Physics, TH Darmstadt, as well as the Institute for High-Frequency Technology under Herbert Döring, TH Aachen.[100]

In addition to the university institutes above mentioned, the following also contributed laser research lectures to the DPG symposia during the 1960s: the Institute of Physics, University of Frankfurt (10 lectures); the Faculty of Physics at the LMU Munich (5), the Institute for High-Frequency Technology, TU Berlin (2), the Department of Applied Physics and Electrical Engineering, University of Saarbrücken (2); the Institute for Applied Physics, TU Hannover (2), the

[96] Proof for the fact that during the early 1960s experimental laser research was also conducted at the Second Physical Institute at the TU Berlin if, however, to a considerably lesser extent, are some lectures held by the institute staff (G. Koppelmann, H. Krebs) on DPG meeting between 1962 and 1964.

[97] Thus a large part of the laser dissertations in Munich in the 1960s dealt with measurement and stimulation issues in plasma physics, like e.g. the work of Franz Hillenkamp (1966), Horst Röhr (1967), Rüdiger Vesper (1968) or Heinz Bernhard Puell (1970). Cf. Jahresverzeichnis Deutscher Hochschulschriften, vol. 82–86, Leipzig 1966–1970.

[98] Toschek habilitated in 1968 with an experimental laser study on atomic collisions in Heidelberg. His staff at that time included Theodor W. Hänsch, who took his doctoral degree in 1969 with a study about the interaction in laser fields in Heidelberg. Hänsch later went into the USA. Upon his return to Germany he was appointed as fourth director of the MPI of Quantum Optics in Garching and a chair in the Physics Department at the LMU Munich.

[99] In addition to Rothe should be mentioned his colleagues O. Aanensen, H. Manger and P. Bauer. Apart from Ruppel, his staff members J. Bille, B.M. Kramer, P. Reimers and R. Stille dealt in the late 1960s with semiconductor lasers and their application in communications engineering.

[100] On behalf of Braunschweig must be mentioned A. Lohmann and D. Gloge; for Darmstadt G. Schaack and E. Bayer; for Aachen R. Hecken, as well as in the late 1960s at the I. Physical Institute J. Czech and W. Gnörich.

Third Physical Institute, University of Göttingen (1); the First Physical Institute, University of Gießen (1); the Institute for Applied Physics, University of Erlangen (1), as well as the Institute for Physical Chemistry, University of Marburg (1). Evaluating the DGaO meetings of the 1960s adds the Institutes A and B for Physics (6 lectures), as well as the Professorship B for theoretical physics (2) at the TH Braunschweig to this list.

However, the actual institutional impact that laser research had within the framework of academic research during the 1960s in the Federal Republic cannot be determined merely from charts of the institutes active in laser research. Only a comparison of all the research capacities available to West German universities in the field of physics and high-frequency technology can provide a more specific picture. In the late 1960s there were 106 chairs for general and experimental physics (including nuclear, plasma and biophysics) at 31 West German universities, 102 chairs for theoretical physics at 30 universities, 49 chairs for applied and technical physics at 22 universities and 11 chairs for high-frequency technology at 9 universities.[101] Basic and applications-oriented laser research was actively conducted within this framework at a minimum of 7 (22.6%) locations and 9 (8.5%) professorships for general and experimental physics; at 6 (27.3%) locations and 6 (12.2%) professorships of applied and technical physics, and at 4 (44.4%) locations and 5 (45.5%) professorships for high-frequency technology. In contrast, theoretical laser research was only represented at 2 (6.7%) locations with altogether 3 (2.9%) professorships.

Taking a look at the University of Hamburg, however, which during the 1960s does not appear in any chart, neither of the VDFL, nor in the dissertations, nor in the DPG/DFG chart, shows that the actual number of German universities with laser research activities was higher than these numbers imply. In October 1964 two papers for qualifying exams (Staatsexamen, for professions like teaching, for example) were produced for the first time at the Institute for Applied Physics in Hamburg. In 1965, during the following summer term, institute director Professor Dr. Heinz Raether offered a one-hour special lecture titled "From the Physics of Lasers".[102] Even though there is no proof for doctoral and professorial dissertations either at Raether's institute or at the Institute for Physics of the University of Hamburg in the field of laser before the end of 1970, a whole series of examination theses as well as several diploma theses on the subject of laser were produced during that time.[103] Though obviously the university of Hamburg did not represent a center of West German laser research during the 1960s,[104] its

---

[101] See Die Lehrstühle an den Wissenschaftlichen Hochschulen in der Bundesrepublik, 15th edition, Göttingen 1969 (Schriften des Hochschulverbandes 9).

[102] Cf. the university calendar of the Hamburg University for the summer term 1965.

[103] Thus at the Physical Institute four Staatsexamen and two diploma theses, as well as ten Staatsexamen and one diploma theses at the Institute for Applied Physics. Cf. Universität Hamburg, Bibliothek des Fachbereichs Physik.

[104] It was not until 1971 that with Franz Lanzl from Darmstadt a laser physicist was called to the Hamburg Institute for Applied Physics, in order to establish there the research field of "Coherent Optics." Cf. Legler, 60 Jahre Angewandte Physik.

example nevertheless proves Wolfgang Kaiser right, for he has argued that at least in the Federal Republic, almost all universities had been concerned with the laser by some form or other.

Only a few West German university institutes emerged during the 1960s as laser research centers, including eight to ten main locations. In particular, institutes at technical universities were able to establish themselves in this field, whether in basic or application-oriented laser research. During this sample period the First Physical Institute of the Technical University Berlin clearly became the leading academic institution in the field of experimental laser research. The same held true for the Institute for Theoretical and Applied Physics of the Technical University Stuttgart with regard to the theory.

CONCLUSION: THE DFG AND WEST GERMAN LASER RESEARCH IN THE 1960s

From 1958 and 1966 a total of about 11 million DM was spent within the framework of the DFG high-frequency physics priority program for the work on "all technical and physical questions regarding molecular amplifiers for microwaves and light frequencies."[105] Unfortunately the exact amount of funds for laser research cannot be deduced from the accessible sources. Yet the increasing number of laser projects and laser researchers that received grants implies (see chart 7, below) that starting in 1961 – that is, the same year in which professor Rothe (TH Karlsruhe) for the first time demonstrated a laser at the annual Research Focus Program Colloquium – an increasing portion of the program's funds went into the laser research.

The lecture programs of the focus program colloquia confirm this assumption. By the fifth Colloquium in 1963 the lectures of laser group clearly outnumbered the lectures of the other seven groups on high-frequency research.[106] Moreover, the 1963 colloquium report indicates an interesting change regarding the disciplinary origin of the laser scientists as well as the content of their research: "[Not only electrical engineering technicians discussed] the optical molecular amplifier (LASER) [this year]", it says there, "but also experimental physicists and theoreticians."[107]

---

[105] Numerical data according to Mitteilungen der DFG, 2, 1966, p. 13.

[106] Cf. the report on the colloquium from October 6 to October 12, 1963, in Hirschegg/Kleinwalsertal, in: Mitteilungen der DFG 4, 1963, pp. 13–15.

[107] Cf. ibid, p. 15.

Chart 7: Laser scientists and projects funded by the DFG priority program "High-Frequency Technology" from 1962 to 1966[108]		
Scientist	Institution	Projects/Subjects (grant period)
Hans Boersch	TU Berlin	– Construction and investigation of different types of lasers (1962) – Applications of lasers (1962)
Hermann Haken	TH Stuttgart	Optical maser properties of the ruby (1962–1965) – Two quantum process (1962–1965) – Theory of the spectral line width of lasers (1966) – Theory of the scattering of photons on photons (1966) – The form of the absorption band of the exciton (1966)
Horst Rothe	TH Karlsruhe	Investigation of molecular amplifiers in the microwave and optical regions (1963–1966)
Karl-Heinz Hellwege	TH Darmstadt	Production of higher light field intensity in solid bodies through powerful laser impulses (1964–1966)
Wolfgang Kaiser	U Munich	– Quantitative investigation of the kinetics of a individual oscillation (1964) – Mode selectivity of a prism reflector for an optical maser (1964) – Development of lasers in the far infrared (1966) – Experiments on producing acoustical phonons in the far infrared (1966) – Resonance spectroscopy in the far infrared (1966)
Friedrich Kirchstein	TH Braunschweig	Diode laser as a light amplifier (1965)
Hans-Georg Unger	TH Braunschweig	Investigations of the bound or directed diffusion of coherent electromagnetic waves at optical frequencies (1965)
Joachim Heintze	U Heidelberg	Optical high frequency spectroscopy (1966)

A compilation of the laser projects that received grants within the framework of the high-frequency physics focus program between 1962 and 1966 shows that this trend continued during the next years (see chart 7). In addition to the studies on molecular amplifiers in the optical sector conducted by Horst Rothe in Karlsruhe, the engineer who studied high frequencies and had initiated the DFG priority program, promotion increasingly focused on the research in laser theory by the Stuttgart theoretical physicist Hermann Haken or the physicist Wolfgang

---

[108] Compiled according to the DFG reports, 1962 ff.

Kaiser's experimental laser work in Munich. In the same way projects on the application of lasers, for example in the field of telecommunication engineering and spectroscopy, received considerably more grants.

The DFG grants for West German laser research within the framework of the High-Frequency Physics priority program during the 1960s were considerably supplemented by the funding of individual laser studies within the DFG standard procedure. Since 1963 there is proof for the funding of laser subjects independent of and parallel to the priority program.[109] The fact, however, that occasionally laser subjects from the standard procedure were after a while included into the priority program's grant catalogue[110] (and vice versa, subjects of the priority program received standard procedure grants after the program expired[111]), proves that there was a close connection between standard procedure and priority program promotion.

By establishing the new priority program "Solid-State Physics" in 1964, the DFG set up another potential promotional focus program in laser research even before the "High-Frequency Physics" program expired. The new program was to stimulate above all "research on the physics of non-metal solids (preferably semiconductors, polar crystals, ceramics) and metal-based materials (especially supra-conductors)", but also had an indirect impact on the promotion of laser research due to the fact that, according to a DFG report on the priority program "Solid-State Physics", "the further development of technical elements, like for instance the development of semiconductor components [...] of materials for microwave and light amplifiers (MASER and LASER)" was immediately linked to this work.[112] From 1964 to 1970 altogether 20.82 million DM was provided for the focus program "Solid-state Research", as it soon was called. During this period the number of proposals increased from 28 to 127 per year, notwithstanding that the program was constantly refocused, leaving in the end only the fields "collective phenomena" as well as "electronic structure" and "interaction with

---

[109] Thus e. g. in 1967 the subject "Kollektives Zusammenwirken vieler Lasermoden (Collective concurrence of many laser modes)" by Wolfgang Weidlich from Stuttgart, in 1968 the subject "Weiterentwicklung von Farbstofflasern (Further development of dye lasers)" by Fritz Peter Schäfer from Marburg or in 1970 the subject "Untersuchung von Plasmen, die durch Laser-Höchstleistung erzeugt werden (Examination of plasmas generated by maximum laser performance) by Wolfgang Kaiser from Munich. Cf. DFG report, 1967, p. 284, resp. 1968, p. 314, as well DFG Programme und Projekte, 1970, p. 249.

[110] An example for this is the subject "Theorie der Linienbreite des Lasers und der Streuung von Photonen an Photonen (Theory on laser linewidth and photon diffusion)" by Hermann Haken, funded in 1963 within the framework of standard procedure, which reappeared in 1966 split into two separate topics in the promotional catalogue of the High-Frequency Physics priority program. Cf. DFG report, 1963, p. 211, as well as chart 3.6.

[111] For instance the subject "Diodenlaser als Lichtverstärker (Diode laser as light amplifier)" by Friedrich Kirchstein from Braunschweig, which was initially funded in 1965 within the priority program and subsequently in 1967 as standard procedure. Cf. chart 3.6, as well as DFG report, 1967, p. 278.

[112] Mitteilungen der DFG (2), 1966, p. 12.

phonons".[113] However, the consequence of this thematic focus for the promotion of laser research was that in 1970 none of the 127 research projects was directly linked to the subject of "laser".

In order to emphasize the increasing significance of solid-state research for science and technology, three years after creating the priority program of the same name, the DFG established a special "Commission for Solid-state Research", which was primarily supposed to elaborate proposals on how the Federal Republic could eventually catch up in this field as well as for the special promotion of important areas in solid-state research. One of the arguments of the commission was that a goal of the to-be funded research projects must be "above all the study [...] of the electronic state in isolating crystals, the knowledge of which is the basis of the laser's light emission."[114] Reverting to plans the two Stuttgart physicists Heinz Pick and Hermann Haken had already developed in the early 1960s, the commission voted at their first meeting to establish a central institute for solid-state research, which was finally set up in 1970 as the Max Planck Institute for Solid-State Research in Stuttgart.[115] Later methodical foci of the MPI for Solid-State Research included Raman and infrared spectroscopy (closely linked to laser research) as well as the development of short pulse spectroscopy.[116]

Finally, in the late 1960s the DFG created two more promotional fields that impacted West German laser research. In 1968 the priority program "Joining by Bonding Solids (Fügen durch Stoffverbinden)" emerged in the engineering sciences, which was designed for a term of six years and received from 1968 to 1970 about 3 million DM in funds.[117] The program aimed to provide bases for the numerous methods of fixed connections in order to get a better understanding of their potential areas of application. The funded work was to concentrate mainly on bonding methods with a high energy density, specifically including "laser welding". In 1970 the program provided the framework for two investigations, one, which conducted basic research on light beam procedures, and another one that dealt with the practical problems of laser beam welding.[118]

And finally in 1969 the DFG established the Special Research Field 65 "Solid-state Spectroscopy" at the TH Darmstadt and Frankfurt University, which was funded during the first two years with 3.1 million DM.[119] The goal of this special research field was to collect experimental material to determine the electronic structure and its interaction in solids. The methods applied also included laser spectroscopy, for which altogether four subprojects were conducted within the

---

[113] See DFG Programme und Projekte, 1970, p. 399.

[114] See Kommission für Festkörperforschung, in: Mitteilungen der DFG (1), 1968, p. 11 ff.

[115] On the founding of the MPI für Festkörperforschung see Eckert, Großes für Kleines, pp. 181–199.

[116] See Gerwin/Holtz, Porträt einer Forschungsorganisation, p. 98.

[117] See DFG Programme und Projekte, 1970, pp. 464 ff.

[118] Prof. Dr.-Ing. Helmut Koch worked on the light beam procedures in Mannheim; Prof. Dr. H. Welling and Dr.-Ing. Gerhard Sepold worked on laser beam welding in Bremen. Cf. ibid., p. 465.

[119] ibid., p. 589 ff.

framework of the special research field.[120] The Special Research Field 65 had two spokesmen, one of which was the director of the Second Physical Institute of the TH Darmstadt, Professor Bruno Elschner, who in 1961 had initiated the first laser research in the German Democratic Republic (Deutsche Demokratische Republik, DDR) with his working group at the University of Jena and had shortly afterwards, just before the construction of the Wall, left for West Germany.[121]

All things considered, the funding the DFG provided for West German laser research within the framework of standard procedures as well as the so-called promotional foci and the "Special Research Field 65" was instrumental in establishing and developing this field of research, especially at West German universities. During the sample period neither the then recently founded Federal Ministry of Education and Research nor the ministries of the German states (Länder) in charge of the advancement of science, were willing to contribute funds worth mentioning to early laser research.

In this context great importance has to be attached to the High-Frequency Physics priority program, because during the crucial early stages it stimulated sustained theoretical and experimental basic research on the laser, and also eased the way for first applications. In particular the fact that by the mid-1960s West German laser theory had not only caught up with the international state of research, but was instrumental in defining it, has to be credited to the DFG as well.

It is striking, however, that the DFG did not establish individual priority programs or special research fields in the mid-1960s for laser physics or laser technology. Yet the reasons behind this will remain guesswork until the files concerning that matter will be disclosed. Still it is safe to assume that one reason was the still rather vague form of the new research field, which only suggested that it might become independent from, for example, high frequency research or solid-state physics. Moreover, the perspectives of the new research field were not easy to assess regarding its technical and economic impacts at that time – at best they only existed as apparently bizarre speculations. At that moment the "laser" was not seen as a key technology. Given the numerous deficits German research showed in international comparison in other fields, like for instance semiconductor technology, nuclear energy or aerospace engineering, West German research policy only attached secondary importance to laser research and laser technology, at least during the 1960s. Hence from the DFG's point of view, and thus, in a manner of speaking, from the perspective of science as such, specific grant programs were not (yet) required.[122]

---

[120] 1. 2 Photon induced luminescence and stimulated Raman effect; 2. 2 Photon absorption measurement in solids; 3. Spectroscopy of color centers in relaxed [excited] states; 4. Luminescence studies in semiconductors. See ibid., p. 589.

[121] Cf. Albrecht, Laserforschung Jena.

[122] That did not change until the 1980s when the Bundesministerium für Forschung und Technologie issued a special grant program for the field of laser research and laser technology, for which altogether 350 million DM in funds were allocated from 1987 to 1993. Cf. Laserforschung und Lasertechnik. Jahresbericht 1990, issued by the VDI Technologiezentrum on behalf of the BMFT. Düsseldorf 1991.

BIBLIOGRAPHY

Albert, W. and Ch. Oehler: Materialien zur Entwicklung der Hochschulen 1950 bis 1967, vol. 1, Hannover 1969 (HIS Hochschulforschung).

Albrecht, Helmuth: Hermann Haken im Gespräch, in: Frieß, Peter and Peter M. Steiner (eds.): Deutsches Museum Bonn. Forschung und Technik in Deutschland seit 1945, Bonn 1995, pp. 250–259.

Albrecht, Helmuth: Laserforschung in Deutschland 1960–1970. Eine vergleichende Studie zur Frühgeschichte von Laserforschung und Lasertechnik in der Bundesrepublik Deutschland und der Deutschen Demokratischen Republik, Habil. Universität Stuttgart 1997.

Albrecht, Helmuth: Alfred Kastler – Nobelpreis für Physik 1966, in: Brockhaus Nobelpreise. Chronik herausragender Leistungen, Mannheim, Leipzig 2001, pp. 612–613.

Albrecht, Helmuth: Laser für den Sozialismus. Der Wettlauf um die Realisierung des ersten Laser-Effekts in der DDR, in: Splinter, Susan, Sybille Gerstengarbe, Horst Remane, and Benno Parthier (eds.): Physica et historia. Festschrift für Andreas Kleinert zum 65. Geburtstag, Halle, Stuttgart 2005 (Acta Historica Leopoldina 45), pp. 471–491.

Albrecht, Helmuth: Laserforschung an der Friedrich-Schiller-Universität in Jena, in: Hoßfeld, Uwe, Tobias Kaiser, and Heinz Mestrup (eds.): Hochschule im Sozialismus. Studien zur Geschichte der Friedrich-Schiller-Universität Jena 1945–1990, Köln, Weimar, Wien 2007, vol. 2, pp. 1436–1469.

Albrecht, Helmuth and Uwe Schulte: Nicolaas Bloembergen, Arthur Leonhard Schawlow, Kai Manne Börje Siegbahn, in: Brockhaus Nobelpreise. Chronik herausragender Leistungen, Mannheim, Leipzig 2001, pp. 784–785.

Anwendungen des Lasers. Heidelberg 1988 (Spektrum Wissenschaft: Verständliche Forschung).

Armstrong, J. A. and A.W. Smith: Intensity Fluctuations in a GaAs Laser, in: Physical Review Letters 14 (3), 1965, pp. 68–70.

Armstrong, J. A. and A.W. Smith: Intensity Fluctuations in GaAs Laser Emission, in: Physical Review 140 (1 A), 1965, pp. A 155-A 164.

Arzt, Volker, Hermann Haken, Hannes Risken, Herwig Sauermann, C. Schmid, and Wolfgang Weidlich: Quantum Theory of Noise in Gas and Solid State Lasers with an Inhomogeneously Broadened Line I, in: Zeitschrift für Physik 197, 1966, pp. 207–227.

Basting, Dirk, Fritz Peter Schäfer, and B. Steyer: A Simple High Power Nitrogen Laser, in: Opto-Electronics 4, 1972, pp. 43–49.

Becker, G. and B. Fischer: Beitrag zum internationalen Wasserstoffmaser-Vergleich mit transportabler Atomuhr. In: PTB-Mitteilungen 78, 1968, pp. 177–184.

Bertolotti, Mario: Masers and Lasers. An Historical Approach, Bristol, Philadelphia 1983.

Boersch, Hans, Hans Eichler, Arno Schmackpfeffer, and Horst Weber: Generation of Short and Intensive Light Pulses by Saturable Absorbers, in: Proceedings of the I. International Conference of Laser Application, Paris 1967.

Brix, Peter: Erinnerungen an die Physik von 1945 bis 1970 – Göttingen, Ottawa, Heidelberg, Darmstadt, in: Marx, Otto M. and Anette Moses (eds.): Emeriti erinnern sich. Rückblicke auf die Lehre und Forschung in Heidelberg. Vol. II: Die Naturwissenschaftlichen Fakultäten, Weinheim, New York, Basel, Cambridge, Tokyo 1994, pp. 9–31.

Bromberg, Joan Lisa: The Laser in America 1950–1970, Cambridge/Mass., London 1991.

Brüche, Ernst (ed.): Physikertagung Wiesbaden. Hauptvorträge der Jahrestagung 1960 des Verbandes Deutscher Physikalischer Gesellschaften, Mosbach 1961.

Brüche, Ernst (ed.): Physikertagung Wien. Hauptvorträge der gemeinsamen Jahrestagung 1961 des Verbandes Deutscher Physikalischer Gesellschaften und der Österreichischen Physikalischen Gesellschaft, Mosbach 1962.

Die Lehrstühle an den Wissenschaftlichen Hochschulen in der Bundesrepublik und Westberlin, 7th edition, Göttingen 1961 (Schriften des Hochschulbundes 9).

Dietrich, Paul: Licht als Informationsträger, in: Jahrbuch für elektrisches Fernmeldewesen 12, 1960/61, pp. 293–323.

Eckert, Michael and Maria Osietzki: Wissenschaft für Macht und Markt. Kernforschung und Mikroelektronik in der Bundesrepublik Deutschland, München 1989.

Eigen, Manfred and R. Winkler-Oswatitsch: Das Spiel, München 1975.

Empfehlungen des Wissenschaftsrates zum Ausbau der wissenschaftlichen Einrichtungen. Teil I: Wissenschaftliche Hochschulen, Tübingen 1960.

Fleck jr., J.A.: Linewidth and Conditions for Steady Oscillation in Single and Multipleelement Lasers, in: Journal of Applied Physics 34 (10), 1963, pp. 2997–3003.

Gerwin, Robert and Barbara Holtz: Porträt einer Forschungsorganisation. Die Max-Planck-Gesellschaft und ihre Institute. Aufgabenstellung, Arbeitsweise, Strukturen, Entwicklung, 3rd edition, München 1983.

Gloge, D.: Gesichtspunkte zum Bau von He-Ne-Lasern für die Nachrichtentechnik, in: Frequenz 16 (5), 1962, p. 196.

Haken, E. and Hermann Haken: Zur Theorie des Halbleiter-Lasers, in: Zeitschrift für Physik 176, 1963, pp. 421–428.

Haken, Hermann: A Nonlinear Theory of Laser Noise and Coherence. Part I and II, in: Zeitschrift für Physik 181, 1964, pp. 96–124, and 182, 1965, pp. 346–359.

Haken, Hermann: Laser Theory, Berlin, New York 1970 (Handbuch der Physik, vol. XXV/2c).

Haken, Hermann: Licht und Materie II: Laser, Mannheim, Vienna, Zurich 1981.

Haken, Hermann: Erfolgsgeheimnisse der Natur. Synergetik: Die Lehre vom Zusammenwirken, 2nd edition, Stuttgart 1981.

Haken, Hermann and Robert Graham: Quantum Theory of Light Propagation in a Fluctuating Laser-Active Medium, in: Zeitschrift für Physik 213, 1968, pp. 420–450.

Haken, Hermann and Robert Graham: Laserlight. First Example of a Second-Order Phase Transition far away from Thermal Equilibrium, in: Zeitschrift für Physik 237, 1970, pp. 31–46.

Haken, Hermann and Robert Graham: Synergetik. Die Lehre vom Zusammenwirken, in: Umschau in Wissenschaft und Technik 71 (6), 1971, pp. 191–195.

Haken, Hermann and Herwig Sauermann: Frequency Shifts of Laser Modes in Solid and Gaseous Systems, in: Zeitschrift für Physik 176 (1), 1963, pp. 47–62.

Haken, Hermann and Herwig Sauermann: Nonlinear Interaction of Laser Modes, in: Zeitschrift für Physik 173 (3), 1963, pp. 261–275.

Haken, Hermann and Wolfgang Weidlich: Quantum Noise Operators for the N-Level System, in: Zeitschrift für Physik 189, 1966, pp. 1–9.

Haug, Hartmut: Multimode-Eigenschaften verschiedener Halbleiterlasermodelle, in: Zeitschrift für Physik 194, 1966, pp. 482–506, and 195, 1966, pp. 74–97.

Head, Peter: German Research in Lasers Independent of US Efforts, in: Electronic News 331 (7), 1962, p. 32.

Herziger, Gerd: Anwendung von Laser-Resonatoren für hochauflösende Interferometrie, Diss. (rer.nat.) TU Berlin 1965.

Herziger, Gerd: Übertragungseigenschaften entdämpfter optischer Resonatoren, Habil. TU Berlin 1967.

Hirsch, Joachim: Wissenschaftlich-technischer Fortschritt und politisches System, 3rd edition, Frankfurt am Main 1973 (edition suhrkamp 437).

Kaiser, Wolfgang: Der optische Maser, in: Physikalische Blätter 17, 1961, pp. 256–262.

Kaiser, Wolfgang: Der optische Maser, in: Brüche, Ernst and H. Franke (eds.): Physikertagung Stuttgart. Hauptvorträge der Jahrestagung 1962 des Verbandes Deutscher Physikalischer Gesellschaften, Mosbach 1963, pp. 30–37.

Kaiser, Wolfgang: Der Laser. Grundlagen und neue Ergebnisse, in: Abhandlungen und Berichte, ed. by Deutsches Museum 35 (2), 1967, pp. 5–26.

Kern, Ulrich: Die Entstehung des Radarverfahrens. Zur Geschichte der Radartechnik bis 1945. Diss. (phil.) Universität Stuttgart 1984.

Kertz, Walter: Forschung in der jüngsten Entwicklungsphase, in: Kertz, Walter et al. (eds.): Technische Universität Braunschweig. Vom Collegium Carolinum zur Technischen Universität 1745–1995, Hildesheim, Zurich, New York 1995, pp. 669–699.

Kinder, Herbert and Walter Stöhr: Die Funkstelle Raisting für Nachrichtenverbindungen über Satelliten, in: Siemens Zeitschrift 38 (10), 1964, pp. 723–733.

Kleen, Werner: Aus der Arbeit des Forschungslaboratoriums von Siemens & Halske, in: Siemens Zeitschrift 36 (11), 1962, pp. 757–771.

Kleen, Werner and Rudolf Müller (eds.): Laser. Verstärkung durch induzierte Emission. Sender optischer Strahlung, hoher Kohärenz und Leistungsdichte, Berlin, Heidelberg, New York 1969.

Krohn, Wolfgang, Günther Küppers, and Rainer Paslack: Selbstorganisation – Zur Genese und Entwicklung einer wissenschaftlichen Revolution, in: Schmidt, Siegfried J. (Hrsg.): Der Diskurs des Radikalen Konstruktivismus, 2nd edition, Frankfurt am Main 1988, pp. 441–465.

Kuhn, Hans W.: The Electron Gastheory of the Colour of Natural and Artificial Dyes: Problems and Principles, in: Zechmeister, D.L.: Fortschritte der Chemie organischer Naturstoffe 16, Wien 1958, pp. 169–205.

Lamb jr., Willis E.: Theory of an Optical Maser, in: Physical Review 134 (6A), 1964, pp. A1429–A1450.

Lamb jr., Willis E.: Laser Theory and Doppler Effects. In: IEEE Journal of Quantum Electronics QE-20, 1984, pp. 551–555.

Landrock, Rudolf: Die Deutsche Akademie der Wissenschaften zu Berlin 1945–1971 – ihre Umwandlung zur sozialistischen Forschungsakademie. Eine Studie zur Wissenschaftspolitik der DDR, 3 vol., Erlangen, Nürnberg 1977 (abg-Analysen und Berichte aus Gesellschaft und Wissenschaft 1–3/1977).

Legler, Werner: 60 Jahre Angewandte Physik in Hamburg, in: uni hh Forschung Nr. XIX/1985: 100 Jahre Physik in Hamburg, pp. 29–34.

Lemmerich, Jost: Zur Geschichte der Entwicklung des Lasers, Berlin 1987 (Berliner Beiträge zur Geschichte der Naturwissenschaften und der Technik 6).

Maiman, Theodore H.: Stimulated Optical Radiation in Ruby, in: Nature 187, 1960, pp. 493–494.

Maiman, Theodore H. and Irnee J. D'Haenens: Stimulated Optical Emission in Fluorescent Solids, in: Physical Review 123 (4), 1961, pp. 1145–1157.

Meinel, Werner, Reinhardt Neubert, and Gerhard Wiederhold: Wirkungsweise der Laser und Demonstration einiger ihrer Eigenschaften, in: Feingerätetechnik 12 (11), 1963, pp. 493–510.

Peterson, O.G., S.A. Tuccio, and B.B. Snavely: CW Operation of an Organic Dye Solution Laser, in: Applied Physics Letters 17, 1970, pp. 245–247.

Pilkuhn, M. and H. Rupprecht: Optical and Electrical Properties of Epitaxial and Diffused GaAs Injection Lasers, in: Journal of Applied Physics 38, 1967, pp. 5–10.

Qing, Lin: Zur Frühgeschichte des Elektronenmikroskops, Stuttgart 1995.

Rexin, Manfred: Die Entwicklung der Wissenschaftspolitik in der DDR, in: Ludz, Peter Christian (ed.): Wissenschaft und Gesellschaft in der DDR, München 1971, pp. 78–121.

Risken, Hannes: Correlation Function of the Amplitude and of the Intensity Fluctuations for a Laser Model near Treshold, in: Zeitschrift für Physik 191, 1966, pp. 302–312.

Risken, Hannes: Zur Statistik des Laserlichts, in: Fortschritte der Physik 16, 1968, pp. 261–308.

Röss, Dieter: Laser: Lichtverstärker und –oszillatoren, Frankfurt am Main 1966 (Technische-Physikalische Sammlung 4).

Rothe, Horst: Bedeutung und Stand der Maser-Entwicklung, in: Umschau 12, 1964, pp. 353–358.

Runge, P.: Ein 1 m langer He-Ne-Laser für 0,63 Fm mit nur einer einzigen Frequenz, in: Archiv für elektrische Übertragung 19, 1965, pp. 573–574.

Schäfer, Fritz Peter: Farbstofflaser – Von der Entdeckung bis zur Erzeugung ultrakurzer Lichtimpulse, in: Jahrbuch 1984 der Max-Planck-Gesellschaft. Göttingen 1984, pp. 46–67.

Schäfer, Fritz Peter: Der Weg zum Farbstofflaser – und was daraus wurde, in: Gerwin, Robert (ed.): Wie die Zukunft Wurzeln schlug. Aus der Forschung der Bundesrepublik Deutschland, Berlin, Heidelberg, New York 1989.

Schäfer, Fritz Peter and Werner Schmidt: Lösungen organischer Farbstoffe als optische Schalter zur Erzeugung von Laser-Riesenimpulsen, in: Zeitschrift für Naturforschung 19a, 1964, pp. 1019–1020.

Schäfer, Fritz Peter, Werner Schmidt, and J. Volze: Organic Dye Solution Laser, in: Applied Physics Letters 9, 1966, pp. 306–309.

Schmidt, Werner: Lichtabsorption in Lösungen organischer Farbstoffe bei hohen Bestrahlungsstärken, Diss. (rer.nat.) Universität Marburg 1966.

Schmidt, Werner and Fritz Peter Schäfer: Blitzlampengepumpte Farbstofflaser, in: Zeitschrift für Naturforschung 22a, 1967, pp. 1563–1566.

Schmidt, Werner and Fritz Peter Schäfer: Self-Mode-Locking of Dye-Lasers with Saturable Absorbers, in: Physical Letters 26A, 1968, pp. 558–559.

Snavely, B.B. and Fritz Peter Schäfer: Feasibility of cw-Operation of Dye Lasers, in: Physical Letters 28A, 1969, pp. 728–729.

Snow, Charles Percy: Die zwei Kulturen. Literarische und naturwissenschaftliche Intelligenz, Stuttgart 1967. (Engl.: Snow Charles Percy: The two cultures and a second look, Cambridge 1963.)

Sorokin, Peter P.: Contributions of IBM to Laser Science 1960 to the Present, in: IBM Journal of Research and Development 23, 1979, pp. 476–488.

Sorokin, Peter P. and Johan R. Lankard: Stimulated Emission observed from an Organic Dye, Chloro-Aluminum Phthalocyanine, in: IBM Journal of Research and Development 10, 1966, p. 162.

Sporleder, F.: 25 Jahre Institut für Hochfrequenztechnik der Technischen Hochschule Braunschweig, in: 25 Jahre Institut für Hochfrequenztechnik, Braunschweig 1985, pp. 15–23.

Szöllözi-Janze, Margit and Helmuth Trischler: Großforschung in Deutschland. Frankfurt am Main, New York 1990 (= Studien zur Geschichte der deutschen Großforschungseinrichtungen 1).

Townes, Charles H. (ed.): Quantum Electronics. A Symposium, New York 1960.

Troe, Jürgen: Laser in der Reaktionskinetik, eine Revolution, in: Gerwin, Robert (ed.): Wie die Zukunft Wurzeln schlug. Aus der Forschung der Bundesrepublik Deutschland, Berlin, Heidelberg, New York 1989, pp. 212–217.

Unger, Hans-Georg: 25 Jahre optische Nachrichtentechnik. Festvortrag, in: 25 Jahre Institut für Hochfrequenztechnik der Technischen Hochschule Braunschweig, Braunschweig 1985, pp. 24–39.

Wagner, W.G. and Groge Birnbaum: Theory of Quantum Oscillators in a Multimode Cavity, in: Journal of Applied Physics 32 (7), 1961, pp. 1185–1194.

Weber, Horst: Theoretische und experimentelle Untersuchungen des zeitlichen Intensitätsverlaufs eines Drei-Niveau-Lasers am Beispiel des Rubinlasers, Diss. (rer.nat.) TU Berlin 1965.

Weber, Horst: Theoretische und experimentelle Untersuchungen des zeitlichen Emissionsverlaufs eines Rubinlasers, in: Zeitschrift für Physik 188, 1966, pp. 444 ff.

Weber, Horst: Das Emissionsverhalten des gepulsten Rubinlasers, Habil. TU Berlin 1967.

Weber, Horst and Gerd Herziger: Laser: Grundlagen und Anwendungen, Weinheim 1972.

Weidlich, Wolfgang and Fritz Haake: Coherence-Properties of the Statistical Operator in a Laser Model, in: Zeitschrift für Physik 185, 1965, pp. 30–47.

Weidlich, Wolfgang and Fritz Haake: Master-Equation for the Statistical Operator of Solid State Laser, in: Zeitschrift für Physik 186, 1965, pp. 203–221.

Weingart, Peter: Wissensproduktion und soziale Struktur, Frankfurt am Main 1976 (stw 155).

SCIENCE AND THE PERIPHERY UNDER STALIN: PHYSICS IN UKRAINE

Paul Josephson, Yury Ranyuk, Ivan Tsekhmistro, and Karl Hall

In February 1952, Kirill Sinelnikov, director of the Ukrainian Physical Technical Institute (UFTI) in Kharkiv, wrote Lavrenty Beria to protest the difficult financial straits of his institute. Sinelnikov was a central figure in the Soviet nuclear program. His brother-in-law was Igor Kurchatov, the head of the atomic bomb program. He had studied with leading scholars in Leningrad, and Cambridge, England, and he led UFTI from its early years in the 1930s, through the turmoil of the purges and the challenges of evacuation during World War II. Although recently founded and located in an agricultural region, UFTI had quickly become a center of Soviet efforts in low temperature and theoretical physics, and a founding European site of nuclear physics. Yet Sinelnikov had become a supplicant before Beria, head of the secret police and government head of the bomb project, to secure additional funding and solve growing personnel problems to maintain the level of scientific quality UFTI had demonstrated from its first days.

Beria used his access to resources and personnel, as well as infamous coercive measures of intimidation, arrest, interrogation and execution, to force the pace of projects large and small. Yet Sinelnikov had the distinct impression that the physicists of UFTI had been slighted by Beria. Accordingly, their efforts to expand research in high energy and nuclear physics, including a new program in fusion (controlled thermonuclear synthesis) lagged. He informed Beria that the funds indicated for building apartments had not be spent, making it impossible to house younger specialists and a part of the senior scholars "who lived under quite difficult circumstances." The Kharkiv City Council had offered the institute space in the central part of the city for construction, but nothing more had happened. Further, the government authorized salaries for staff of UFTI at levels significantly lower than in other Academy of Sciences institutes, in particular those in Moscow and Leningrad. Both the housing and salary shortfalls made it hard to attract and keep the best researchers. In as much as UFTI scientists undertook research of central importance to the government during the Cold War on various nuclear topics, Sinelnikov asked Beria to ensure that the 1952–53 capital budget included funds to build another thirty apartments, and that his staff receive salaries commensurate with scientists in other nuclear institutes.[1]

---

[1] Letter from K. Sinelnikov to L. Beria of 29 February 1952, Archive of UFTI (hereafter A UFTI), f. 1, op. 1/c, ed. khr. 78.

In secret documents officials referred to UFTI as Laboratory No. 1. Kurchatov's institute in Moscow was Laboratory No. 2. How did it happen that Laboratory 1, UFTI, the center of Ukrainian physics, a birthplace of nuclear and low temperature physics, a major site for the development of particle accelerator technology, whose physicists were both national and world leaders in a variety of fields, had fallen apparently on hard times? This was not the first time. The Ukrainian physics community, perhaps more than any other scientific community in the USSR, suffered through the Stalinist era. The authorities imposed autarky on the scientific community, while ideological and bureaucratic strictures handicapped their research. In Kharkiv, the authorities delayed the reestablishment of Kharkiv State University (KhGU) until 1934, when they were confident they controlled the processes of "Sovietization" of Ukrainian higher education and culture. The physicists faced close scrutiny for alleged philosophical and political transgressions that were manifested in "anti-Soviet," Trotskyite, and other alleged tendencies. It lost a number of leading specialists to the Great Terror; the secret police carried out a series of purges that led to the arrest and execution of a number of leading physicists from 1936 to 1938.

Yet the damage to Ukrainian physics was also the result of Soviet policies directed specifically toward Ukraine. One of these policies was the effort to "russify" or "sovietize" the nation and to destroy Ukrainian (and other) nationalism. Another concerned the violence of collectivization in Ukraine. On top of this, Ukraine was then decimated by the Germany invasion during World War II. In the rush to establish UFTI as part of rapid industrialization, in the violence of the effort to establish ideological control over its scientists, in the brilliance of its personnel, and in the transformation of UFTI into an institution of Soviet power, the experience of UFTI reflected both national trends and regional and local concerns, geopolitical desiderata and Ukrainian realities.

STALINIST SCIENCE POLICY: THE CASE OF PHYSICS

The expansion of the physics enterprise in Eastern Ukraine in Kharkiv took place against the backdrop of Stalinist programs in the 1930s for rapid industrialization, collectivization of agriculture, cultural revolution in art, literature, film, and higher education, and new policies for science. The imposition of Stalinist science policies involved the centralization of policy making in such Moscow-based government, party and scientific bureaucracies as the Ministry of Heavy Industry or the Presidium of the Academy of Sciences. Centralization enabled officials to enforce an emphasis on applied science at the expense of basic research. In general, the five-year plans indicated that the periphery – any periphery – would serve the center, Moscow. Stalin left no doubt about the centralization of science policy when he ordered the presidium of the Soviet Academy of Sciences to be moved from Leningrad to Moscow in 1934.

Simultaneously, the authorities introduced a new system of planning of scientific activity. They required that institutes produce one-year and five-year plans of research that could be audited, verified and enforced. Of course, specialists found it challenging to predict their discoveries, a strangely impossible task even in Stalin's heroic world, and worried about having their hands tied to completing one project when another avenue of research might turn out to be more promising. The costs of failure grew throughout the 1930s to include firing, arrest and in some cases even execution.

Another crucial aspect of Stalinist science policy was ideological control, which had both practical/immediate and epistemological/long term implications. Fearing contamination by Western ideas, ideologues, officials and their scientific allies undertook a series of interventionist measures to secure Soviet science. They believed that deviation from philosophical norms would lead to a weakening of research, if not the outright production of results harmful to the Soviet state and the proletariat. They tied their assertions to an ongoing discussion about what the Soviet philosophy of science, dialectical materialism (diamat), meant for each science. In physics this reflected disputes over whether and how relativity theory and quantum mechanics were commensurate with diamat.[2] We will not spend much time on this subject.

Marx's theory of history, historical materialism, was more closely related to some of the disputes over the scientific enterprise that embroiled physicists. It was related to two concepts: class war and the construction of socialism in one country. Regarding the former, ideologues asserted that "ivory tower reasoning" prevalent in capitalist societies, what might be called "science for the sake of science," produced little of value for workers; workers and scientists alike toiled for their capitalist masters to make profit. How awareness of class struggle would show physicists the paths to take in quantum mechanics, low temperature or nuclear physics, or what dangers to the proletariat lurked in these fields remain difficult to explain. Regarding the second, the matter is clear: planners and builders established economic autarky in response to Stalin's effort to create a socialist fortress of advanced industry and agriculture. Autarky extended to science, with Soviet scholars increasingly isolated from international happenings, colleagues, and trends. Officials sought control over such fields as sociology, economics, and, infamously, Lysenkoist biology.

Physicists became increasingly isolated, were largely unable to attend conferences abroad, entertain their colleagues at conferences in the USSR, or exchange reprints with them freely. Journals were censored, with libraries putting current issues out – with suspicious articles often excised – after months of delay. Autarky made provincial scientists feel even more provincial. Ideological control extended to personnel policy. Class, political reliability and other factors played a role in who was admitted to university or graduate school, or accepted for employment. Working class origins often dictated advancement and in some cases rivaled merit as a criterion of success. Knowledge of Party history and Marxist-Leninist phi-

---

[2] On philosophical controversies and Soviet science, see Joravsky, Marxism, and Graham, Science.

losophy was often as important as knowledge of science for advancement. The philosophy of science became a quicksand of ideological pitfalls.

In many respects, challenges facing UFTI as a new institute were like those anywhere. The new directors needed to define the direction of research, simultaneously organizing laboratories, hiring personnel, financing research and securing facilities that, in the USSR, included gaining housing and a dining hall for employees. Yet the physicists faced problems unique to the USSR. They organized UFTI during Stalin's self-proclaimed Great Break, which included more than rapid economic change. It involved cultural revolution and the efforts to advance young personnel of working class origin and communists into positions of responsibility throughout the economy. In research institutes this was a difficult process that created tension between scientists, a large number of whom had been educated in the Tsarist era, and radical communists who lacked scientific skills but believed they ought to supervise scientists. At UFTI this was not as great a problem as in some other institutes because of the young age of the staff. Indeed the chair of the theoretical department, future Nobel laureate Lev Landau, was only twenty-three years old when assigned to the institute in 1931.

Toward the ends of cultural revolution, Party officials constantly organized verification commissions to check the work of the scientists. Many of them did not trust the scientists whom they considered in the worst case "Tsarist remnants." They feared ideological deviation from materialist philosophy and in extreme cases allegiance to foreign enemies. They organized show trials of mining engineers and other specialists in the late 1920s and early 1930s, many of whom were foreigners, to signal their intention vigilantly to supervise research, ensure its consonance with plans, and indicate their ability to identify and destroy "wreckers."[3] In Ukraine all of this meant additionally that individuals who were devoted to Soviet causes understood that "Ukrainian" attitudes, beliefs and even language and customs were to be subordinated to the Soviet worldview. "Bourgeois nationalist" views were anathema to Soviet power.

Yet even if the struggle against "bourgeois nationalism" accompanied the centralization of planning, economic and scientific organizations, many other forces worked against centralization. These include the very scale of industrialization, the inability of planning organizations to follow, comprehend and control unfolding processes involving millions of people across a huge territory and the understandable persistence of local and regional interests.[4] Local and regional party and economic personnel naturally wished to see investment increase in their region, demanded greater resources because of what they maintained was the crucial nature of their economic, educational and scientific institutions to the ongoing campaign. In this case, UFTI therefore was both a product of the center and a centripetal force against it. Yet the centripetal forces were not solely "national" (i.e., Ukrainian). With respect to science the categories of "national"

---

[3] See Bailes, Technology, for a discussion of the difficult relationship between engineers and communist party officials as background to the show trials of specialists.

[4] For example, Hughes, Stalin.

vs. "international" did not reflect a good dichotomy, especially when "national" prestige was not a valid category in Soviet discourse if the referent was Ukrainian science.

Still, under Stalin, Ukrainian science became a scientific colony of Russian science, no less than Ukraine itself was subjugated to Soviet interests. The highly centralized system allowed administrators in Moscow to skim the best scientific talent available for their institutes. Their physical and political proximity to Communist Party and ministerial officials enabled them to force many of the major applied tasks of science onto Ukrainian science, for example designing and producing equipment, which enabled institutes in Moscow and Leningrad to maintain their positions as leading institutes of basic research.

The case of Ukrainian physics reveals that central planning of scientific research and development was no more efficient than decentralized Western systems of administration in managing science. In fact, a myth exists that centralized command and control would enable the USSR more efficiently to allocate personnel, machinery, equipment, financial and other resources to the rapid and successful completion of projects. Instead, the successful completion of projects owed much more to the ingenuity of Soviet scientists – and to their remarkable, dangerous and often fortuitous avoidance of the pitfalls of Stalinist science policy. They managed to maintain a modicum of autonomy by paying lip service to ideological instructions in part through participation in Marxist study circles, and through publication of philosophical essays which staked out safe positions. They secured sufficient funding within the windfall of investment given to institutes that had connections with the industrialization effort, as physics did through metallurgy, heat engineering and solid state physics. At UFTI they had strong, capable leadership in the persons of Sinelnikov, Alexander Leipusnkii, Anton Valter, Lev Landau and others. Yet their precarious position indicates that Stalinist science policy, like Stalinism itself, was more than anything an arbitrary and capricious doctrine that might be used against alleged enemies, including physicists. These factors explain the glorious, painful history of Ukrainian physics.

UFTI IN THE 1930s

UFTI was founded in October 1928, and was based on Leningrad physicists, programs and facilities. Ivan Obreimov was appointed director of UFTI on December 12, 1928, although the institute did not open its doors officially until two years later. UFTI grew out of the plan of Abram Ioffe, head of the Leningrad Physical Technical Institute, the so-called cradle of Soviet physics, to establish a network of research centers throughout the country. Ioffe first announced this plan publicly in a slogan to the sixth congress of Russian physicists in Moscow in 1928: "Physics to the Provinces!" The idea was that using core LFTI personnel as the kernel of a new research center, they would attract promising young local talent to a series of physical technical institutes established throughout the country. In each case, these institutes would be located in or near new or planned industrial centers, and in each case the research focus would be connected with the major industrial tasks of that city. Quickly physics centers were founded in Siberia (in Sverdlovsk and Tomsk), and in Dnepropetrovsk and Kharkiv, Ukraine, and later in other cities.

While the initiative for and shape of the new physics institutes owed much to Ioffe and his colleagues, the responsibility for setting national science policy, approving personnel, allowing the formation of new centers and providing funding for research fell to Glavnauka (the Main Scientific Administration of the Commissariat of Enlightenment and NTU, the Scientific Technical Administration of the Supreme Economic Council). Once the nation embarked on Stalin's programs for industrialization and collectivization, the government transferred jurisdiction for most institutes under Glavnauka and NTU to the Commissariat of Heavy Industry. This jurisdictional decision left no doubt about the expected role that scientists were to play in industrialization; the Ukrainian Academy of Sciences voted to transfer UFTI into the Ukrainian Academy of Sciences, a move that reflected both the institute's position as a center of fundamental research and as the leading physics center in Ukraine, only in 1939.[5]

At a meeting of the collegium of the NTU early in 1929, Ioffe discussed how he saw LFTI taking a leading role in the creation of this network. Regarding UFTI, he explained how UFTI must be an independent center of research, not a branch of his own institute, and pointed to the fact that such capable physicists as Sinelnikov and Obreimov were available to staff it. Scientists, drawn largely from LFTI at the start, would serve as core personnel and provide a focus of research until local cadres were trained to fill out positions. Ioffe pointed out initial costs would be significant because of new equipment needed to facilitate a focus on solid state and low temperature physics, but they would not be excessive. He promised close ties with factory laboratories – as his own institute had achieved with Leningrad factories. He worried about ministerial barriers to the growth of the institute in the way higher education, research and development had been spread across institutes of Narkompros, Glavnauka and NTU. But he suggested that UFTI would thrive on personnel drawn from Kharkiv University, as soon as it was restored and strengthened.[6]

On the whole, the members of the collegium "welcomed the initiative of Abram Fedorovich [Ioffe]" since scientific research would be of crucial importance in the national republics in the five-year plans. One official criticized a number of institutes in Moscow and Leningrad for opposing Ioffe's plan because they feared diminution of their own positions. Others warned that the new institutes would duplicate the effort of institutes in the center, producing just that kind of "parallelism" that planned Soviet science was supposed to avoid. But the

[5] A UFTI, f. 39, op. 13, d. 1620.
[6] A UFTI, f. 39, op. 13, ed. khr. 1620, ll. 48–62.

NTU collegium unanimously endorsed Ioffe's effort and encouraged NTU to provide the 125,000 rubles requested to open UFTI.[7]

Several factors interfered with the establishment of such new facilities as UFTI. Primary among them was the complex and frustrating task of establishing physical plant, acquiring equipment and embarking on research without wilting under incessant pressures to meet extremely high economic and other targets. Finding housing was as much a problem as securing research facilities. For this reason, it was hard to attract young, promising scientists to go to the provinces, and the brightest physicists wanted to be nearer the cutting edge of research in Moscow and Leningrad and dreaded daily life in provincial backwaters. It was also risky to subordinate new institutes to applied tasks because fundamental research suffered, innovation always seemed to lag, and local party and economic bosses constantly harped on the need for results without delay.

Yet UFTI would be connected to Kharkiv as an engine of Ukraine's industrialization, no matter the worries of physicists. Machine building, electrical and farm equipment industries were located in the city. Officials viewed them as crucial to the nation's electrification effort and to the modernization of agriculture. Two major facilities that grew to have thousands of employees were the Kharkiv Tractor Factory and the Kharkiv Turbogenerator Factory. They and other important industrial giants garnered personnel, financial and other resources – and the attention of Party bosses who felt pressure to meet outrageous targets for increasing production of tractors, harvesters, turbines and the like. This made it difficult for the authorities to pay attention to the physicists' requests for capital construction, apartments and sufficient funds to purchase equipment and magnets to build cyclotrons and other increasingly expensive devices needed for research on the cutting edge of nuclear physics.

The creation of UFTI took place in the center of one of the most disturbing and inhuman events of Soviet history. Kharkiv sits in the center of an agricultural region. During forced collectivization and "de-kulakization," millions of peasants were forced to join collective farms. The somewhat wealthier among them, or those with more political acumen, were dubbed "kulaks," and faced even harsher measures. They had their private property, homes and equipment confiscated. Many were forced to march hundreds of kilometers without food, shelter or adequate clothing to new state farms that were little more than undeveloped fields in the Urals Mountain region or Siberia. The kulaks were largely an object of fiction of the political authorities who feared any political opposition to collectivization. The peasantry opposed the process both directly and indirectly. They slaughtered their livestock rather than see the state confiscate them; roughly half of the nation's pigs, cattle and horses were killed. Enough was planted for subsistence, not to serve the Bolsheviks, their industrialization plans, or the workers in the cities. The Bolsheviks sent armed detachments to confiscate any grain they could find. They affixed the label of "wrecker" to peasants who sought keep even seed grain. In the ensuing battle against the countryside millions of Ukrainians – some "ku-

---

[7] Ibid.

laks," but mostly simple peasants – perished of starvation. The Kharkiv region was one of the worst hit. As food disappeared from stores, starvation hit the city as well.[8] Some peasants had left the collective farms in search of food in the cities and died on the streets outside of the walls of UFTI. Surely the physicists noticed all of this and found it hard to concentrate on their work.

Hence, the creation of UFTI was part of the process of establishment of industry to serve the center and subjugation of the peasantry, in a word, the colonization of periphery of the national republics by the Bolsheviks. This colonization represented reconstruction of the Tsarist Empire. It involved military conquest during the Civil War (1918–1920), and the forced assimilation of Marxist ideas, notions of nation, class and family throughout the USSR. It included the transformation of lands and indigenous peoples in those lands.[9] As part of colonization, party officials and Marxist ideologues, engineers, scientists and other intellectuals from Moscow, Leningrad and other urban centers took up leading positions in various bureaucracies to ensure allegiance of local and regional officials to central plans and ideals. They were emissaries of an urban, industrial worldview. UFTI was to be a major success story of this effort.

Yet UFTI was not merely a scientific colony. Although built from scratch, the physicists managed rapidly to establish a world class research program. Although the institute's basic research program was tied to Kharkiv's intended status as a big machine-building center, the level and quality of physicists, and the panache and originality of their research were world-class from the very start. Sinelnikov, Anton Valter, Ilya and Evgenii Lifshits, Lev Landau, Obreimov, Lev Shubnikov, and Aleksandr Leipunskii gathered in Kharkiv. Viktor Weisskopf passed through, and Niels Bohr, Paul Dirac, Paul Langevin, P. M. S. Blackett, Robert Van de Graaff, George Placzek, Paul Ehrenfest, and Boris Podolsky visited. The foreign physicists Alexander Weissberg, Fritz Houtermanns and Laszlo Tisza joined the institute staff (the first two would experience the purges of the Great Terror firsthand). And UFTI physicists became pioneers in theoretical physics, in nuclear physics, in fusion, in the design of the Romashka nuclear battery, and in linear accelerators. They published extensively, won Lenin and Stalin prizes, and the institute earned a coveted Order of Lenin. Very few of them became full members of the more prestigious Soviet Academy of Sciences, however; normally they were appointed to the Ukrainian Academy of Sciences.

It helped that, before the xenophobic isolation of Soviet society under Stalin in the name of "socialism in one country," Ioffe and other leading scientists had succeeded in establishing fairly regular scientific contacts with leading Western scholars that ensured they remained active in ongoing developments. The contacts included opportunities for publication and travel abroad. Even the Rockefeller Foundation underwrote sabbaticals of Soviet scientists in Europe and America.[10] Foreign contacts served future UFTI personnel well. In 1928 Ioffe

---

[8]　Conquest, Harvest.
[9]　Slezkine, Arctic.
[10]　V. Ia. Frenkel', Josephson, Sovetskie.

suggested that Sinelnikov join Kapitza, Obreimov, and George Gamow, already at Cavendish Laboratory, in Cambridge, England, with a Supreme Economic Council Fellowship, to work with Ernest Rutherford. Kapitza spent each academic year in Cambridge, returning to Leningrad, where he had been affiliated with Ioffe's institute, each summer. In 1934, Stalin refused to let him return to England, setting off a chain of events that saw Kapitza face house arrest, nearly abandon physics, finally gain his own Institute of Physical Problems in Moscow, and ultimately earn a Nobel Prize in 1978.[11] In the event, Sinelnikov remained at Cavendish for two years. In 1930, when Ioffe and others determined to establish institutes in other cities with the personnel and equipment from Leningrad, they sent Valter, Obreimov and Sinelnikov to Kharkiv to establish UFTI.

In the first years of the institute's existence, cooperation with foreign specialists and publication in foreign languages was a sign of success, not yet a sign of espionage, treason, or another threat to the prestige and safety of the nation. The need for qualified researchers led physicists to search for employees in Europe, especially in Holland, Germany and Austria. The rise of National Socialism eventually led left-leaning and Jewish physicists to seek employment outside of Central Europe. At this stage, socialism looked attractive to many people. There seemed to be full employment, and the Soviet authorities successfully controlled the media to create the image that they really were building a worker's paradise.[12] Lev Shubnikov arrived from Leiden where he had worked with Paul Ehrenfest, a leading specialist in low temperature physics and a friend of Ioffe's. Alexander Weissberg, a German Communist who worked in Kharkiv until being accused of being an agent of foreign powers, arrested, and eventually handed over to the Gestapo, recruited Martin Ruhemann. By 1932 specialists at the institute proudly proclaimed that they had published twenty articles on quantum electrodynamics, photo-elements, solar physics, quantum theory, bombardment of nuclei, and other subjects, including as co-authors with such physicists as Paul Dirac and Boris Podolsky. Hence, in spite of indications to the contrary, Kharkiv turned into a place to be for young Soviet and European physicists.

One case of recruitment provides tenor of the atmosphere in Central Europe. Weissberg went to Germany to recruit for UFTI and sent a letter to Ruhemann in Stuttgart about the great opportunity Kharkiv provided for young specialists. Ruhemann and his wife agreed that Germany offered little hope in the near future because of a poor economic situation, inflation and lack of jobs. (Ruhemann did not foresee the rise of the Nazis that became a reason to emigrate for many others.). A day long visit with Weissberg in Munich that included a trip to the theater convinced Ruhemann to accept a job in Kharkiv. When lengthy contract negotiations were completed some eight months later, Ruhemann and his wife left on a forty-hour train journey to Kharkiv. Weissberg met him at the station in

[11] On Kapitza, see Badash, Kapitsza; Boag/Rubinin/Shoenberg, Kapitza; Kapitsa/Rubinin, Kapitsa; and Rubinin, Nauke. See also Josephson, Kapitsa.

[12] On the fascination of American scientists with socialism, see for example Kuznick, Laboratory.

horse-drawn buggy, and they went directly to the institute where he received a flat attached to the laboratory.[13]

Ruhemann arrived at the beginning of the second five-year plan. Ukraine was in the throes of collectivization, the peasants fought the government, the Party ordered grain confiscation, and the countryside descended into famine. One scientist reported of a shocking visit to collective farm that convinced him the cities ought to send bread to the peasants to prevent them from starving. In Kharkiv they had bread but little else. Ruhemann recalled that everything was rationed, but that rations were seldom available. At least Ruhemann and other foreigners had access to special foreigners' shop that was slightly better stocked. Life became more normal in 1934. Ruhemann enjoyed Kharkiv. He praised Ioffe for his efforts to advance technology founded on fundamental research. "Technology, as far as it existed," Ruhemann observed, "was harnessed to cope with immediate needs. There was no time for science."[14]

Ruhemann noticed purges in 1935, but these took mostly big wigs, not small fry like Ruhemann. Then in 1936 "madness began." "All kinds of people, Party members, intellectuals, but also ordinary people" disappeared. Shubnikov, Fomin, Weissberg and others were arrested, most of them never heard from again. One of the charges these physicists faced was collaboration with foreigners, a real problem at UFTI given the make-up of the staff. Leipunsky called Ruhemann into his office in the summer 1937 to tell him his contract would not be renewed. Ruhemann was not surprised, but not happy since he had nowhere to go, and certainly not to Nazi Germany. Having had a British passport in the past, he was able to get a visa from the British Consul and moved to England in the nick of time.[15]

INSTITUTIONAL GROWTH, RESEARCH PROGRAM, PERSONALIA

Ioffe's solid state physics program dominated research decisions in Leningrad, so the opportunity to move to Kharkiv and build on the *annus mirabilis* of 1932 with one's own research on nuclear structure attracted many promising young researchers. As with many recently founded institutes, there was great intellectual excitement that accompanied the drudgeries of putting an institute together, equipping it, and dealing with government officials and funding agencies. Until equipment was in place, daring theoretical pronouncements might dominate the landscape. This was certainly true with Landau's presence. Away from Moscow and Leningrad, UFTI had openings for émigrés from the anti-Semitic and Red Scare politics that forced Jewish and socialist physicists hastily to abandon their homelands. Weissberg arrived from Austria, Rudolph Peierls and Houtermans from Germany, and Weisskopf for a short time from Austria. They got enough

---

[13] Ruhemann, Kharkiv.
[14] Ibid.
[15] Ibid.

money to publish a German language physics journal, Physikalische Zeitschrift der Sowjetunion, through which they might compete with European physicists for priority. Physikalische Zeitschrift der Sowjetunion appeared in twelve volumes of nearly ten thousand pages over six years, and contained first-rate contributions to world physics by all of the major Soviet physicists and by a number of foreigners who happened to be in Kharkiv – Boris Podolsky, Weissberg, and others.

In the environment of industrialization campaigns, the tension between the pressure to conduct research with immediate applications and the desire of physicists to engage in path-breaking research with unclear economic benefit played out. The physicists were able to gain support for nuclear and low temperature physics research programs by pointing to their connection to electrification and other areas of industry. It helped that the institutes under the jurisdiction of the Commissariat of Heavy Industry received large increases in their budgets. This provided physicists with the leeway to move ahead into such areas as nuclear physics. They built a series of expensive devices such as Van de Graaff generators in the pursuit of nuclear knowledge.

Experiment and theory eventually went hand in hand in Kharkiv. Like their colleagues in Cambridge, England, Rome, Italy, and Pasadena and Berkeley, California, Sinelnikov and others pushed accelerator technology to penetrate deeper into atomic structure, at first using charged ions, and then, following the work of Enrico Fermi in Rome, irradiating various elements with neutrons. These elements absorbed neutrons and emitted alpha particles, gaining a unit of atomic weight. Lithium became beryllium, beryllium boron, and so on. By the beginning of 1932, UFTI's research plans included a new theme, "the study of the atomic nucleus with the help of collisions of fast particles," achieved through two high voltage machines and a Tesla transformer. In 1933 the institute added the theme "research on the neutron – a new kind of matter." As a source, they bombarded beryllium with artificially produced alpha particles or deuterons. Leipunskii led the group investigating the scattering and capture of neutrons in a large group of elements.[16] Physicists defended their work as vital to industrialization.[17] In 1935, in Zeitschrift für Physik der Sowjet Union, Sinelnikov, Valter, and two other colleagues published two works on the absorption of neutrons in iron and selective absorption of neutrons, and proposed looking at absorption over a range of energy. Eventually Leipunskii took over direction of this direction of research, before leaving in the late 1940s to establish the Soviet breeder reactor program.

Even if officials initially did not understand the significance of nuclear studies, the institute's physicists exceeded all of their own expectations in this area within a few years. They had built a first generation of particle accelerators to bombard nuclei with high-energy particles. This was a "central area in contemporary physics and the institute is the only one in the USSR which has overcome all difficulties in preliminary work and has begun directly to attack the nucleus,"

[16] Trofimenko, Istoriia, pp. 77, 98, citing TsGAOR, f. 34, op. 16, ed. khr. 611, ll. 6–12, and f. 806, op. 1, ed. khr. 1041, l. 16, ed. khr. 2196, l. 6, and ed. khr. 3095, l. 3.
[17] Bunimovich, Atoma.

they reported. In solid-state physics, photo-elements, magnetism, theoretical physics and instrument building they had other successes. All of these efforts, they assured readers of their reports, involved no duplication of efforts with other institutes. They complained about tight financing, but noted a series of contracts with institutes and factories to build specialized equipment earned extra funds. Low temperature physics and materials science also developed rapidly. Under Lev Shubnikov the physicists founded the first cryogenics laboratory in the USSR in 1931, working with liquid hydrogen, and soon helium;[18] the highly talented thirty-six-year-old Shubnikov, was arrested and executed in 1937. Studies of super-conductivity logically followed, with Landau creating the theoretical foundation for a series of advances in the mid-1930s. On the basis of this laboratory, they established the world-renowned Institute of Low Temperature Physics in Kharkiv (in 1960) where physicists studied electronic, magnetic, and thermal properties of metals, later the physics of isotopes of condensed matter, and pursued a series of related research agendas.[19]

Yet, as the institute grew, so did a range of problems. For example, the high voltage and cryogenic facilities lagged considerably behind in construction, even for several simple apparatuses. Housing was another dreadful spot: of seventeen graduate students, only seven were housed at UFTI. Institutions took on the responsibility for industrial feeding of the masses in the USSR with the essential elimination of private cafes and restaurants, and the food in the UFTI canteen was miserable. Finally, the authorities were worried about the low representation of communists among researchers. In 1931 of forty-three scientific workers, of whom eleven were leading scientists, only four belonged to the Communist Party (and six of forty-three in 1932).[20] Then there was the matter of the Great Terror.

PARTY PENETRATION OF UFTI: THE STALINIST PURGES IN SCIENCE

Generally, membership in the Communist Party of the Soviet Union grew throughout the 1920s. Higher rates of growth occurred during periods of crisis when the Party needed new recruits to bolster its strength, for example after the Krondstadt rebellion of formerly loyal soldiers and sailors and the Workers' Opposition to Bolshevik rule, both early in 1921. Periodic purges of party rolls then followed. In the 1930s, the purges followed no rhyme or rule except for the arbitrary behavior of local officials determined to demonstrate their vigilance against perceived enemies. Toward that end they produced lists of Trotskyites, counter revolutionaries, enemies of the people and other criminals.[21] They appearance on a list might lead to the destruction of an individual or his career.

---

[18] Shubnikov, Izbrannye.
[19] http://www.ilt.kharkov.ua/bvi/general/history_e.html (accessed on 26 June 2009).
[20] A UFTI, f. 806, op. 1, d. 574, ll. 16–25.
[21] Shapiro, Communist.

We tend to view science as the disinterested pursuit of the absolute truth in which there can be no politics. In the Soviet circumstances, analysts have considered officials of the Communist Party who seek control and orthodoxy as being in conflict with the values of science. Yet there was no inherent conflict between scientists and party officials in Ukraine; perhaps the most important conflict was over resources. Scientists wished for funding to do their work. In Ukraine, an increasing number of scientists joined the party in the 1930s, a number of them to advance their careers, a number because of sincere commitment to communist ideals. Many scientists indeed believed that they might contribute to industrialization and the building of an unassailable communist fortress. They engaged in such Soviet rituals as "socialist competitions" to demonstrate their superior contributions to the construction of socialism in one country. As a result, not only opportunistic or second-rate physicists benefited from party membership. Many UFTI physicists spent a good deal of their time in several world crossing boundaries between scientific, economic and party institutions. UFTI initially thrived because its staff was so well integrated into local and central party-state structures. UFTI directors Kirill Sinelnikov and Alexander Leipunskii were devoted and sincere members of the Communist Party, but they could not forestall the purges in Ukrainian science.

Throughout the mid-1930s, as momentum built toward the Great Terror, the provincial, district and institute party committees such as those in Kharkiv engaged in periodic purges of communists from their ranks. There were three levels of communist: full member, candidate member and sympathizer. For various crimes of commission or omission, lack of vigilance, alleged infractions, and, increasingly, rumors, hearsay and fabricated evidence, one could be censured or strongly censured, demoted to another level or expelled from the party entirely, with the added embarrassment of having shamefully to give up his party membership card. The earliest purges centered on inappropriate behaviors and activities that were vaguely defined, and did not necessarily mean arrest would follow. The result might be a reprimand, a strong reprimand with a notation in one's record, expulsion from the party, and loss of one's job. The loss of a job became problematic, because the individual might face the charge of being a social "parasite."

Issues concerning the behavior of a communist that, to the contemporary reader seem relatively minor, might draw the attention of local, regional and even national party organs. The focus might be an agitation group that operated without a plan and met irregularly, or had a plan but failed to follow it, a poor attitude toward updating the bulletin board, poor political education in schools and collective farms, or insufficiencies regarding recruitment of party members and sympathizers. Officials passed dozens of resolutions at meeting after party cell meeting calling for redoubled efforts to avoid these mistakes and to ensure vigilance.[22] Each effort in the 1930s to recruit new, more militant members into the party apparatus meant almost inevitably another purge was soon to follow.

[22] GAKhO, f. 12–23, op. 1, del. 2, str. 132.

A devoted communist would successfully identify wreckers and other enemies. But soon he himself might fall under suspicion and into the maelstrom. This happened at all levels of the party apparatus from the provinces to the Central Committee.

At the beginning of the Great Terror, the party faithful increased their vigilance by expanding the categories of enemies and potential enemies. Within each instance of the party apparatus, party members worked diligently to identify tendencies, behaviors, and other faulty actions. They held constant meetings, evaluations, re-evaluations and more meetings to be certain that everything was in concert with higher directives. This meant that local officials often outdid the higher authorities in vigilant enforcement, seeking to demonstrate preeminent diligence in identifying wreckers and enemies of the people, and in setting the purges into an ever advancing cycle of crushing of hostile elements, including party officials themselves. New, ever-more vigilant officials moved into vacated spots from which they identified hostile elements to crush.

The party organization of UFTI was actively engaged in these kinds of activities, as well as in monitoring scientific, social and other areas of interest. In spite of the growing political tensions, scientists were very productive. In 1934 institute scholars published or prepared to publish fifty-six works. The party committee noted such achievements as the placing of most employees in apartments and improvement in the quality of the dining hall. They had found the wherewithal to give each graduate student a book allowance and all workers a two-month vacation. They had expanded the library holdings, including foreign journal subscriptions, and saw to it that all worker communists had learned German and were learning English. They discussed the concern that several employees still lived in the kindergarten in the absence of housing.[23]

Regarding science, party verification commissions considered such broad issues as the extent to which an institute had effectively tied its research program to the needs of industry (demonstrating the crucial "tie between theory and practice"), and its success at providing communist upbringing of young scientists by demonstrating a Marxist-Leninist approach toward science. But more mundane issues might dominate some meetings. At one UFTI meeting, a commission reported that the library had no catalog, books had not been shelved, and people borrowed books without checking them out. Even in the growing terror, several scientists courageously rejected ludicrous accusations or pressure to identify still other enemies of the people among their colleagues. They tired of the insistence that they take on obligations for party activities that interfered with research. Lev Shubnikov called for releasing qualified workers from their burdensome party responsibilities that slowed research and prevented younger scientists from raising their qualifications. They also requested to be freed from the requirement of universal military service since the institute itself had defense contracts.[24]

---

[23] GAKhO, f. p-99, op. 1, d. 10, str. 83–85.
[24] A UFTI, f. 806, op. 1, d. 1091, ll. 30–39, and GAKhO, f. p-99, op. 1, d. 10, str. 83–85.

At the beginning of the Great Terror the party faithful ratcheted up their vigilance. In addition to the foregoing transgressions, they now sought out enemies in their midst, enemies who somehow had shielded their anti-Soviet inclinations for years, enemies who yesterday had been loyal party members, enemies who were enemies by virtue of close blood ties with other enemies or by virtue of distant circumstantial contacts, usually fabricated. They identified counterrevolutionary and Trotskyite tendencies. They unmasked class enemies. A resolution identifying someone as an enemy of the people usually meant he had already been arrested. Constant meetings, evaluations and meetings followed these meetings and resolutions to be certain that everything was in concert with higher directives. Local officials sought to outdo the higher authorities in rooting out danger, in branding wreckers and enemies of the people, in setting the terror into an ever-advancing cycle of crushing hostile elements, including previously trusted but now suddenly criminal colleagues, putting newer, more vigilant officials in their spots, and identifying more hostile elements to crush, which might eventually include themselves. In other words, identification of enemies was no guarantee that you yourself would not be identified as an enemy.

As early as October 14, 1935, Ivan Andreevich Musul'bas, secretary of the Kharkiv regional party committee, called for a purge of UFTI because of the presence of hostile class and counterrevolutionary elements.[25] The Komosomol of UFTI was, according to party officials, an especially weak point. It had too few propagandists who only half-heartedly engaged audiences both on the grounds of the institute and at other higher educational institutions in the region.[26] Musul'bas, who served as second secretary of the Kiev and Kharkov provincial party committees, a specialist apparently in industrial policy, was himself arrested in October of 1936 and shot in May of 1937.

Purges occurred with greater frequency at factories, schools, theaters and institutes of all sorts. A behavior that escaped notice in 1934 might become a mortal crime in 1937; those who carried out purges in 1934 were often the focus of "cleansing" in 1937; and those whose long-term service saved them from expulsion from the party or death sentence in 1934 found that long-term service a mine field of missteps, impolitic associations and crimes against the state. Abram Ruvim-Osifovich Garber, a member of the UFTI party committee in 1934, a member of the Komsomol from 1923, with an unimpressive employment record until 1925 when he commenced work in a series of factories and then entered UFTI as a graduate student in 1931, apparently had lied about an illness to avoid mobilization into the Military Academy in 1931. At an investigative meeting in 1934 he answered political questions satisfactorily, and his party affiliation remained intact. But by 1935, the Kharkiv provincial Party committee often heard evidence of the presence of class enemies and counterrevolutionaries in the institute who would be purged,[27] and Garber was apparently among them.

[25] GAKhO, f. 17–2, op. 1, d. 325, str. 52.
[26] GAKhO, f. 17–23, op. 1, del. 3, str. 141.
[27] GAKhO, f. 17–2, no. 1, d. 326, l. 52.

No one was immune from the danger, and certainly not leading personnel. Well over half the members elected to the Central Committee of the Communist Party in 1934 were purged and executed by the late 1930s.[28] Similarly, such loyal communist scientists at the local level as Leipunskii faced the constant threat of arrest. Leipunskii was in and out of trouble in the late 1930s. On one occasion, he faced a party interrogation. As usual for party members who no doubt hesitantly engaged in various forms of public self-criticism, Leipunskii first offered a statement about his suspect middle class origins, his parents' occupations, and his party and professional careers. He noted his increasingly involvement in party activity from 1928; for several years he had been a member of UFTI's party committee. Yet he admitted that his record was not unblemished. He had been reprimanded during his career for the fact that he had "in several cases reacted 'liberally' to hostile presentations of several scientific workers." He had belatedly fired one of the foreign specialists at UFTI who was consequently arrested by the NKVD, yet had given the man a good recommendation.[29]

The members of the institute's party committee then delved deeper into his "liberalism." Leipunskii admitted that Landau had fired one employee who had spoken up at public meeting unfavorably about another employee. Leipunskii defended that employee and reprimanded Landau rather than present the matter before the party cell for proper adjudication. In another case, a young communist engineer who worked in very poor conditions in a laboratory complained to Leipunskii, who called the laboratory director and told him to rectify the situation. The charge this time was that Leipunskii failed to follow up to see what had been done. Leipunskii's acknowledgement of these mistakes – his self-criticism – seems to have done the trick, for the UFTI party committee voted to let Leipunskii remain in the party without reprimand.[30]

A scandal involving Lev Landau and a strike by professors at Kharkiv State University triggered recriminations and arrests that also touched Leipunskii. Leipunskii admitted not having properly dealt with this "strike." As we discuss more fully in the next section, the university administration had fired Landau for his suspect political views, his disregard for students who did not measure up to his standards, and his haphazard attitude toward reporting for work. In solidarity, Landau's university colleagues presented a bundle of letters in protest, an act that was taken to be an illegal strike, not merely a protest against Landau's firing. When the matter came up at a party meeting, according to criticism leveled at him, Leipunskii had not spoken in defense of anyone who correctly criticized Landau nor against those who defended Landau.[31]

In 1937–38, UFTI fell directly into the Stalinist maelstrom. Party officials of the Kaganovich district where UFTI was located found themselves under great pressure to demonstrate vigilance against enemies. The tenor of institute meet-

[28] Conquest, Terror.
[29] GAKhO, f. p-23, op. 1, d. 7, str. 10–12.
[30] Ibid.
[31] Ibid.

ings changed fundamentally, especially after NKVD officials identified a nest of anti-Soviet vipers in UFTI. They singled out a series of party members, many of whom were Jewish, who were "careerists," foreign elements, members of the Trotskyite opposition, and counterrevolutionaries. Arrested party members faced the charges that they had failed in efforts to unmask foreign enemies, although the arrests of such individuals as Weissberg and Davidovich left no doubt about the need to unmask them. Other charges included the accusation that they had not adopted a Bolshevik attitude to the struggle against enemies, instead merely gossiping; they tolerated the liberalism of Leipunskii; they ignored theft of state property; and they avoided criticism and self-criticism. On top of this, the investigators alleged, a number of departments performed poorly. Tkachuk and Demidov engaged in shenanigans with women, while Tsevetkov, Pavlenko, Romanov and Demidov again told anti-party counterrevolutionary jokes and lies about Stalin.[32]

The purges of enemies from the institute – and from important positions in the Kharkiv economy generally – took on a ferocity that seems to have exceeded that in most other regions of the country. We can only speculate at the reasons for this ferocity. Perhaps the desire to end Ukrainian nationalism or to destroy the Ukrainian peasantry played a role. Another factor seems to be the effort of party and secret police officials on the periphery to demonstrate their unquestioned devotion to the cause. This led them to exceed whatever unspoken targets that Moscow had in mind. Yet it must be noted that physicists who suffered from the purges at later stages could benefit from them at earlier ones. Outcomes were not pre-ordained, whether one takes the "good" scientists as innocents led to the slaughter, or as fully "creatures of the system." In any event, Kharkiv NKVD operatives were vigilant investigators, and UFTI and its physicists fell squarely within their sites, in part because scientists were accustomed to intellectual autonomy that the Party officials feared, in part because of the presence of foreign scientists, and in part because there truly were Trotskyites like Lev Landau on the staff.

PURGES AND ARREST: THE CASE OF LANDAU

During his training in Leningrad at Ioffe's institute, Landau fell in with the "Musketeers" and their "jazz band" in Leningrad. The Jazz Band embraced the new physics of quantum and relativity theory. The members included the Odessa native George Gamow, or "Johnny," who as an expatriate in America worked on the hydrogen bomb and developed the Big Bang theory; the brilliant theoretician Matvei Bronshtein, "the Abbot", and "Jimmy" – Dmitrii Ivanenko, well-known for his work on quantum field theory, nuclear theory, and synchrotron radiation. Each of the members of the Jazz band paid for real and imagined transgressions against the state. Gamow was able to emigrate, ending up at the University of

[32] GAKhO, f. p. 23, op. 1, ed. khr. 10, str. 146–147.

Colorado; Ivanenko was briefly arrested, Bronshtein was shot in 1937, and Landau was in prison for a year before being released.

Landau's reputation for breadth of physical knowledge and penetrating insights was such that he was often invited to the European centers of physics for study. Landau spent eighteen months in Cambridge, Copenhagen, and Zurich, before moving to Kharkiv in 1932 as head of the theoretical department at UFTI where he remained until 1936. Indeed, the standard of scientific excellence at UFTI was set in large measure by the Landau school of theoretical physics. Like UFTI itself, this school was formed as the result of the joining of a scientific diaspora of local, regional and national talent, a chance coming-together possible perhaps only in the USSR, for in other countries written and unwritten prohibitions against Jews and foreigners would have prevented it. The ongoing industrialization campaign contributed to this center of excellence on the periphery of the scientific enterprise, for millions of rubles were pumped into the economy to build an industrial superpower. The authorities may have had a mercenary and mechanical understanding how scientific research operated, but they supported the expansion of the enterprise. The industrial significance of Kharkiv for machine-building, agricultural machinery and the burgeoning electrical power industry in Ukraine meant that the authorities ordered funding and personnel to the city's industry, and to its educational and scientific institutions be made available.

Perhaps the best known physicist to emerge from the ranks of UFTI, Landau (1908–1968) established a legendary school of theoretical physics. Dozens of aspiring scientists, each hoping to study with the impetuous, self-assured, difficult, yet unquestionably talented scholar attempted to pass Dau's "theoretical minimum" to study with him. Landau contributed widely to quantum mechanics, diamagnetism, phase transitions, weak interaction theory and superconductivity, and his development of a mathematical theory for superfluidity of liquid helium won him a Nobel Prize in 1962. The areas of interest of the talented specialists who gathered around Landau, many of whom would lead Soviet physics into the 1960s and beyond, included solid state and low temperature physics, the theory of conductivity of metals, quantum electrodynamics, and neutron physics. These theorists ultimately contributed significantly to the nuclear enterprise, not only proposing improvements in weaponry but also in such peaceful applications as fission and fusion reactors and accelerators.

Though he began working in Kharkiv in August 1931 as an emissary of Abram Ioffe's physics empire based in Leningrad, settled there a year later, and departed in haste early in 1937, Landau quickly made his mark. The political and social dynamics of Kharkiv contributed the initial impetus to the launching of the Landau school. In the past twenty years several of his students from the Kharkiv days have penned memoirs describing Landau's early exploits. Central to these narratives is the figure of the prophet without honor in his own land, a man despised by unworthy opponents, rejected for speaking truth to power, and persecuted for defending professional solidarity. The tension between Landau and several UFTI physicists, and between him and the local authorities, contributed to the atmosphere in which Landau wrote his world-famous Course of Theoretical Physics. Another important dynamic was between the pursuit of the theoretical advances and pressures for immediate applications from science in industry, manifested in Kharkiv in such "hero" projects of Stalin's industrialization campaigns as the Kirov Kharkiv Tractor Factory, and factories manufacturing machine tools, cranes, pistons, pneumatic equipment and so on.

Having already achieved youthful European renown well beyond the ambitions of his Soviet peers, Landau also developed a reputation among party officials and the secret police for his anti-Soviet ruminations. In fact, Landau embraced Marxist ideas but was a self-proclaimed Trotskyist, disliked Stalin and Stalinism intensely, and was unable to control his very sharp tongue about his political orientation. He engaged in public and private protests against a regime that had no tolerance for heterodoxy in any way. During the Great Terror the regime swept up hundreds of thousands of innocent victims, having manufactured charges of "anti-Soviet" activity against them. They did not need to manufacture evidence that Landau rejected Stalinist rule. To make matters more difficult, Landau and his colleague Lev Shubnikov let it be known that they disliked life and work in Kharkiv. They considered it to be a provincial outpost of industry, not science, in the center of an agricultural region. They desired to be in Leningrad or Moscow, perhaps even Kiev, but not Kharkiv. Disliking provincial Kharkiv, and sensing that the NKVD would arrest him during the purges of UFTI, Landau fled to the Peter Kapitza's Institute of Physical Problems in Moscow. The secret police caught up with him in April 1938, arrested, interrogated and tortured him, freeing him at near death only after the personal intervention of Peter Kapitza after one year in prison.

Nationality politics were another aspect of the history of Soviet physics on the (Ukrainian) periphery. UFTI as an institution contributed crucially to the reconstitution of the post-revolutionary university in Kharkiv, a "sovietizing" process that had lasted much longer than in the Russian capitals. The Ukrainian universities were abolished after the Bolsheviks took power, when they were "transformed into pedagogical institutes with bizarre programs" to promote the proper world view and proletarian attitudes. A distressed observer wrote in 1921, "There is no physics, and none foreseeable," and physics instruction and research foundered during the remainder of the decade.[33] As late as 1930, outside of the nascent processes of Stalinist cultural revolution to replace "bourgeois specialists" with working class if not communist scientists, party and university officials were expelling faculty members for suspected Ukrainian nationalist sympathies,

---

[33] So testified the mathematician and mechanical engineer A. P. Psheborskii, who for a time headed an abortive "Academy of Theoretical Sciences," founded in Kharkiv after the revolution, though it came to nothing and he subsequently emigrated to Lithuania. See Archive Russian Academy of Sciences, St. Petersburg Division, f. 162, op. 2, d. 365, as cited in Ermolaeva, Russkoe. Prompted by UFTI staffers, Meissner judged physics at the university to be "scarcely represented" in 1932.

and the ranks of the Ukrainian intelligentsia remained greatly depleted by these repressions.[34]

Not only the struggle against nationalism halted the effort to establish Kharkiv State University; famine in which millions of peasants died obviously diverted attention from education to survival. Finally, during the 1933–34 academic year the authorities approved the reconstitution of KhGU, although its various orphaned scientific "institutes" and sectors did not even include one devoted solely to physics. The university's first rector, Ia. S. Bludov, appointed in September 1, 1933, was arrested in 1934, convicted of anti-Soviet crimes, and sent to a gulag camp in Arkhangelsk province. A philosopher who had been director of the hybrid institute of physics, chemistry, and mathematics that had provided the core faculty for the reconstituted university, Bludov may have been soiled by his allegiance to several of the older Marxist intellectuals, especially to Semen Semkovskii (a pseudonym for "Bronshtein," like Trotsky), a Jew and prominent Kharkov philosopher of science who defended the dialecticians' viewpoint of the new physics. Semkovskii's Teoriia Otnositel'nosti i Materializm (Kharkov, 1924) had fallen out of favor. Semkovskii had long been associated with Russian Marxist causes but with their Menshevik wing. He joined Anatoli Lunacharskii, latter Commissar of Enlightenment, Alexander Bogdanov, Lenin's main epistemological foil, Alexandra Kollontai, a member of Workers' Opposition in the early 1920s and others at playwright Maxim Gorky's villa in Capri, Italy, in conducting a party "propaganda" school for workers in 1909.[35] Bludov, freed from the gulag in 1939, was denied the right to vote or live in a large city, but after rehabilitation in 1956 under Khrushchev returned to the university as a professor of philosophy.

Before the reconstitution of the university, several UFTI physicists hosted a regular seminar on experimental physics that served in place of a graduate school program.[36] Several staff members then returned to part-time teaching at the university, mostly in applied domains on the borderland between physics and engineering. What it meant to be trained as a professional physicist in the Soviet Union, let alone in Ukraine, remained far from clear given the ongoing industrialization, collectivization and other campaigns. The experienced party apparatchik, Aleksei Neforosnyi, assumed the rectorship of KhGU in December 1934, in no small part precisely because his notions of pedagogy and curriculum were consistent with those of the dominant UFTI physicists. Neforosnyi worked throughout 1933 and the first half of 1934 as head of department of agitprop of the Kharkov City Communist Party Committee. He knew some of them quite

[34] Kostiuk, Stalinist, p. 88; Kas'ianov/Danilenko, Stalinizm, pp. 53–54.

[35] See papers of Grigorii Aleksinskii, Houghton Library, Harvard College Library, Harvard University, bMS Russ 73, and Papers of S.IU. Semkovskii, 1912–1917 [Box Boxes 185–186] in Register of the Boris I. Nicolaevsky Collection, 1801–1982, University of California.

[36] Butiagin/Saltanov, Universitetskoe, p. 159; Khar'kovskii, pp. 265–298; Pavlenko et al., UFTI, pp. 94–106. This all-too-brief characterization of non-UFTI physics institutions before 1934 should not obscure the fact that research in a broad variety of topics was sustained in the 1920s, as Pavlenko et al. make clear.

well, for example jointly serving with institute director Alexander Leipunskii on the municipal party committee,[37] and through him Neforosnyi likely learned of Shubnikov and Landau. At Neforosnyi's bidding, the UFTI physicists expanded their roles in the development of curricula at the university precisely when nearly everyone agreed that physics ought to have a central place in the university. On their side, Landau and Shubnikov recognized the larger pedagogical context. In their teaching they played on the theme of applications to their student audiences.

The party purge that led to Neforosnyi's appointment as rector had faulted his predecessor with the familiar refrain of standing by as groups of Trotskyites and nationalists were recruited into the ranks of the university's instructors. These travails had not touched physics at the university, where the head of the party committee was Lev Piatigorskii, a graduate student of Landau's at UFTI.[38] University physicists apparently avoided active criticism in the municipal party committee during the 1934 purge, and Piatigorskii himself came in for favorable mention as a good communist. Once Neforosnyi had consolidated his position as rector, he persuaded both Shubnikov and Landau with the backing of the department's party committee to take up permanent teaching duties at the university in the fall of 1935; Shubnikov's position opened up following the death of D. S. Shteinberg (1888–1934), a specialist in magnetic phenomena.

This development was not simply an instance of first-rate scientists catching politically motivated second-raters in a moment of weakness and insisting on the prerogatives of "best science" in the face of obscurantist opposition. Whatever their relative status as physicists, the university contingent had not been momentarily weakened by the 1934 purge, because they did not have much political authority at their disposal in the first place. For its part, the first administration of the revived university had been hesitant or unable to initiate significant new agendas for the department. The 1935 expansion of the physics curriculum came about both because Shubnikov and Landau were physicists at the cutting edge of their fields and crucially aware of how to train more physicists, and because they understood how party-state administrative structures operated to utilize them better for what they regarded as the best interests of science.

Perhaps the real problem for Landau was his personality. Landau's confrontational teaching style revealed intolerance of the slightest weakness. His brash personality and open Trotskyism certainly generated dislike. The final rupture between him and university administrators and party officials came in the wake of the usual tired accusations of "idealism" by a university rival in 1936, with Landau unceremoniously dismissed by the Communist rector. His flight to Moscow may thus be seen as an effort to escape Ukraine, where the purges had taken

---

[37] Much of the information in this section is drawn from Vorob'ev, Landau, and Ranyuk/ Shevchenko/Josephson, Eshche. For biographical information on Neforosnyi, see Vorob'ev, Neforosnogo.

[38] Davydov, Pro stan vykladovs'kikh kadriv v KhDU, 26 October 1934, and O sostoianii kadrov KhGU, DAKhO, f. 69, op. 1, d. 175, l. 17 ff.

on an intense level of energy, for purportedly organizing an instructors' "strike" in response to his firing.[39]

Yet if the pressures of industrialization and the campaign to root out idealism at Kharkiv University were the ultimate cause of Landau's downfall, the paths to his defeat were much more roundabout than the received view would have it. For example, one cannot simply pin the blame for Landau's dismissal on a feckless Communist rector, when that same rector recruited him and Shubnikov to the university in the first place. Landau in particular was scarcely orthodox in political terms, but one thing that made Kharkiv so volatile was the continual disagreement among diverse provincial constituencies about what constituted orthodoxy at any given moment in time. A figure like Landau yielded to no one in his confidence that a more effective Soviet system lay within reach, and that his professional agendas could serve that end. Heterodox he may have been – ecumenical he was not. "At the time Landau was 'red,'" UFTI theoretician Alexander Akhiezer further reminds us, "and he didn't tolerate the expression of any anti-Soviet opinion."[40] Although notoriously impatient with the pervasive cant of Stalinist political discourse, Landau was fully capable of adopting "more Soviet than thou" stances and defending them with a fierceness that many of his contemporaries would have found quite familiar outside the hollowed precincts of science.[41]

At the university Landau was initially enrolled as an instructor in theoretical physics, despite the fact that the formal head of the theory group was Piatigorskii. Piatigorskii's predecessor in the theoretical chair, Lev Rozenkevich, had also worked as in-house theorist at UFTI, but Landau had been recruited in part because Rozenkevich was not meeting the broader needs of the institute's experimentalists. (Rozenkevich studied with Iakov Frenkel in Leningrad, arrived in Kharkiv in 1930 to head the theoretical department at UFTI, and then moved to a laboratory connected with Leipunskii's work on nuclear physics as part of the effort to retrain as an experimentalist. Along with other talented physicists, Rozenkevich was arrested and executed in 1937, as yet another needless loss to Soviet and world science.)

Neforosnyi had a more convoluted agenda for Landau, however. He forced the Ukrainian dean of the physics faculty, Andrei Zhelekhovskii, to yield the chair of experimental physics to Landau, with the thought that the young theorist could thus formally take charge of the freshman and sophomore general physics lectures for the entire department, which counted some 800 students at all levels. (Zhelekhovskii himself worked part-time at UFTI, once heading the "electro-vacuum brigade" involved in instrument building. By 1935 all of the brigades were folded back into single, larger laboratories.) Initially, at least, Landau appears to have been very persuasive in pushing for curricular revisions. His success was

---

[39] Akhiezer, Ocherki; Akhiezer, Recollections; Akhiezer, Uchitel; Kikoin, Kak, p. 161. See also interview with A. I. Akhiezer by Yu. M. Raniuk, Niels Bohr Library, AIP.

[40] Akhiezer, Recollections, p. 38.

[41] Hall, Purely, chap. 6–11.

not necessarily tied to the fact that the physics students were being held to lower standards before his arrival. In the previous semester the physics department had awarded fewer high grades and more unsatisfactory marks than any other department. In the spring 1935 semester only 16.5% of physics students received the highest mark, while in the economics department the figure was 32.2%. Fully 10.9% of physics students received poor marks, well above the next strictest department, mathematics, at 4.6%.[42] But this state of affairs enabled Neforosnyi to force through rapid changes.

If anything, the fact that he awarded higher grades on average may well have made Landau's job more difficult. Consider first the social dynamic of the Soviet classroom. Attendance at lectures was optional, and Zhelekhovskii's audiences for the introductory courses had generally been dismal.[43] Landau's unusual lecture style initially piqued the interest of students, but he flatly insisted that the few textbooks available were entirely unsuitable for his purposes – including Zhelekhovskii's own Ukrainian-language textbook. This left the students in a bind, because no matter how well-disposed they may have been toward the animated young lecturer, with his resolute pronouncements on Hamiltonian coordinates, current affairs, and the idiocy of most scholarship outside the natural sciences, the vast majority of them were ill prepared for the experience. Upsetting the old Newtonian order, as it were, Landau also insisted on teaching mechanics to physics students using the Principle of Least Action as his point of departure. He compounded this challenge to the standard historical approach by actively competing for students from the analogous courses offered by the stronger joint faculty of mathematics and mechanics.[44]

The intensity of class warfare that reflected debates about the new physics also played a role in the escalating disputes over science and politics in Kharkiv (and elsewhere). In this case, fully one quarter of the students were older than the twenty-seven-year-old Landau, and those who had arrived with the proper social credentials were not inclined to show deference to their intellectual superior. Many students had come to the university via remedial "workers' faculties" (*rabfak*, abolished in the late 1930s), and few came with previous instruction in physics or higher mathematics. Though they were far less evident in physics than in other departments, a sizeable number of students had politically-mandated extracurricular activities. These distracted them from their studies, if providing them with a vital source of social identity. Most physics students were well attuned to invidious social distinctions within Soviet academia, without necessarily being very politically engaged. In many instances uncertain or insecure about their own labile social identities, but more often confident of their political importance to

---

[42] See the headline Fizmat ta khemfak – u khvosti naividstalishikh. Likviduvati vidstavannia, vidminno sklasti chervnevi sessii – obov'iazok reshti kursiv tsykh fakul'tetiv, Za Novi Kadry 31 (148), 14 June 1935. Za Novi Kadry was the newspaper of the university's party committee.

[43] This is according to the recollections of Ia. B. Fainberg, then a student in the department, as related in Vorob'ev, Landau, p. 95.

[44] Kikoin, Kak, p. 161.

the administration vis-à-vis the faculty, students did not always know what to make of Landau, who was uncowed by any intimations that his own social background in the middle class might be suspect.[45] Landau's father, an engineer and former manager in the Baku oil industry, had briefly been arrested in 1930 at the time of the organization of the Industrial Party and "Shakhty" affairs that led to show trials of engineers, many of them foreign, on charges of wrecking.

Older "bourgeois" colleagues may have been intimidated at times by the more militant students, but Landau did not hesitate to put students in their place while flanking from the left. They in turn took to referring to him among themselves as "Levko Durkovich" (loosely, "Lefty Dolty"). Few niceties were observed in the classroom, by student or by teacher, and Piatigorskii occasionally had to report to the university party committee to persuade them that the latest Landau escapade had come in the midst of a brilliant lecture that was a credit to the university.[46] The pro-rector for academic affairs, one Rakhunov, complicated matters by promising to arrange printing of Landau's lectures in mimeograph form but never delivering.[47] At first the unhappiness of the students was not channeled at Landau so much as it was at the pro-rector's office, widely understood to be the source of the problem. Students even attempted to print lecture outlines "by Prof. Landau's method" on their own, without much success.[48]

To make matters worse, Landau publicly criticized Marxist philosophy as an impediment for doing good physics. He triggered a student protest against him for his strict – and impetuous – behavior concerning oral examinations that were the basis of grades. On one occasion, his students refused to show up for exams. He also declared that "theoretical physics is not an activity for Slavs." The high number of Jewish theoreticians at UFTI and KhGU was visible enough without these kinds of comments, and given latent anti-Semitism in the USSR would only engender bitterness. Soon Landau faced attacks as being an idealist and a Trotskyist. He was, in fact, the latter. NKVD officials framed the attacks within a larger narrative about the existence of a hostile network of anti-soviet physicists in Kharkiv. As noted, there was no such thing since Neforosnyi was in the city party committee, had close connections in the NKVD, and wouldn't have invited suspect individuals to the university in the first place.

But Stalinism created an atmosphere of constant battle against enemies, real and imagined, internal and external, and as purges gained momentum in Leningrad and Moscow, they gained momentum throughout the country. Stalin ordered attorney general Andrei Vyshinsky to conduct investigations and public show trials of leading Bolsheviks as the purges began to consume even loyal

---

[45] On Soviet students, see Konecny, Builders, and on Soviet youth more generally, see Gorsuch, Youth.

[46] Lengthy excerpts from Piatigorskii's recollections may be found in Ranyuk, Landau.

[47] An undated excerpt from Landau's memo, written sometime in 1936, is quoted in Vorob'ev, Landau, p. 95.

[48] Under the heading, Nalagoditi uchbovyi protses, one finds the brief account of Student Egupov, Koly zh matimeto [sic?] pidruchnyk?, in: Za bil'shovishch'ki kadry 5–6 (178), 17 February 1936.

Old Bolsheviks. In August 1936 Vyshinsky prosecuted the so-called Trotskyite-Zinovievite Terrorist Center that led to the conviction and execution of Grigory Zinoviev and Lev Kamenev who had served in a triumvirate with Stalin against Trotsky after Lenin's death. A second show trial followed in January 1937 that finished off many of the Old Bolsheviks, including Nikolai Bukharin. The trials apparently convinced those in the periphery of the need to be vigilant and to identify enemies. Neforosnyi determined to rid of Trotskyites in university. Almost in parallel, in October 1935 an extensive purge of personnel began at UFTI soon after the institute's physicists began undertaking research with more direct military significance. Two months later, in December, Piatagorskii gave testimony to the NKVD that both Landau and Shubnikov were class hostile elements and counterrevolutionaries.[49]

One explanation is that Neforosnyi fired Landau from KhGU on the testimony of the dean of the physics department, Zhelekhovskii, for being an idealist who did not accept the law of the conservation of energy. The philosophical disputes of the 1930s meant that many scientists were fired or arrested for alleged ideological errors or, worse still, anti-Soviet activity include "wrecking," spying, and engagement in various counterrevolutionary conspiracies involving "great power chauvenists," "nationalists," Trotskyites and others. In any event, Neforosnyi and Landau had a private meeting in 1936 after which Landau was fired. We do not know the substance of their conversation. We know that Landau's fellow physicists were incensed. Shubnikov and others in theoretical department voluntarily offered their resignations. In the tense atmosphere of purges, the authorities considered this act a strike; strikes were illegal in the workers' paradise. The Kharkiv officials demand that the physicists appear before the Commissar of Enlightenment of Ukraine, Vladimir Zatonskii, a devoted Bolshevik official with training in the sciences, commissar from 1933 to 1938, and a member of the Central Committee. Of course, no one was immune from the Great Terror. Neforosnyi himself was arrested in July of 1937 and shot in December. Zatonskii disappeared in 1938. Landau was arrested at the Institute of Physical Problems in Moscow on April 27, 1938, and was interrogated and tortured for a year while institute director Peter Kapitza and Niels Bohr, among others, tried to save the talented young theoretician. The NKVD released Landau to Kapitza's recognizance on April 28, 1939. Landau remained under constant surveillance into the late 1950s.

THE DESTRUCTION OF UKRAINE, THE DESTRUCTION OF KHARKIV PHYSICS

World War II dealt another major blow to the position of the institute as a center for Ukrainian physics. When the Nazis invaded, the Kharkiv physicists evacuated what they could just ahead of the onslaught. German bombers hit apartment

---

[49] Weissberg, Accused, pp. 80–81.

buildings visible from the institute grounds. The physicists were able to disassemble some of the equipment, and cart it off to Almaty, Kazakhstan, where they set up quarters in a few rooms of the university. Here, on the orders of Kurchatov, they eventually contributed to the initial steps of the atomic bomb project. They studied the theory of moderation and scattering of neutrons in crystals, especially in graphite, with Alexander Akhiezer and Isaak Pomeranchuk, two other leading theoreticians connected with Landau, completing the first Soviet monograph on nuclear theory, and investigated the interaction of particles with matter.

When Kharkiv was freed from Nazi rule in 1943, physicists returned to find their city and institute in rubble. The institute's main building had been bombed. All the windows were broken. The building was ransacked. There was no heating fuel. The plumbing and sewer systems were destroyed. Sinelnikov found his flat absolutely empty and filthy. His beautiful Steinway lay on the road near the garage, having been used by the Germans as a platform for washing lorries. Luckily, Sinelnikov and Valter reported to Kurchatov that the electrostatic generator at UFTI was only slightly damaged and could be repaired with materials at hand. The vacuum system seemed to be in full order, with electrical equipment of the generator practically in working condition. Sinelnikov asked Kurchatov to authorize funding and materials needed to reestablish a fully operable workshop to repair or replace this equipment; provide orders to reestablish normal supply of gas; and repair a neutron generator. Problems in securing the early release of several physicists from service in the Red Army along with the special rations for them delayed the completion of repairs for some months.[50]

As if the difficulties of recovering from the war were not bad enough, the authorities chose the postwar years as the time to renew the call for ideological vigilance in science. In every scientific discipline, in every region of the country, Marxist philosophers and scientists of firm Stalinist conviction asserted their authority in science. In physics, this meant interest in how relativity theory and quantum mechanics were commensurate with the Soviet philosophy of science, dialectical materialism. Sinelnikov assisted Akhiezer and Valter in developing readings in the Marxist philosophy of science for young students preparing for qualifying examinations for the candidate of science degree. But as in Leningrad and Moscow, many scientists' careers were damaged. In biology, Trofim Lysenko, who rejected genetics, gained the authority to ban genetics from books and banish researchers from their jobs. In physics, only by the grace of Kurchatov was the enterprise saved from a similar outcome.

Pressure intensified on physicists, biologists and chemists to identify the correct ideological positions and be prepared to marshal the appropriate citations from writings of Marx, Lenin, Engels and Stalin to demonstrate they had not engaged in anti-Soviet behavior. A party cell meeting on November 30, 1949, considered the on-goings and on-comings in the Physical Mathematical department of Kharkiv University. Self-criticism was the rule. While the faculty lectured on Marxism and Soviet power, conducted workshops on Marxism and Soviet

---

[50] 48 KIAE, f. 2, op. 1, ed. khr. 71/8, ll. 1–4.

power, and published the newspaper *Vektor* regularly (with articles on Marxism and Soviet power), idealist tendencies were present. While the department organized three special evenings for veterans of the war, a traditional evening of meetings of teachers and students, and a third evening on the institute's history, and while seven (!) student scientific circles operated at which attendees had heard fully 35 (!!) speeches on the priority of Soviet science, and finally, while the physicists organized a successful scientific conference "On the Struggle of Materialism With Idealism in Physics," a radical improvement of work against idealist tendencies in the department was required, loyal communist party scientists demanded. Terekhov observed furthermore that attendance at meetings to discuss the foundations of Marxism-Leninism was inadequate, while others criticized the quality of methodological efforts. In a final resolution, the party bureau of the department criticized the role of the faculty in the ideological upbringing of the students, and admonishing such professors as Akhiezer, Pines and others as "apolitical." Clearly, they concluded, the major questions of the struggle with idealist tendencies in physics that had been highlighted in the party press had not become a subject of central importance.[51]

Most observers assume that the centralized nature of the Stalinist system enabled economic planners to allocate resources in a rational fashion. But the physicists at UFTI knew first hand that Party officials played favorites in rationing personnel and equipment among the growing number of research institutes of the scientific establishment after the war. Even UFTI's position as a central institute of the atomic bomb project did not guarantee access to those resources.

In January 1951, Anton Valter addressed Beria about the unsatisfactory level of support for his sector that had led to continuing difficulties in resurrecting research and development of electrostatic generators. He asserted that UFTI should be entitled to keep some of its own equipment for research programs, not serve as a production bureau for other institutes of the nuclear enterprise that needed generators. He pointed out how the institute had suffered terribly during evacuation; the Nazis gutted he facilities. Having returned to Kharkiv in March 1944 after the Nazi withdrawal, Valter and his associates set to work, getting some of the equipment back on line by 1946, including two electrostatic generators that were, to his understanding, the only such functioning devices in the USSR. Yet they were insufficient to expand crucial research, and the institute required 3 million rubles to build new ones. Valter pointed out that UFTI was a pioneer in linear accelerator technology and other projects important to the nation's atomic bomb programs. Yet Valter's laboratory had not received sufficient resources to build new, more powerful electrostatic generators. To make matters worse, of the twenty-nine specialists recently graduated from UFTI programs on nuclear themes, not one gained employment in the institute; they were all transferred to Moscow. Valter put it simply: "From my point of view it would be significantly undesirable to transform the role of the department that I direct at the institute

---

[51] GAKhO, f. 854, op. 1, d. 68, ll. 116–120.

into the role of a design production outfit for the production ... of electrostatic generators for other organizations."[52]

In autumn 1949 Kurchatov reminded Sinelnikov of the contract between laboratories one and two for Sinelnikov's physicists to provide him with a new generator in the first quarter of 1950. Sinelnikov responded that without the specifications for the generator his hands were tied. Further, UFTI required the requisite materials and equipment, more workers and technical personal, and the right to pay overtime, without all of which there was no way to meet the contract terms.[53] Because of the demands on the institute, Sinelnikov and Valter simply began turning down requests to build smaller electrostatic generators (on the order of 4 MeV), and gained permission to involve other institutes and laboratories to build them instead. UFTI would focus instead on developing a 20 MeV electrostatic generator to become operational in 1952. The authorities established a special institute, the Efremov Institute of Electrophysical Apparatuses outside of Leningrad, to meet the growing demand for magnets, chambers, accelerator components and so on,[54] thus saving UFTI from merely serving Moscow physicists as a supply depot.

In the 1950s the institute developed alternative approaches in fusion research, high energy physics, solid state and nuclear physics. While UFTI's research programs rapidly improved with the infusions of funding from the nuclear weapons program, however, the institute had a relatively marginal status compared to institutes in Moscow and Leningrad. Those institutes benefited as product of larger physics community's commitment to strong centralization over the long run. "Regionalism" among physicists and efforts to promote science in the periphery would be revived in another setting again only later in the 1950s, for example, through the founding of Akademgorodok near Novosibirsk as part of the rapid expansion of the scientific enterprise. For the majority of scientists in Kharkiv, and not only the Landaus of the 1930s but the promising young graduates of the 1950s and 1960s, Ukrainian science remained peripheral science. The essential processes of Sovietization, centralization, central planning and ideological control ensured that outcome.

[52] A KIAE, f. 1, op. 1/c, ed. Khr. 63.
[53] A KIAE, f. 1, op. 1/c, ed. Khr. 43, str. 82–83.
[54] A KIAE, f. 1, op. 1/c, ed. Khr. 78.

BIBLIOGRAPHY

Akhiezer, A. I.: Ocherki i vospominaniia, Kharkiv 2003.

Akhiezer, A. I.: Recollections of Lev Davidovich Landau, in: Physics Today 47 (6), 1994, p. 38.

Akhiezer, A. I.: Uchitel' i drug, in: Khalatnikov, I.M. (ed.): Vospominaniia o L.D. Landau, Moscow 1988.

Badash, Lawrence: Kapitsza, Rutherford and the Kremlin, New Haven 1985.

Bailes, Kendall: Technology and Society Under Lenin and Stalin, Princeton 1978.

Boag, J. W., P. E. Rubinin, and D. Shoenberg: Kapitza in Cambridge and Moscow, Amsterdam 1990.

Bunimovich, V. I.: Opyty po razrusheniiu atoma, in: Elektrichestvo (8), 1933, pp. 4-9.

Butiagin, A. S. and Iu. A. Saltanov: Universitetskoe obrazovanie v SSSR, Moscow 1957.

Conquest, Robert: The Great Terror, London 1973.

Conquest, Robert: Harvest of Sorrow, London 1986.

Ermolaeva, N. S.: Russkoe matematicheskoe zarubezh'e, in: Priroda (11), 1994, p. 84.

Gorsuch, Anne E.: Youth in Revolutionary Russia: Enthusiasts, Bohemians, Delinquents, Bloomington 2000.

Graham, Loren: Science, Philosophy, and Human Behavior, New York 1987.

Hall, Karl: Purely Practical Revolutionaries: Soviet Theoretical Physics in Stalin's Day, Ph.D. dissertation, Harvard University, Cambridge, Mass. 1999.

Hughes, James: Stalin, Siberia, and the Crisis of the New Economic Policy, Cambridge 1991.

Joravsky, David: Soviet Marxism and Natural Science, 1917-1932, London 1961.

Josephson, Paul: Kapitsa, Petr, in: New Dictionary of Scientific Biography 4, Detroit 2008, pp. 425-430.

Josephson, Paul: Sovetskie fiziki – stipendiaty rokfellerovskogo fonda, in: Uspekhi fizicheskikh nauk 160 (11), 1990, pp. 103-134.

Kapitsa, E. L. and P. E. Rubinin: Vospominaniia, Pis'ma, Dokumenty, comp. by Petr Leonidovich Kapitsa, Moscow 1994.

Kas'ianov, G. V. and V. M. Danilenko: Stalinizm i ukrains'ka inteligentsiia, Kiev 1991.

Khar'kovskii gosudarstvennyi universitet za 150 let, Kharkiv 1955.

Kikoin, A. K.: Kak ia prepodaval v Khar'kovskom universitete, in: Khalatnikov, I.M. (ed.): Vospominaniia o L.D. Landau, Moscow 1988, p. 161.

Konecny, Peter: Builders and Deserters: Students, State, and Community in Leningrad, 1917-1941, Montreal 1999.

Kostiuk, Hryhory: Stalinist Rule in the Ukraine, Munich 1960.

Kuznick, Peter: Beyond the Laboratory, Chicago 1987.

Pavlenko, Yu. V., Yu. A. Ranyuk and Yu. A. Khramov: "Delo" UFTI 1935-1937, Kiev 1998.

Ranyuk, Iu. N.: L. D. Landau i L. M. Piatigorskii, in: VIET (4), 1999, pp. 92-101.

Ranyuk, Yuri, Oksana Shevchenko, and Paul Josephson: Eshche raz ob 'Antisovetskoi zabostovke Khar'kovskikh fizikov', in: Voprosy Istorii Estestvoznaniia i Tekhniki (3), 2007, pp. 69-81.

Rubinin, P. E.: Pis'ma o Nauke, Moscow 1989.

Ruhemann, M.: Kharkiv in the Thirties (from Recollections), in: Soviet Journal of Low Temperature Physics 18 (1), 1992, pp. 50-56.

Shapiro, Leonard: The Communist Party of the Soviet Union, New York 1960.

Shubnikov, Lev: Izbrannye Trudy. Vospominaniia, Kyiv 1990.

Slezkine, Yuri: Arctic Mirrors, Ithaca 1994.

Trofimenko, A. P.: Istoriia razvitiia iadernykh issledovanii na Ukraine, candidate of science dissertation, Moscow (Institute of History of Science and Technology) 1974.

Vorob'ev, V. V.: Lev Landau i 'antisovetskaia zabastovka fizikov', in: VIET (4), 1999, pp. 92-101.

Vorob'ev, V. V.: Vy znaete, iakoiu vin buv liudinoiu? Pravda pro rektora Oleksiia Neforosnogo, in: Kharkivs'kii universytet, 1994.

Weissberg, Alexander: The Accused, New York 1951.

PHYSICISTS AS POLICYMAKERS IN POSTWAR JAPAN: THE RISE OF JOINT-USE UNIVERSITY RESEARCH INSTITUTES

Morris Low

INTRODUCTION

How did the availability (or lack of) research support systems for science in Japan influence the activities of physicists after World War II? To understand how physicists negotiated the landscape of postwar Japan, we need to examine their wartime experience. Only then can we understand the general disdain on the part of many scientists (and many Japanese people) for university-industry cooperation in service to the state. In postwar Japan, any research that could be construed as having possible military uses was considered almost taboo.

This paper consists of two major parts. We firstly examine sources of research funding from 1920 through to 1970, and then explore how physicists organized themselves, especially after the war, to promote experimental physics and the development of nuclear power. We will see that physicists were clever actors who made the most of funds that were available to them. At a time of economic difficulties, kôza-based (see below) core funding to universities, and the establishment of many joint-use research institutes attached to national universities, provided ways of making the most of what meagre funds were available.

I RESEARCH FUNDING FROM 1920 TO 1970

During and after World War I, a number of research facilities were established including the Institute of Physical and Chemical Research (abbreviated in Japanese as "Riken") in Tokyo, the largest pre-World War II scientific research organization in Japan. It was analogous to the Imperial Institute of Physics and Technology in Germany (established in 1887).[1] Riken was formed in 1917 with funds from government and the private sector. As its name suggests, the Institute was, at least until 1921, divided into two divisions: physics and chemistry. The head of physics was Hantarô Nagaoka, a highly influential scientist who was a key player in policymaking up until the end of World War II. He will be discussed in some detail in the second major part of this paper.

---

[1] For further details of the Imperial Institute see Cahan, An Institute.

At Riken, a laboratory system was established from 1922 that enhanced the institute's research. The man who was responsible for this structure was Masatoshi Ôkôchi, director of Riken from 1921 to 1946. He abolished the divisions of physics and chemistry, setting up a system of chief-researchers who would be in charge of their own laboratories. Riken attracted many promising young researchers, many of whom used the facilities there for research towards a doctoral degree.[2]

Despite the wide interest in science, there were only three schools of science in Japanese universities in 1926, the first year of the Showa period. The school at Tokyo dated back to 1877; Kyoto to 1897; and Tohoku to 1911. (Kyushu Imperial University was established in 1911 with programs only in medicine and engineering.) Before long however, science schools were created at the Hokkaido (1930), Osaka (1933), Kyushu (1939), and Nagoya (1942) imperial universities. Tokyo Bunrika University and Hiroshima Bunrika University also had a major science focus,[3] but the imperial universities dominated scientific research.

The formal teaching of physics largely rested with the seven imperial universities: Tokyo, Osaka, Nagoya, Kyoto, Hokkaido, Kyushu and Tohoku. Some of the physics graduates from these institutions went on to become fully-fledged researchers at institutions such as Riken. By 1940, university science graduates would be over double the number in the first year of the Showa period, 1926. Five years later it would be over double that again.[4]

MOBILIZATION OF SCIENCE

Despite the existence of Riken and the establishment of several academic research institutes, funding remained scarce in the 1920s. Most government funding was allocated on a per capita basis according to the kôza system of chairs where full professors controlled funds and the associate professors, research associates and graduate students in the same research unit were subordinate. In the 1930s, when Japan was increasingly placed on a wartime footing, there was an increase in government funds for research. From 1932, there were grants from the newly-established Japan Society for the Promotion of Science (Nihon Gakujutsu Shinkôkai, JSPS).[5] The JSPS grants encouraged university-industry collaboration.

It was only after full-scale war with China broke out at Shanghai on 13 August 1937 that more attempts were made to mobilize science for the war effort. In October, a Science Division (Kagakuka) was created within the newly-established Cabinet Planning Board (Kikakuin) to integrate research activities into the wartime economy. Following major military setbacks, further attempts were made

---

[2] Shimao, Some Aspects, pp. 84–85.
[3] Hirosige, Reevaluating, pp. 181–183.
[4] Shimao, Some Aspects, p. 86.
[5] This account is largely derived from three sources: Hirosige, Science and History, pp. 169–184; Hirosige, Social History, pp. 200–222; Kamatani, History.

in 1938 to mobilize science for military purposes with the promulgation of the National Mobilization Law in April of that year. The government immediately established the Science Council (Kagaku Shingikai) within Cabinet, virtually incorporating JSPS. Given that the role of Cabinet's Science Division had now changed to the mobilization of research, it was upgraded to Science Department (Kagakubu). These organizations would later be incorporated into the powerful Agency of Science and Technology.

On 26 May 1938, Cabinet was reorganized and General Sadao Araki was appointed Minister of Education. Despite his reputation as a right-wing ultranationalist who dismissed and imprisoned progressive-minded teachers and students,[6] he had a positive impact on science policy. This is one of the ironies of the wartime period. The war effort put into play processes of modernization and rationalization that might have otherwise been much delayed.[7] One of the first things that Araki did as minister was to establish an Association for the Advancement and Investigation of Science (Kagaku Shinkô Chôsakai), the forerunner of the Science and Technology Council. The Association recommended the establishment of a Science Division (Kagakuka) within the Ministry of Education. The Division controlled university and college research, and the funding of other Ministry of Education research organizations. Sectionalism within government thus led to a disastrous, two-tier approach to the mobilization of science.[8]

On 15 March 1939, while Araki was still minister, Cabinet approved a new budgetary allocation of 3 million yen per year for grants-in-aid for scientific research (kagaku kenkyûhi or shortened as kakenhi), to be distributed by the Ministry of Education. This was a major policy achievement and testimony to Araki's forceful personality. Araki's brief stint as minister ended on 30 August 1939, but his belief in the importance of basic research continued. On 19 November 1940, Araki gave an invited lecture before the Imperial Institute of Invention and Innovation where he emphasized the importance of basic science to Japan's industrial might and future. He also considered museums important, praising the Deutsches Museum in Munich and contrasting it to the poor state of Japan's Science Museum.[9]

Infrastructure was important. Attempts to mobilize scientists included the establishment of research laboratories, funded by the government and often attached to universities. Liaison organizations such as the Japan Federation of Science and Technology Organizations (Zen-Nippon Kagaku-Gijutsu-sha Dantai Rengô), which was established on 2 August 1940, sought to improve lines of communication between institutions and prevent wasteful overlap in research.[10] The Federation was linked with the Ministry of Education's Science Division through the National Research Council. Its efforts towards rapid rationalization

[6] Kurihara, Japan's Educational System, p. 36.
[7] Shillony, Universities, p. 769.
[8] Kamatani, History, pp. 40–41, 48–49.
[9] Hirosige, Social History, pp. 153–155.
[10] Hirosige, Science and History, pp. 169, 175.

of science expectedly met with a great deal of opposition from the scientific community.[11] The Cabinet Planning Board's Science Department set up an auxiliary organization to do this as well. On 8 December 1940 the Science Mobilization Association (Kagaku Dôin Kyôkai) was created with the aim of bringing together scientists of different fields and affiliations.[12]

The war effort provided the opportunity for a concerted attack on the feudalistic manner in which research had been conducted in Japan. Scientists were blamed for creating a gulf between basic and applied research, and of encouraging sectionalism. In order to overcome these problems, a "General Plan for the Establishment of a New Scientific and Technological Structure" ("Kagaku Gijutsu Shintaisei Kakuritsu Yôkô") was passed by Cabinet on 27 May 1941.[13] In January 1942, the Cabinet Science Department was expanded to become the Agency of Science and Technology (Gijutsuin), Japanese closest equivalent to the US Office of Scientific Research and Development, of which the Japanese are unlikely to have known about.

The Agency was founded to oversee the general administration of the new structure, and to pay special attention to aeronautical engineering. Responsibility for the Central Aeronautic Laboratory was transferred from the Ministry of Communications to this Agency.[14] Later that year the Science and Technology Council (Kagaku Gijutsu Shingikai) was established by the government to deliberate on policy and to provide central support for mobilization. The Ministry of Education, not wishing to lose ground, expanded its former Science Division to Science Department in 1942. This inter-ministry antagonism extended to the Army and Navy.[15]

In August 1943 the "Emergency Plan for Scientific Research" ("Kagaku Kenkyû Kinkyû Seibi Hôsaku Yôkô") was conceived in a desperate attempt to mobilize science. The plan involved the establishment, by the National Research Council (Gakujutsu Kenkyû Kaigi) of a Research Mobilization Committee (Kenkyû Dôin Iinkai) with liaison committees in each university.[16] In October 1943, Cabinet decided to reorganize the overall mobilization system again. A Research Mobilization Council (Kenkyû Dôin Kaigi) was established to decide important areas of research. The government's "Integrated Plan for the Mobilization of Science and Technology" ("Kagaku Gijutsu Dôin Sôgô Hôsaku") prioritised aeronautic research and the development of new scientific weapons.[17] In the closing stages of the war, there were other attempts to muster science and technology to avert

---

[11] Ibid., p. 178.

[12] Ibid.

[13] Ibid., p. 175.

[14] "Gijutsuin" is sometimes translated as "Board of Technology" or "Board of Technics". For further details, see Yamazaki, Mobilization.

[15] Kamatani, History, pp. 40–41, 48–49.

[16] Hirosige, Science and History, p. 178.

[17] Hirosige, Social History, p. 211.

defeat. In January 1945 there was a structural reform of the National Research Council, to little avail.[18]

It is clear that the wartime mobilization of science was a case of too little too late,[19] but a major pillar of research funding, the Ministry of Education's grants-in-aid for scientific research, were put into place at this time, under the watch of General Araki. Araki was put on trial after the war as a Class A war criminal and sentenced to life imprisonment but was released early in 1955 for health reasons, only to die much later. In Japan, like Germany, there was a need to take a pacifist stance and to embrace science and technology. The dropping of the atomic bomb and Japan's defeat showed that Japan was clearly behind – something that Araki had been all too aware of even during the war.

Soon after the end of World War II, the Scientific and Technical Division of the Economic and Scientific Section of the Supreme Commander of the Allied Powers estimated that between 1931 and 1940, around 1159 doctorates in science, technology, agriculture and forestry were awarded in Japan.[20] As for the US during the same period, there were 11,443 Ph.D. degrees. This means that the rate of production of highly trained scientists and engineers, per head of population, was about one-fifth that in the United States.[21] It was still a substantial figure, although no match for the Americans.

During the Allied Occupation (1945–52), there were calls for changes to the kôza system and a shift to research policy determined by the academic community rather than the state. At the same time, however, given the need to rebuild Japan, there were those who felt that research should be more responsive to social and economic needs. As Japan rebuilt and recovered from the war, in the late 1940s and 1950s, scientists became more emboldened to call for academic autonomy. At times, this took on an anti-government character.[22]

Despite a desire to change the kôza-based system under which each kôza receives research funding via general university funds (kôhi) provided by the government, the situation remained unchanged. What's more, this system was expanded to include all national universities, not just the former imperial universities. Given the financial difficulties that universities experienced after the war, these funds came to be used for purposes other than research, such as administration and student-related expenses, albeit with the permission of each chair or full professor. These monies constituted core funding for institutions and were distributed rather mechanically according to the number of faculty in each field. They preserved the hierarchy of kôza, and did not reward research excellence as kôza received equal allocations.[23]

---

[18] Ibid., p. 214.

[19] See Grunden et al., Laying the Foundation, pp. 93–98.

[20] See Table 115 "Estimated Trained Personnel in Selected Scientific and Technical Professions, 1947", in: General Headquarters, Japanese Natural Resources, p. 519.

[21] General Headquarters, Japanese Natural Resources, p. 506.

[22] Asonuma, Funding, p. 153.

[23] For a discussion of the problems of the kôza system, see Coleman, Japanese Science, pp. 19–21, 41.

The Ministry of Education, Science and Culture[24] (hereafter referred to as the Ministry of Education) continued the grants-in-aid program for scientific research, but it was the newly-formed Science Council of Japan (JSC) (established in 1948), which was elected by the academic community, that had a great influence on the distribution of funds. These grants, combined with funds from the JSPS, served to augment in a small way, the main, core funding for kôza. The autonomy of the academic community increased, free from the constraints of working with the military or industry.

Amidst the postwar reorganization of the administrative structure, responsibility for distribution of research funds came to rest with the Ministry of Education. This resulted in an outflow of funds to other ministries but not the reverse. As a result of the apportionment of responsibilities, applied R&D was consigned to other ministries and the grants-in-aid for scientific research focused on basic research, especially conducted at universities.[25]

The JSC provided a forum for scientists to discuss policy and make recommendations to government. The JSC did not consist of government appointees. Rather, members of the JSC were elected. The government was not obliged to take its advice and was frequently in conflict with it over the importance of basic research to, for example, the development of nuclear power. Government and industry preferred to fast-track implementation by importing nuclear technology from the USA under the Atoms for Peace plan rather than develop its own.

In 1956, the Japanese government established the Science and Technology Agency to manage Japanese nuclear R&D and in 1959 created the Science and Technology Council as the highest government advisory board on science and technology policy. The latter was designed, in part, to bypass the JSC. We shall see in the second part of this paper that physicists who cooperated with government were indeed able to influence the distribution of support for Japanese science. The JSC did have some influence but the strategy of establishing joint-use research institutes meant that scientists were able to secure some government funding of basic science and at the same time achieve the autonomy that they desired.

RISE OF JOINT-USE RESEARCH INSTITUTES

One solution to the meager funds available for research has been a policy of concentrating funds in major scientific facilities at inter-university research institutes and joint-use laboratories attached to national universities.[26] While the idea of

[24] The Ministry is now known as the Ministry of Education, Culture, Sports, Science and Technology after the Ministry merged with the Science and Technology Agency as of 6 January 2001. The Ministry oversees approximately 75 per cent of government research funding. See Normile, Japan.

[25] Asonuma, Funding, p. 155.

[26] Irvine et al., Investing, pp. 171–172.

attaching research institutes to universities was not new, with many having been established during World War II,[27] promoting joint-use was novel and suited the economic situation and the emphasis on promoting wide access to scientific facilities. There was also a trend after the war, to attach national research facilities to universities to ensure their relative autonomy, free from the interference of politicians and bureaucrats.

The Research Institute for Fundamental Physics at Kyoto University, which was established in 1953, was the first major joint-use institute to be created after the war. It was followed soon after by the opening of the Cosmic Ray Observatory, University of Tokyo, located at Mt. Norikura. The 1950s and 1960s saw a succession of institutes being established, taking advantage of the funding environment: The Institute for Nuclear Study, University of Tokyo (1955), the Institute for Protein Research, Osaka University (1958), Institute of Plasma Physics, Nagoya University (1961), Ocean Research Institute, University of Tokyo (1962), Institute for Mathematical Analysis and Reactor Laboratory, both at Kyoto University (1963), and the Institute of Space and Aeronautical Science, University of Tokyo (1964). The Research Institute of Languages and Cultures of Asia and Africa, Tokyo University of Foreign Studies, was also established in 1964. Yet another institute, the Primate Research Institute, was established in 1967 at Kyoto University.

The institutes are the legacy of the Laboratory Democracy Movement in the 1950s and 1960s, in which young academic scientists promoted a network of common-use university-affiliated research institutes. These young researchers found it impossible, however, to conduct big academic science within the constraints of limited university budgets. Since the 1970s, research institutes have been reorganized as national research institutes without university affiliation and directly administered by the Ministry of Education. A good example is the Institute of Space and Astronautical Science (formerly known as the Institute of Space and Aeronautical Science, University of Tokyo). The end-result has been that for much of the postwar period, the kôza-based funds, the grants-in-aid for scientific research, and funding for the research institutes attached to universities, have been the three main pillars of research funding.

We have seen how the number of research institutes linked to universities grew quickly. In 1960, there were 57 research institutes attached to universities but by 1967, 13 more had been established. Seven of these new institutes were for joint-use. From 1968, the funding of the attached research institutes began to stagnate and the grants-in-aid for scientific research at last began to increase.

How do we account for this growth? The launch of the first artificial satellite Sputnik I by the Soviet Union on 4 October 1957 marked the beginning of the space race with the US and as a result Japan dramatically increased its levels of scientific manpower. To this end, the Ministry of Education announced a plan that same year to increase the number of students admitted into science and engineering programs. The aim was to increase the number of admissions from

---

[27] Shillony, Universities, p. 773.

22,000 to 30,000 by 1960. The Ministry launched a further plan in 1961 to increase the number of new student places by a further 20,000 places by 1964.[28]

It was against the background of a massive increase in the number of students majoring in science that we see rapid growth in the number of students studying mathematics and physics. Numbers increased 5.5 and 4.3 times respectively over the period 1958–1970. Mathematics and physics majors found ready jobs in electronics, space research, nuclear power, and information technology.[29]

In the 1950s, university budgets began to increase, as did the kôza-based funding, but it was in the 1960s, in particular, that there was rapid growth. Over the years from 1960 to 1967, kôza-based faculty research funding (kyôkan kenkyûhi) almost quadrupled, going from 62 billion yen to 237 billion yen. These changes reflected the rapid economic growth that Japan was enjoying in the 1960s.

Although the Japanese government sought to increase support for applied research in the 1960s and 1970s, this push for socially and economically relevant research was resisted by scientists, especially the Science Council of Japan. This led the government to channel funding to the Science and Technology Agency for key areas such as materials science and the life sciences. For such reasons, the Ministry of Education's share of the government R&D budget declined from 1965.[30]

THE PRIVATIZATION OF KNOWLEDGE

It is difficult to compare national R&D expenditure for various countries, as its definition varies from one place to another. It is, however, clear that as far as the research budget is concerned, Japan is strong in the "private" science of company laboratories, and much weaker in academic science, as practiced in the academic sector of universities and affiliated institutes. By the mid-1960s, over three quarters of Japanese spending on scientific research came from the private sector.[31] Things have not changed very much. In 2005, the private sector accounted for 80.7% of total R&D expenditure. Government spending accounted for 19%. It is not surprising to learn that 77.3% of R&D funds for the natural sciences went to industry, 12.9% to universities and colleges, 8.1% to government research institutions, and a small 1.7% to private research institutions. Over half of government funding (51%) goes to universities. A little less (40%) goes to government research institutions, and about 9% goes to the private sector. Most private R&D expenditure (i.e. 98.9%) goes to the private sector.[32]

---

[28] Nakayama, Science, pp. 61–62.

[29] Hirosige, Social History, p. 295.

[30] Irvine et al., Investing, pp. 169–170.

[31] Cummings, Culture, p. 432.

[32] All figures are full-time equivalent values which take into consideration the proportion of time that researchers actually devote to research activities, rather than teaching and research which a simple head count would reflect. Japanese Government, White Paper, pp. 95, 98.

Let us compare Japan to other countries. In Germany in 2004, government expenditure accounted for 30.4%, the private sector 67.2%, and foreign sources the remaining 2.5%. In the United States in 2004, the government share of R&D expenditure was almost the same: 31% and the private sector about 69%. In contrast, government funding in France in 2004 was a high 37.6% of total R&D expenditure. What is clear is that although private sector funding of R&D expenditure dominates, Japan has the lowest level of government funding of R&D amongst industrially-advanced nations.[33]

The privatization of knowledge has led to the relative impoverishment of Japanese academic science. Rather than the privatization of academic science, this refers to the much better-funded state of private science. The Japanese have often been criticized for the lack of government funding of basic science, compared to the private sector. The budgets of national universities (former imperial universities) funded by the central government and other public universities funded by prefectural governments, have failed to keep up with inflation. And it is these universities where most government-funded research occurs.

Japanese corporations have long regarded the country's universities as providers of human resources for their own in-house research centers rather than as places to go to have research done. The tendency to oppose any research that might be construed as linking research to private profit was the product of student unrest in the late 1960s and Japan's postwar academic culture. This resulted in an over-reliance on the Ministry of Education for its general university funding based on kôza and competitive grants-in-aid for scientific research.

The reliance on kôza-based, core-funding to institutions may seem "equitable" in that it results in the equal distribution of resources, but it does prevent highly-talented researchers from greatly expanding their projects, for they are not considered independent researchers but a member of a larger research unit. Though efforts have been made to overcome the constraints of the kôza, it continues to hamper research activity, especially interdisciplinary cooperation. The combination of the chair system and life-long employment has meant that the mobility of researchers is severely hampered and fast-promotion is difficult to achieve.

II PHYSICISTS AS POLICYMAKERS FROM 1920 TO 1970

Nagaoka and Nishina

In the 1920s and 1930s, few Japanese physicists were prominent as policymakers. Hantarô Nagaoka (1865–1950) and Yoshio Nishina (1890–1951) were two major exceptions. Nagaoka was not only a distinguished physicist in his own right. He first proposed a Saturnian model of the atom at a meeting of the Physico-Mathematical Society in Tokyo in 1903 and published it in the Philosophical Magazine

---

[33] Japanese Government, White Paper, p. 95.

in 1904.[34] He suggested that the atom was like the planet Saturn with a positively-charged particle in its center and electrons orbiting around it.

Nagaoka married into a prominent family of scholars. Nagaoka's wife Misako was one of the three daughters of the legal scholar Rinshô Mitsukuri (1846–1897). Mitsukuri was, in turn, the brother-in-law of the mathematician Dairoku Kikuchi (1855–1917). Kikuchi had, in addition to being professor and President of Tokyo Imperial University and Kyoto Imperial University, also served as Minister of Education, member of the House of Peers of the Imperial Diet, and President of the Imperial Academy of Japan. He was made a Baron in 1902, and appointed the inaugural director of Riken when it was established in 1917, the year of his death. There were other famous scholars in the extended Mitsukuri family, too numerous to name.[35]

Nagaoka had studied at Tokyo Imperial University, later becoming a professor there and chief researcher at Riken. He seems to have followed in Kikuchi's footsteps not only in terms of the Riken connection but also in university administration, as the first president of Osaka Imperial University. Nagaoka had the distinction of being the first recipient of the Order of Cultural Merit, President of the Imperial Academy of Japan, Chairman of the Japan Society for the Promotion of Science, and member of the House of Peers.

Nagaoka became one of Nishina's mentors. It is noteworthy that Nishina attended university lectures given by Nagaoka at Tokyo Imperial University and developed a strong interest in physics despite a background in electrical engineering. Upon Nishina's return in 1928 from a long stint in Europe that included research with Niels Bohr in Copenhagen, Nishina was given a position in the Nagaoka laboratory at Riken. In July 1931, Nishina was appointed chief researcher, with strong backing from Nagaoka. In this way, Nishina became the youngest person to hold such a position that came complete (in 1932) with his own laboratory.

Nagaoka belonged to the Science Deliberative Council (Kagaku Kyôgikai), which was first established in 1921 to improve liaison between scientists and the military. This Council became the Society for Science for National Defense (Kokubô Kagaku Kyôkai) in 1934, with Nagaoka at one stage being a director. Nagaoka was also Chairman of the National Research Council Committee for Radio Research, and his roles as Division Head and later Chairman of Directors of the Japan Society for the Promotion of Science meant that he would become heavily involved with the military on matters of science.[36]

Nagaoka was a central figure in the mobilization of Japanese science. It appears that he enjoyed a special relationship with the Army and Navy. In 1940 he became a consultant to the Army Technical Headquarters and councillor of the Army Ordnance Department. Nagaoka belonged to many military and Agency of Science and Technology committees. He held membership in the Navy's own Science and Technology Council (Kaigun Kagaku Gijutsu Shingikai) (1942)

---

[34] Nagaoka, Kinetics of a System.
[35] Koyama, Japanese Students, pp. 2–3.
[36] Itakura et al., Nagaoka, pp. 636–637.

and the Army and Navy Committee for Radio Technology (Rikukaigun Denpa Gijutsu Iinkai) (1943). He was also a councillor of the Agency of Science and Technology (1944). Around 1943–1944 he was employed as a part-time consultant by the Naval Technical Research Institute.[37]

As for Nishina, he had a key role in Japan's atomic bomb project. This required soliciting funds from the military and is testimony to his ability to negotiate and his perceived credibility. The research at Riken was dubbed the "Ni-Project" ("Ni-gô Kenkyû") after Nishina, its head.[38] It is claimed that up to two million yen (around $US 500,000)[39] was allocated by the Army to the Ni-project. Meanwhile, the Navy provided funding, around 600,000 yen for the Kyoto-based "F-Project" ("F-gô Kenkyû") from around May 1943. The "F-Project" was headed by Professor Bunsaku Arakatsu of Kyoto University; the letter "F" standing for "fission".[40]

While some left-wing physicists such as Mituo Taketani were interrogated and jailed during the war, and others were deployed in military-related research, it can be said that Japanese scientific institutions such as the Riken, where Nishina was based, did indeed prosper. The immediate challenge after the war for senior physicists such as Nishina was to save the institution. He did so by cutting staff and promoting the production of penicillin[41] and liquid oxygen.[42]

The dropping of the atomic bomb elevated physicists to a prominent role after the war as public men of science, but despite their increased prestige, nuclear research was banned by the Occupation authorities and there was little in the way of funding for science. When surveyed after the war, Japanese scientists viewed the war years in a positive light, in stark contrast to the conditions they found themselves in after the war when there was little to eat and next to no funding of research to speak of. Indeed, the war years were remembered as the "good old days," when funds and research were the least restricted.[43] In a survey conducted by the Science Council in 1951, a number of scientists replied the Pacific War period was the time of their greatest freedom in research.[44]

There are interesting parallels in this regard between Japan and Germany. Although the poor treatment of Jewish scientists in Germany is well known, it can be argued that many other scientists did well under the Nazis.[45] Nishina has even

[37] Itakura et al., Nagaoka, pp. 636–637.

[38] A detailed account can be found in Yomiuri Shinbun, Shôwashi, pp. 78–229.

[39] This figure uses the 1930s exchange rate of 4 yen per US dollar, given in Alletshauser, The House, appendix 6.

[40] Dower, Science, p. 51; Dower, Japan in War, pp. 55–100.

[41] Coleman, Riken.

[42] Riken, Stories.

[43] For example see Anon., Nihon.

[44] Reference is made to this survey in Hirosige, Kagaku to rekishi, p. 180. Yamamoto writes that the most frequent reply to the question asking the period of greatest intellectual freedom during the past decades, was the Pacific War period. See Yamamoto, Nihon, p. 19.

[45] See Beyler, Maintaining Discipline.

been portrayed as an "ardent nationalist".[46] However, whereas Japan's defeat was seen by the Japanese in terms of superior Western (especially American) science and technology, this appears not to have been the case in Germany.[47] In Japan, there was a consensus among politicians that there needed to be a policy for science whereas in Germany, this was not the case.

After the war, German scientists claimed that the Kaiser Wilhelm Society, and science in general, had maintained its independence during the National Socialist regime. The line was that it had been politically innocent, and any instances that showed it to be otherwise were examples of the misuse of science.[48] The postwar discourse of Otto Hahn (1879–1968), President of the Kaiser Wilhelm Society and its successor, the Max Planck Society, from 1946 through to 1960, fits into this overall pattern of denial.

Nishina was busy promoting ties with industry in order to save Kaken, Riken's successor. Rather than promote the idea of pure science, he strived to use science for the economic reconstruction of Japan. Although he was a senior physicist, he had been trained in electrical engineering, and his entrepreneurial approach to science differed from other physicists based in universities. It was the latter who tended to promote the view remarkably close to that of German scientists, that much of what was carried out under the name of "wartime research" was related to pure science. Involvement in Japan's atomic bomb project and military research was to help save the lives of young Japanese who might otherwise be sent to war.[49]

In postwar, pacifist Japan, academic scientists sought to distance themselves from industry in case there was any doubt.[50] Scientists promoted the idea of the autonomy of science in ways similar to the utterances of their colleagues in Germany. Such rhetorical strategies helped to ensure that science in both Japan and Germany could recover and eventually prosper. But it also meant that a research institution like the former Riken, funded by both government and the private sector, was not an option. Instead, the establishment of joint-use research institutes attached to national universities provided a solution for the perceived need to be independent of industry and spread what research funds were available to the maximum number of scientists.

There are many parallels between postwar Japan and Germany. Both were defeated Axis nations and both were occupied by the United States and the Allied Powers. Both went on to become major economic powers. We can also point to considerable scientific exchange between Japan and Germany in the prewar period. In what ways was the German experience important to the Japanese?

The young physicists Seishi Kikuchi, Sin-itirô Tomonaga, and Satoshi Watanabe all visited and studied under Werner Heisenberg at his Institute for Theoreti-

---

[46] Ito, Values, p. 63.
[47] Carson/Gubser, Science Advising, p. 177.
[48] See Beyler, Maintaining Discipline.
[49] For Otto Hahn, see Sime, The Politics of Memory, p. 41; Walker, Otto Hahn, p. 137.
[50] Hicks, University-Industry, p. 373.

cal Physics, University of Leipzig at various times from the late 1920s through to the 1930s. In the postwar period, Japanese intellectuals who had traveled to Europe during these years reflected on their experiences and compared Japan and Germany. Both "were modernizing societies that had slipped back into irrationality and repression".[51]

While it is tempting to compare Japan and Germany as if they were on a par with each other after the war, General Douglas MacArthur himself suggested otherwise. On his return to the US after having overseen the Allied Occupation of Japan and served as commander of the United Nations forces in Korea, he addressed a joint session of Congress on 19 April 1951. He argued that

> Well, the German problem is a completely and entirely different one from the Japanese problem. The German people were a mature race. … The Japanese, however, in spite of their antiquity measured by time, were in a very tuitionary condition. Measured by the standards of modern civilization, they would be like a boy of twelve as compared with our development of 45 years …
> But the Japanese were entirely different. There is no similarity. One of the great mistakes that was made was to try to apply the same policies which were so successful in Japan to Germany, where they were not quite so successful, to say the least. They were working on a different level.[52]

There are, nevertheless, interesting similarities between the German and Japanese postwar experience.[53] Japan and the US are also sometimes compared and depicted as competitors in high-energy physics,[54] but Japan and the Japanese – MacArthur's twelve-year old "children" as it were – had much catching up to do.

What became of the Japanese physicists who had ties with Germany and who seemed so promising before the war? Did they contribute to the reconstruction of Japan? In the book Werner Heisenberg in Leipzig, 1927–1942,[55] there is a photograph dated 1951 of Tomonaga, Kikuchi and a third unnamed physicist, Ryôkichi Sagane. Both Kikuchi and Tomonaga had visited Heisenberg's Institute during the periods 1929–30, and 1937–39 respectively. While the Nobel Prize-winner Tomonaga is well-known, this paper also throws light on the activities of Kikuchi and Sagane who arguably were just as important as policymakers in the mid-1950s. It can be difficult reconstructing their activities after so much time has passed, but we are fortunate that transcripts of deliberations in two key national parliament committees are still available.

In the lower house, House of Representatives, we will examine the Special Committee for Policies for the Promotion of Science and Technology (Kagaku Gijutsu Shinkô Taisaku Tokubetsu Iinkai) meeting on 11 April 1956 attended by Kikuchi and Sagane, and the 19 February 1958 meeting attended by Kikuchi. In

---

[51] Hein, Reasonable Men, p. 14.
[52] Cited in Dower, Embracing Defeat, pp. 550–551.
[53] See for example, Beyler/Low, Science Policy; Inoguchi/Jackson, Memories.
[54] Hoddeson, Establishing K.E.K.
[55] Kleint/Wiemers, Werner Heisenberg, p. 151.

the upper house of parliament, the House of Councilors convened the Committee for Science, Education and Culture (Bunkyô Kagaku Iinkai), which dealt with matters affecting the Ministry of Education. The committee consisted of around twenty members. We will examine discussions at the meeting on 16 June 1955 which Tomonaga and Kikuchi were called on to attend as experts. Coming a few years after the end of the Allied Occupation of Japan, these committees provide a snapshot of the exciting times of the mid-1950s that physicists were immersed in. What is most fascinating is how many physicists sought to promote a science free of government and private enterprise, and how others like Sagane, sought to engage them.

Tomonaga

Sin-itirô Tomonaga was born in 1906, the eldest son of Sanjurô Tomonaga, a professor of philosophy at Kyoto Imperial University who had spent time at Heidelberg University. Sin-itirô majored in physics at Kyoto, graduating in 1929. He joined the laboratory of Yoshio Nishina at Riken in 1932. He studied nuclear theory at Leipzig University under Werner Heisenberg from June 1937 to August 1939, thanks to a researcher-exchange agreement between Riken and Leipzig University. Other Japanese who had preceded him at Leipzig included Seishi Kikuchi. Upon his return to Japan, Tomonaga received his D.Sc. from Tokyo Imperial University. From June 1940, Tomonaga was a Riken research fellow, as well as being a professor at the Tokyo University of Science and Literature from 1941.[56]

During the war, Tomonaga experienced mobilization for the war effort firsthand. From around mid-1943, he was ordered to carry out research on magnetrons and ultra-shortwave circuits at the Naval Research Institute laboratory at Shimada.[57] In 1944, he was appointed to a lectureship at Tokyo Imperial University. From August 1944 to July 1945, his work also included a temporary assignment to the Tama Army Technical Institute.[58]

After the war, Tomonaga was elected to the Science Council of Japan in 1948, and in 1949 was re-appointed professor to what had become the Tokyo University of Education. On 9 August 1949, he left Japan to accept an invitation from Robert Oppenheimer to visit the Institute for Advanced Study, Princeton.[59] Tomonaga was impressed by the intellectual environment of the Institute for Advanced Study and hoped to recreate that environment back in Japan. The award of the 1949 Nobel Prize for physics to Hideki Yukawa provided the catalyst for the Japanese government to establish Yukawa Hall at Kyoto University.[60] This would become the Research Institute for Fundamental Physics. Tomonaga's

---

[56] Anon., Biography, pp. 137–139.
[57] Oda, Sin-itirô Tomonaga, pp. 109–110.
[58] Tomonaga, Personal History.
[59] Anon., Japan's Hidden Physicists.
[60] Matsui, Reminiscences, pp. 392–394; Anon., Biography, pp. 137–139.

own Nobel Prize would come much later in 1965. Tomonaga returned to Japan in 1950. With the death of Nishina in 1951, Tomonaga became highly involved in science policy out of a sense of duty and gratitude to Nishina.[61] He lobbied for the establishment of the Cosmic Ray Observatory at Mt Norikura as a joint-use facility (but formally attached to Tokyo University), and sought to revive cyclotron research. The Research Institute for Fundamental Physics and the Mt. Norikura Cosmic Ray Observatory were viewed as inexpensive ways in which the joint-use research institute concept might be started.

In 1953, Tomonaga was involved in discussions with Seishi Kikuchi, Yoshio Fujioka and others regarding the establishment of an Institute for Nuclear Study.[62] The Institute for Nuclear Study, with Seishi Kikuchi as its inaugural director, was important in helping to shape a new kind of research system.

Kikuchi and Sagane

Kikuchi and Sagane are noteworthy for many reasons. They were particularly well-connected. Seishi Kikuchi (1902–1974) was born in Tokyo, the fourth son of the distinguished mathematician Dairoku Kikuchi. Dairoku was the second son of Shûhei Mitsukuri, whose original name had been Shirô Kikuchi. Shûhei had been adopted into the Mitsukuri family, a remarkable family of scholars into which Nagaoka had married.[63] Dairoku, in turn, was adopted back into the Kikuchi family by his uncle.

Younger sons in Japan have long been adopted into families in order to preserve the family line, and as a way of finding them a home. Eldest sons traditionally inherited the family wealth. In families where there were no sons, adoption provided a ready solution, a way of preserving an elite class. Adoption was also a way of moving talented people into positions of influence, a form of social mobility in a society in which status had traditionally been inherited. It is thus fascinating to see that amongst the families of physicists, who tended to come from samurai backgrounds, such practices were also prevalent.[64] Most of the faculty at Tokyo Imperial University in the 1880s and 1890s came from former samurai families although there were few who were particularly wealthy. As the abovementioned Mitsukuri family suggests, many university professors came from families in which there had been a tradition of scholarship.

Seishi graduated from Tokyo Imperial University in 1926, and worked at Riken where his research earned him international recognition. He studied in Germany from 1929 to 1930, and a few years after his return, was appointed professor at the newly-established Osaka Imperial University whose president was Hantarô Nagaoka, Ryôkichi Sagane's father. Nagaoka had been associated with Riken and attracted some of its best young talent to Osaka. It is remarkable that

[61] Hayakawa, Prof. Tomonaga, p. 799.
[62] For details of its history see Institute for Nuclear Study, Twenty Year History.
[63] Fushimi, Seishi Kikuchi, p. 197.
[64] Moore, Adoption and Samurai, Mobility.

Seishi Kikuchi and Ryôkichi Sagane were distant relatives, but perhaps not surprising given the tendency for sons to follow their fathers in terms of occupation,
and for academics to choose promising scholars as husbands for their daughters.
In this way, many academics during the Meiji period (1868–1912) were related
by birth, adoption or marriage.[65] To what extent family lineage helped Seishi
Kikuchi's career is difficult to say. Both Kikuchi and Sagane did come from families in which overseas study was usual, and it is clear that the opportunity to do
so was a turning point in their careers.

During the years 1950–1952, Seishi Kikuchi visited Cornell University. The
Newman Laboratory of Nuclear Studies at Cornell had been dedicated in late
1948 and Robert R. Wilson who had been chief of experimental physics at Los
Alamos was director. Renowned physicists such as Hans Bethe and Richard
Feynman were conducting research at Cornell. Kikuchi was affiliated with Wilson
whose group had recently completed a 300 MeV synchrotron with funding from
the Office of Naval Research. In 1953, Kikuchi was appointed the first director of
the Institute for Nuclear Study, attached to Tokyo University.[66]

Sagane, too, was born in Tokyo on 27 November 1905, the fourth son of
Hantarô Nagaoka. After encouragement from his father, he decided to pursue a
career in physics.[67] In 1914, he was adopted by Mrs. Chiyo Sagane and thence-
forth took on her family name.[68] His education followed the prestigious academic route, graduating from the First Middle School in 1923, the First Higher
School in 1926, and the Department of Physics, Tokyo Imperial University in
1929 where he would later become professor. Sagane entered graduate school in
April 1929. He finished postgraduate studies in June 1931, upon which time he
took on the status of research student at Riken. The following month he became
one of the first members of Yoshio Nishina's lab when Nishina became a chief
researcher at Riken. In April 1933, he was appointed lecturer in the Faculty of Science at Tokyo Imperial University, but nevertheless remained busy at the Nishina
lab. From August 1935 to November 1936, Sagane conducted research at Ernest
Lawrence's Radiation Laboratory, after which he visited the Cavendish Laboratory, Cambridge and other research institutes in Europe. In March 1936, during
his absence, Sagane was promoted to Associate Professor at Tokyo.

After visiting Europe, Sagane returned to Berkeley in June 1937, staying there
until February 1938.[69] The worsening relationship between the US and Japan
meant that he had little alternative but to return to Japan. From January 1939, he
was employed part-time at Riken, receiving his D. Sc. in December 1939 with a
dissertation entitled "On Artificial Radioactivity".[70] In May 1942, he was made

[65] Marshall, Academic Freedom, pp. 39–41.
[66] Kodansha, p. 207; Itô et al, Historical Dictionary, p. 245.
[67] Aizu, Recollections, pp. 395–398.
[68] Translation of extract of Census Register, Ômura city, Nagasaki prefecture, 17 September
1949; National Diet Library, Tokyo (NDL), GHQ/SCAP records, ESS (B) 11649, "Travel Abroad
of Dr. Sagane" file, p. 9.
[69] Background resumé of Ryôkichi Sagane, NDL, GHQ/SCAP records, ESS (B) 11649.
[70] Fushimi, Six Aspects, p. 121.

a full research member of Riken. This acknowledged Sagane's central role in nuclear physics experiments and in the construction of the prewar cyclotron there. In March 1943 he was promoted to Professor at Tokyo Imperial University.

Sagane was one of a number of outstanding physicists such as Hideki Yukawa, Tomonaga, Kikuchi, Satio Hayakawa and Yôichirô Nambu who went to the US after the war. In December 1949, Sagane was given permission to leave the country to accept a Visiting Professorship in the Department of Physics, Iowa State College, from 1 January to 1 July 1950.[71] On his way there, he stopped at Berkeley. Lawrence drew on a "small private fund" to support visits by Sagane and Kikuchi. Sagane extended his period overseas until 30 April 1955. On 28 February 1955, Sagane resigned from Tokyo University because of his protracted stay in the US. Out of a job, the prospect of the development of atomic energy in Japan suggested exciting work opportunities.

On 16 March 1954, the Yomiuri shinbun newspaper reported that twenty-three Japanese on a tuna fishing boat called "The Lucky Dragon" had been contaminated by radioactive fallout near the Bikini Atoll during hydrogen bomb testing about two weeks earlier. This had inauspiciously occurred on almost the same day that it was decided to present Japan's first reactor budget before parliament so the tensions between those fearful of atomic energy and those keen to embrace it were apparent from early on. On 9 June, the Federation of Economic Organizations nevertheless made a submission to the government urging the establishment of a science and technology agency, for the promotion of science and technology, and to help formulate and administer government policy, especially for nuclear power.

The Japanese rush to establish an atomic energy program was partly a result of a desire to take full advantage of the possibilities of transfer of nuclear technology from the US under the Atoms for Peace plan. In 1954 and 1955, there was considerable support and interest in atomic energy across the political spectrum by politicians, bureaucrats, business leaders and academics. In May 1954, Cabinet established the Preparatory Council for the Peaceful Uses of Atomic Energy. It was hoped that atomic energy would eventually provide the Japanese with an independent energy source. Scientists argued the need to conduct basic research in atomic energy. The concept of energy "self-sufficiency" was used by many proponents of nuclear power.[72]

Sagane and Kikuchi facilitated the growing technological ties between the US and Japan. Americans had been eager to include Japan in the Atoms for Peace nuclear power program. On 14 November 1955, the US-Japan Atomic Energy Agreement was signed, only a little more than a decade after the atomic bomb

---

[71] Letter G.W. Fox to H.C. Kelly, 27 October 1949, NDL, GHQ/SCAP records, ESS (B) 11649.

[72] Hein, Energy.

was dropped on Japan and three years after the end of the Occupation.[73] The agreement enabled Japan to buy nuclear fuel and purchase nuclear technology.[74]

In the wake of defeat, the old military elite was discredited. Scientists, engineers, and economists were seen as being the only ones to really offer "viable strategies for starting over".[75] Kikuchi and Sagane were among the small number of physicists who were part of Japan's new technocratic elite. With their hands-on knowledge of experimental physics, overseas study experience, and family connections, both Kikuchi and Sagane had a special authority. Both were relatively apolitical and willing to work with the government, shaping science policy for the public good. Both were sons of university presidents. There was a sense of public duty as physicists.

In 1950, Kikuchi published a short history of fifty years of nuclear physics in the Japanese science magazine Kagaku (Science).[76] The most recent concerns included new theory, nuclear theory, cosmic rays, mesons and elementary particles, nuclear reactions, radioactivity and the uses of atomic energy. For the latter, there was a direct line drawn to the possible development of nuclear reactors, and studies of the atomic bomb and nuclear fission.

There was a yawning gap between the concerns of those wishing to promote pure science and those involved in the development of nuclear power. While Sagane's involvement in the push to promote nuclear power precedes Kikuchi's participation, Kikuchi did eventually become a director of the Japan Atomic Energy Research Institute. Both were, importantly, experimental physicists and it is their hands-on experience, which sets them apart from theoretical physicists such as Hideki Yukawa, Sin-itirô Tomonaga, Shôichi Sakata, and Mituo Taketani.

Sagane and Kikuchi's involvement in public science (as opposed to academic science) was not a case of second-rate or "bad" physicists choosing to work with the government and private enterprise, but more a situation in which the postwar academic environment made it difficult to choose this as an option. Crossing boundaries required hard work and special, interpersonal skills. Both Kikuchi and Sagane took leadership positions by heading various institutes for nuclear research and the development of nuclear power. For example, Kikuchi was the inaugural director of the Institute for Nuclear Study at Tokyo University. Sagane returned from the United States to spearhead the establishment of civilian nuclear power infrastructure. Minutes of the various committees they were subsequently involved in reveal technocratic cooperation between bureaucrats, scientists, and business leaders and how these different groups were able to arrive at a type of consensus about the direction of science and technology in postwar Japan.

How did Kikuchi and Sagane seek to influence others? How did they assert their authority and negotiate the political and scientific terrain before them? While it is difficult to provide answers, we can say that the deliberations on record

---

[73] Levey, Nuclear Reactor, cited in Hein, Energy, pp. 470–471, 490.
[74] Hein, Energy, pp. 470–471.
[75] Hein, Reasonable Men, p. 8.
[76] Kikuchi, Recollections, p. 109.

are testimony to how decision making was effectively removed from the realm of politics and transplanted to committees where technical specialists could be consulted and a type of consensus achieved.

COMMITTEE FOR SCIENCE, EDUCATION AND CULTURE

Tomonaga and Kikuchi testified as experts at the 16 June 1955 meeting of the Committee for Science, Education and Culture where they were careful to distance the proposed Institute for Nuclear Study from research into nuclear power, military-related research and links with private enterprise. At the committee meeting, Tomonaga highlighted the need for Japan to make progress in experimental physics. He pointed out the difficulty of sometimes distinguishing between fundamental studies in nuclear physics and nuclear power-related research, but given the intensity of feelings of his fellow scientists and members of the public, it was felt that there was a need to be seen as separating the two and for the Institute for Nuclear Study to pursue pure physics and investigate the depths of the structure of matter. The best way to do this was through a joint-use laboratory attached to a university, Tokyo University, the facilities of which would be open to researchers throughout the country.[77]

A committee member Nahoko Takada echoed the concerns of townsfolk when he discussed a scenario whereby researchers might be forced to turn their efforts towards atomic energy rather than fundamental studies in nuclear physics. Takada pointed out the possible production of plutonium through reactors and how atomic energy research could give birth to a nuclear weapons program. He wondered how scientists would be able to stop this from happening. He noted how under the US-led Marshall Plan (1947–51) to rebuild Europe after World War II, scientists who had pursued pure research had become "goyôgakusha" or "servants of the state", at the beck and call of the government, at a time of growing Cold War tensions. Tomonaga responded that it was for such reasons, to avoid external influence, that it had been decided to establish the Institute for Nuclear Study as an institute attached to a university, the University of Tokyo.

One of the committee directors, Manji Yoshida, wondered whether collaborative research would discourage in-depth, Nobel Prize-winning research from being carried out, given the need to share facilities and spread resources around. This was a very perceptive question and comment. Tomonaga responded by pointing out that collaborative research had become a common feature of research, and that times had changed.

[77] Japanese Government, House of Councilors, Minutes of the Bunkyô Kagaku Iinkai (The Committee for Science, Education and Culture) meeting, 16 June 1955, NDL.

SPECIAL COMMITTEE FOR POLICIES FOR THE PROMOTION
OF SCIENCE AND TECHNOLOGY

At the 11 April 1956 meeting of the Special Committee for Policies for the Promotion of Science and Technology, Sagane spoke with authority, referring to the American physicist Ernest Lawrence throughout his testimony, as if to shore up his academic credibility. However, despite these mentions of Lawrence, the discussion largely revolved around how to most quickly develop nuclear power in Japan. This was in stark contrast with the discussions of the meeting of the Committee for Science, Education and Culture on 16 June 1955, where any possible connection between such ambitions and the aims of the Institute for Nuclear Study were denied.

Sagane pointed out at the meeting of the Special Committee that Japan was imitating other countries in terms of policies for atomic energy, however for almost all of those countries, atomic energy had a role in national defense. In contrast, Japan had a declared position of atomic energy for peaceful uses only.[78]

Another problem that Sagane referred to was the question of privatization and when the responsibility for nuclear power could be shifted to private enterprise. In connection with this, Sagane was very worried about the lack of Japanese patent activity. Japan would soon be introducing nuclear technology on a large scale, much of which had been patented by American companies. And Japan would have to learn as it went. Sagane likened policymaking in the nuclear age to a game of rugby. There was bound to be some futile running around. There were no shortcuts, and one had to learn by trying.[79]

Less then two years later, Kikuchi spoke to the same Committee in his capacity as director of the Institute for Nuclear Study, University of Tokyo. He argued for the importance of basic research. Even in the case of nuclear reactors, it was not completely the domain of engineers. There were still many areas where basic research could contribute, and universities were where most basic research was conducted. Despite this, the funding of basic research in Japan was embarrassingly poor. This was in contrast to the large budget allocation for atomic energy. Unfortunately, there was next to no pathway for some of this largesse to flow to university laboratories. Kikuchi expressed his frustration at the situation and called for reforms, especially given the interest in the development of nuclear fusion in which universities could play a part.[80]

The gap in concerns and conflicting views about matters such as public-private cooperation in research, and the need for separation of curiosity-driven research from research associated with the practical problems posed by the de-

[78] Japanese Government, House of Representatives, Minutes of the *Kagaku Gijutsu Shinkô Taisaku Tokubetsu Iinkai* (Special Committee for Policies for the Promotion of Science and Technology) meeting, 11 April 1956, NDL.

[79] Ibid.

[80] Japanese Government, House of Representatives, Minutes of the *Kagaku Gijutsu Shinkô Taisaku Tokubetsu Iinkai* (Special Committee for Policies for the Promotion of Science and Technology) meeting, 19 February 1958, NDL.

velopment of nuclear power – created a longstanding schism among Japanese scientists which is apparent from reading the written records of committees. For better or for worse, the theoretical physicist Tomonaga represented the interests of academic science. The experimental physicists Sagane and Kikuchi were more accommodating to the concerns of government and took up positions in public science.

In 1958, Kikuchi became a part-time member of the Japan Atomic Energy Commission that had been formed a couple of years earlier. His role was to assist in deliberations relating to nuclear fusion. He went on to become Chairman of the Directors of the Japan Atomic Energy Research Institute in September 1959 upon which time he resigned as Director of the Institute for Nuclear Study. By doing so, he moved loyalties and affiliation from the Ministry of Education to the Science and Technology Agency (STA), which had been created in May 1956. This signaled a switch from academic science to the public science of the STA, which oversaw the development of nuclear power and also aerospace technology. Despite Kikuchi's embrace of public science, he maintained his ties with academia, including his friendship with Heisenberg who visited Japan in 1967. By this time, Japan had rebuilt, played host to the 1964 Olympics, and enjoyed high economic growth. As we saw in the first half of this paper, Japan was indeed catching up, and Heisenberg was no doubt impressed.

CONCLUSION

In this paper, we traced the emergence of physicists such as Nagaoka and Nishina as policymakers who took advantage of what funding opportunities existed before and during World War II. With Nishina's death and Nagaoka's retirement from active involvement in science after the war, a national system of innovation emerged that was shaped by actors such as Kikuchi, Sagane, and Tomonaga, all of whom had links to Nagaoka and Nishina. In space research and the development of nuclear power, responsibility has been shared by different government agencies with different responsibilities. The influence of the Science Council of Japan was limited, but the physicists discussed in this paper cooperated with government. What is fascinating is that the language of physicists can change, depending upon the context. Physicists had differing perceptions of what constituted a culture of innovation. For some, it meant self-sufficiency through borrowed technology. For others it was by developing a domestic science infrastructure in which Japan could nurture its own Nobel Prize winners such as Tomonaga.

Japan had limited financial resources after the war and it is not surprising that for some physicists, it was a question of choosing academic research conducted at joint-use research institutes such as the Institute for Nuclear Study. Such institutes provided an acceptable way for some scientists to receive government funding and at the same time maintain their autonomy. Others such as Sagane and later Kikuchi turned to government-led projects such as the development of nuclear power in which the private sector had a major stake. Given that the private funding of R & D is so dominant in Japan, it may appear odd that physicists were reluctant to establish ties. But we need to remember the context of postwar science. Still in the shadow of a disastrous war and the atomic bomb itself, many scientists felt compelled to argue for a pure science, untainted by ties with the military or industry that had been co-opted during the war. In this regard, Japanese physicists had much in common with their colleagues in Germany.

BIBLIOGRAPHY

Aizu, Akira: Hito gakusei no omoide (The Recollections of One Student), in: Publication Committee (ed.): Sagane Ryôkichi kinen bunshû (Collection of Writings to Commemorate Ryôkichi Sagane), Tokyo 1981, pp. 395–398.

Alletshauser, Al: The House of Nomura, London 1990.

Anon.: Biography [Sin-itirô Tomonaga], in: Nobel Lectures, Physics: 1963–1970, Amsterdam 1972, pp. 137–139.

Anon.: Japan's Hidden Physicists: Men Who Will Follow in Yukawa's Footsteps, in: Nippon Times, 22 November 1949, NDL, GHQ./SCAP records, ESS (E) 06365.

Anon.: Translation of extract of Census Register, Ômura city, Nagasaki prefecture, 17 September 1949, NDL, GHQ/SCAP records, ESS (B) 11649, Travel Abroad of Dr. Sagane file, p. 9.

Anon.: Nihon no butsurigaku (Japanese Physics), in: Shizen, 18 Jan. 1963, cited in History of Science Society, 13: Physics, p. 418.

Asonuma, Akihiro: Sengo kokuritsu daigaku ni okeru kenkyûhi hojo (The Funding of Research in Postwar National Universities), Tokyo 2003.

Beyler, Richard H. and Morris F. Low: Science Policy in Post-1945 West Germany and Japan: Between Ideology and Economics, in: Walker, Mark (ed.): Science and Ideology: A Comparative History, London 2003, pp. 97–123.

Beyler, Richard H.: Maintaining Discipline in the Kaiser Wilhelm Society during the National Socialist Regime, in: Minerva 44, 2006, pp. 251–266.

Cahan, David: An Institute for an Empire: The Physikalisch-Technische Reichsanstalt, 1871–1918, Cambridge 1989.

Carson, Cathryn and Michael Gubser: Science Advising and Science Policy in Post-War West Germany: The Example of the Deutscher Forschungsrat, in: Minerva 40, 2002, pp. 149–179.

Coleman, Samuel K.: Riken from 1945 to 1948: The Reorganization of Japan's Physical and Chemical Research Institute under the American Occupation, in: Technology and Culture 31 (2), 1990, pp. 228–250.

Coleman, Samuel: Japanese Science: From the Inside, London 1999.

Cummings, William K.: The Culture of Effective Science: Japan and the United States, in: Minerva 28 (4), 1990, pp. 426–445.

Dower, John W.: Science, Society, and the Japanese Atomic-Bomb Project During World War Two, in: Bulletin of Concerned Asian Scholars 10 (2), 1978, pp. 41–54.

Dower, John W.: Japan in War and Peace: Essays on History, Culture and Race, London 1995.

Dower, John W.: Embracing Defeat: Japan in the Wake of World War II, New York 1999.

Fox, G.W.: Letter to H.C. Kelly, 27 October 1949, NDL, GHQ/SCAP records, ESS (B) 11649.

Fushimi, Kôji: Kikuchi Seishi: Denshi no hadôsei o kakuninshita kagakusha (S. Kikuchi: The Scientist Who Confirmed the Wave-Like Nature of Electrons), in: Kagaku no jikken (Science Experiments) 1979. Reprinted in Fushimi, Kôji: Watashi no kenkyû henreki (My Life in Research, Tokyo 1988, pp. 197–209).

Fushimi, Kôji: Sagane Ryôkichi sensei no muttsu no katsudô sô (Six Aspects of the Activities of Prof. Ryôkichi Sagane), in: Publication Committee, Ryôkichi Sagane, pp. 121–124.

General Headquarters, Supreme Commander of the Allied Powers: Japanese Natural Resources: A Comprehensive Survey, Tokyo 1949.

Grunden, Walter E., Yutaka Kawamura, Eduard Kolchinsky, Helmut Maier and Masakatsu Yamazaki: Laying the Foundation for Wartime Research: A Comparative Overview of Science Mobilization in National Socialist Germany, Japan, and the Soviet Union, in: Osiris 20, 2005, pp. 79–106.

Hayakawa, Satio: Kenkyû shinkô ni tsutometa Tomonaga sensei (Prof. Tomonaga: A Man Who Strived for the Advancement of Research), in: Kagaku (Science) 49 (12), 1979, pp. 799–803.

Hein, Laura Elizabeth: Energy and Economic Policy in Postwar Japan, unpublished Ph. D. dissertation, University of Wisconsin, Madison 1986.

Hein, Laura Elisabeth: Reasonable Men, Powerful Words: Political Culture and Expertise in Twentieth-Century Japan, Washington, D.C., Berkeley 2004.

Hicks, Diana: University-Industry Research Links in Japan, in: Policy Sciences 26, 1993, pp. 361–395.

Hirosige, Tetu: Kagaku no shakaishi: Kindai Nihon no kagaku taisei (The Social History of Science: The Organization of Science in Modern Japan), Tokyo 1973.

Hirosige, Tetu: Kagaku to rekishi (Science and History), Tokyo 1965.

Hirosige, Tetu: Kindai kagaku saikô (Reevaluating Modern Science), Tokyo 1979.

History of Science Society of Japan (ed.): Nihon kagaku gijutsushi taikei, 13: butsurigaku (An Outline of the History of Science and Technology in Japan, Vol. 13: Physics), Tokyo 1970.

Hoddeson, Lillian: Establishing K.E.K. in Japan and Fermilab in the U.S.: Internationalism, Nationalism and High Energy Accelerators, in: Social Studies of Science 13, 1983, pp. 1-48.

Inoguchi, Takashi and Lyn Jackson (eds.): Memories of War: The Second World War and Japanese Historical Memory in Comparative Perspective, Tokyo 1998.

Institute for Nuclear Study, Tokyo University: Kakken nijû nen shi (The Twenty Year History of the Institute for Nuclear Study), Tanashi, Tokyo 1978.

Irvine, John, Ben R. Martin, and Phoebe Isard: Investing in the Future: An International Comparison of Government Funding of Academic and Related Research, Aldershot 1990.

Itakura, Kiyonobu, Tôsaku Kimura, and Eri Yagi: Nagaoka Hantarô den (A Biography of Hantarô Nagaoka), Tokyo 1973.

Ito, Kenji: Values of "Pure Science": Nishina Yoshio's Wartime Discourse between Nationalism and Physics, 1940–1945, in: Historical Studies in the Physical and Biological Sciences 33 (1), 2002, pp. 61–86.

Itô, Shuntarô et al (eds.): Kagakushi gijutsushi jiten (Historical Dictionary of Science and Technology), Tokyo 1983.

Japanese Government, House of Councilors: Minutes of the Bunkyô Kagaku Iinkai (The Committee for Science, Education and Culture) meeting, 16 June 1955, NDL.

Japanese Government, House of Representatives: Minutes of the Kagaku Gijutsu Shinkô Taisaku Tokubetsu Iinkai (Special Committee for Policies for the Promotion of Science and Technology) meeting, 11 April 1956, 19 February 1958, NDL.

Japanese Government, Ministry of Education, Culture, Sports, Science and Technology: White Paper on Science and Technology 2007: Results of Promotion of Science and Technology – Creation, Utilization and Succession of Knowledge, Tokyo 2007.

Kamatani, Chikayoshi: The History of Research Organization in Japan, in: Japanese Studies in the History of Science (2), 1963, pp. 1–79.

Kikuchi, Seishi: Genshikaku butsuri 50 nen no kaiko (Recollections of 50 Years of Nuclear Physics), in: Kagaku (Science) 20 (3), 1950, pp. 107–112.

Kleint, Christian and Gerald Wiemers (eds.): Werner Heisenberg in Leipzig, 1927–1942, Berlin 1993.

Kodansha Encyclopedia of Japan 4, Tokyo 1983.

Koyama, Noboru: Japanese Students at Cambridge University in the Meiji Era, 1868–1912: Pioneers for the Modernization of Japan, transl. by Ian Ruxton, Morrisville 2004.

Levey, Stanley: Nuclear Reactor Urged for Japan, in: New York Times, 22 September 1954, p. 14.

Kurihara, Kenneth K.: Japan's Educational System, in: Far Eastern Survey 13 (4), 1944, pp. 35–38.

Marshall, Byron K.: Academic Freedom and the Japanese Imperial University, 1868–1939, Berkeley 1992.

Matsui, Makinosuke (ed.): Kaisô no Tomonaga Sin-Itirô Tomonaga (Reminiscences of Sin-itirô), Tokyo 1980.

Moore, Ray A.: Adoption and Samurai Mobility in Tokugawa Japan, in: The Journal of Asian Studies 29 (3), 1970, pp. 617–632.

Nagaoka, Hantaro: Kinetics of a System of Particles illustrating the Line and the Band Spectrum and the Phenomena of Radioactivity, in: Philosophical Magazine Sixth Series 7, 1904, pp. 445–455.

Nakayama, Shigeru: Science, Technology, and Society in Postwar Japan, London 1991.

Normile, Dennis: Japan: Superagency Seeks to Reconcile Two Cultures, in: Science 291 (5501), 2001, p. 29.

Oda, Minoru: Bannen no Tomonaga Sin-Itirô sensei (Sin-Itirô Tomonaga during His Later Years), in: Itô, Daisuke (ed.): Tsuisô Tomonaga Sin-Itirô (Reminiscences of Sin-Itirô Tomonaga), Tokyo 1981, pp. 99–115.

Riken, Stories from Riken's 88 Years (13), online at www.riken.go.jp/r-world/info/release/riken88/text/image/pdf/no13e.pdf (accessed on 22 January 2007).

Sagane, Ryôkichi: Background resumé, NDL, GHQ/SCAP records, ESS (B) 11649.

Sime, Ruth Lewin: The Politics of Memory: Otto Hahn and the Third Reich, in: Physics in Perspective 8, 2006, pp. 3–51.

Tomonaga, Sin-Itirô: Personal History Statement, NDL, GHQ/SCAP records, ESS (B) 11647.

Shillony, Ben-Ami: Universities and Students in Wartime Japan, in: The Journal of Asian Studies 45 (4), 1986, pp. 769–787.

Shimao, Eikoh: Some Aspects of Japanese Science, 1868-1945, in: Annals of Science 46, 1989, pp. 69-91.

Walker, Mark: Otto Hahn: Responsibility and Repression, in: Physics in Perspective 8, 2006, pp. 116–163.

Yamamoto, Yôichi: Nihon genbaku no shinsô (The Truth about Japan's Atomic Bomb), in: Daihôrin 20 (8), 1953, pp. 6–40.

Yamazaki, Masakatsu: The Mobilization of Science and Technology during the Second World War in Japan: A Historical Study of the Activities of the Technology Board Based Upon the Files of Tadashiro Inoue, in: Historia Scientiarum 5 (2), 1995, pp. 167–181.

Yomiuri Shinbun, Shôwashi no tennô 4 (The Emperor and Showa History 4), Tokyo 1968.

PHYSICS IN CHINA IN THE CONTEXT OF THE COLD WAR, 1949–1976

Zuoyue Wang

In April 1952, just days after he was appointed the associate director of the Institute of Modern Physics of the Chinese Academy of Sciences in Beijing, the Chinese nuclear physicist Wang Ganchang was called into the headquarters of the academy. A secret mission awaited him at the battle front of the Korean War: The Chinese forces suspected the US had used "atomic shells" and wanted Wang to investigate the matter. Wang, a physicist who had gone to Berlin University to study with Hans Geiger but ended up receiving his Ph.D. with Lise Meitner at the Kaiser Wilhelm Institute in 1933, was well-qualified for the job. With a primitive but effective Geiger counter, Wang checked fragments of the suspect shells and found that there was no increase in radioactivity. He concluded that they were not mini atomic bombs, but perhaps a new type of conventional weapon.[1] A relatively minor incident in the Korean War, it nevertheless marked a milestone toward the beginning of an era when the Cold War and nuclear weapons increasingly and decisively shaped the context within which physics was practiced in the People's Republic of China under Mao Zedong. A decade later, Wang emerged as a major architect of the Chinese nuclear weapons project and his whole institute and most of the leading Chinese physicists devoted themselves to it.

In recent decades historians of science have become increasingly interested in the interactions of science and state during the Cold War. In his influential paper on quantum electronics Paul Forman, for example, has asked us to examine to what extent the needs of the national security state drove the development of American physics. Other scholars have since painted a picture of the science/state relationship where American scientists embraced the national security state, with its objective of containing Soviet expansionism, much more readily than previously thought. All, however, seem to agree that the Cold War transformed American science, especially American physics. While this debate has generated fruitful discussions on the interactions between science and state during the Cold War in the US and, to a lesser extent, in Europe and the Soviet Union, little has been known about the experiences of Chinese scientists.[2]

What drove the dynamics of the relationship between scientists and the state in the Chinese context during the Cold War? To what extent was the science-state

---

[1] Li et al., He wulixuejia, pp. 108–112. Wang later recognized that if it had been an atomic bomb, there would not have been fragments left.

[2] Forman, Behind Quantum Electronics; Kevles, Cold War; Forman/Sánchez-Ron, National Military Establishments.

interaction mediated by the particular political and social environment of Maoist China? In this paper, I explore the forces and circumstances, especially their sense of nationalism, that led Chinese physicists and other scientists to cast their lot with the Communists, to survive the harsh political and ideological purges, and eventually to participate in the nuclear weapons program. As in Stalinist Russia, Mao and other Communist leaders needed to soften their mistrust of western-trained scientists for their service to the Chinese national security state. The Cold War, especially the pursuit of nuclear weapons and the accompanying political confrontations between China, the Soviet Union, and the United States, pro-foundly affected the funding and direction of Chinese science, especially phys-ics. Ultimately, the experiences of Wang Ganchang and other Chinese physicists indicate that a complex web of interactions that was built on Chinese national-ism, scientific professional autonomy, Communist ideology, and the geopolitical strategies of the Chinese party-state shaped the social/political environment for the practice of physics during the Cold War.

WAITING FOR MAO

In order to understand the experience of Chinese physicists during the Cold War, it is necessary to examine the background of modern science in China. Perhaps more than their counterparts in other countries, a sense of scientific nationalism motivated most of the first generation of modern Chinese scientists to pursue sci-ence and technology in the early twentieth century. Their aim was not so much to make China into a leader in world science as to use science to help make China a prosperous and strong country, or "saving China through science." Born in the late 19th and early 20th century, many of these first generation of modern Chinese scientists were trained in the West, especially in the US, in the 1910s and 1920s, and went back to China to establish its earliest educational, research, and industrial institutions in science.

Organizationally, the Chinese scientific community cohered around two main groups during the Republican period (1911–1949): the older and privately-run Science Society of China and the newer, official Academia Sinica. Prominent Chinese physicists played an active role in both institutions. Founded by Chinese students at Cornell University in 1914, the Science Society can be seen in many ways as an attempt by Chinese scientists to build a civil society and public sphere, especially through their publication of the influential Kexue (science) magazine, in China. Together the society and the journal provided Chinese scientists with a forum where they could both critique the government's public policy and, as a departure from civil society institutions in the West in general but in some ways analogous to the German Research Foundation (Deutsche Forschungsgemein-schaft, DFG), collaborate with the government in furthering a form of scientific nationalism that sought to strengthen Chinese national development through science and technology. Receiving funding from private donations, local govern-ments, and the semi-public China Foundation for the Promotion of Education and Culture, the society in turn supported research at its own institutes as well as that carried out by scientists elsewhere. In contrast, the Academia Sinica was established by the new Nationalist government under Jiang Jieshi (Chiang Kai-shiek) in 1927 as an official entity to help unify the country politically as well as scientifically. As a step forward in the struggle for Chinese scientific nationalism, the Academia, modeled after the Soviet and French academies, won overwhelming support of the leaders of the Science Society of China. Indeed, many of them became directors of institutes in the new academy.[3]

Given such close ties between the scientific leaders and the Nationalists and the West, it was striking that few of them moved with the US-backed government to Taiwan when it lost the civil war to the Communists in 1949, one of the signal events of the Cold War. The estrangement between the scientists and the Nationalist regime derived largely from the latter's reputation for corruption, but also partly from a resentment over a perceived American and European indifference toward Japanese aggression in China before the Pearl Harbor attack in 1941. For example, Zhu Kezhen, a Harvard-trained meteorologist who was a leader in both Academia Sinica and the Science Society and, in 1936, hand-picked by Jiang as president of Zhejiang University, grew critical of the Nationalist and US policy in the late 1930s. In 1938, while leading the university on a perilous journey inland to evade the advancing Japanese army, which eventually claimed the lives of his wife and a son, Zhu indignantly wrote in his diary that "the Americans, the British, and the French were all helping the [Japanese] invaders [by refusing to cut off its oil supplies]."[4] Two years later, he added angrily that the Japanese used newly-designed airplanes purchased from the US to shoot down Chinese airplanes.[5] Another entry, in 1944, expressed his dissatisfaction with the Nationalists:

> I am not opposed to the Nationalists per se, but was extremely reluctant toward joining the party. As to the various behaviors of the Nationalists, I view them with complete disgust and abhorrence. Recently, I have become even more horrified and outraged by individuals within the party who became followers of the German Nazis ... If we are forbidden from criticizing those embezzling officials such as Kong [one of the Jiang's brothers-in-law], where is the freedom of speech?[6]

Wang Ganchang, likewise, quickly became disillusioned with the Nationalists when he returned to China from Germany and was recruited by Zhu Kezhen to Zhejiang University. He sympathized with his students, such as Xu Liangying, who had joined the underground Communist Party in opposition to the Nationalist policy. He openly approved, for example, of Xu's couplet "Science Is Supreme; Physics Comes First" on his (Wang's) make-shift laboratory, which was a direct challenge to Jiang Jieshi's edict that "The State is Supreme; The Military Comes

---

[3] Wang, Saving China.

[4] Zhu Kezhen diary for April 29, 1938, in: Zhu, Zhu Kezhen riji, vol. 1, p. 227.

[5] Zhu Kezhen diary for October 18, 1940, ibid., p. 461. Two years later, his wife and one of his sons died from disease contracted on the road while escaping from the Japanese army.

[6] Zhu Kezhen diary for July 13, 1944, in Zhu, Zhu Kezhen riji, vol. 2, p. 768.

First." [7] Qian Sanqiang, another leading Chinese physicist who was trained in the labs of Irene and Frédéric Joliet-Curie in Paris in the 1940s, returned to China in 1948 with the hope of setting up a government-sponsored nuclear research center in Beijing under his leadership. Unbeknownst to Qian, however, intervention from the US embassy led to a cancellation of the center.[8] The American action likely derived from its Cold War concerns: Beijing was on the verge of being overtaken by the Communists and Qian had been closely associated with Frédéric Joliet-Curie, one of the best known French Communists. The apparent sudden change of plans left Qian frustrated; he became disappointed with what he interpreted as a sign of the Nationalists' lack of interest in science. Thus when the government sent for him to retreat to Taiwan, he declined and instead waited for the Communists to take over the city.[9]

Despair with the Nationalist and a hope, though an uneasy one, for the Communists made Qian, Wang, Zhu, and most other scientists decide to ride out the coming storm by remaining on the mainland. As Zhu told Wu Youxun, a physicist trained in the US with Arthur Compton in the 1920s and another leader of the Science Society:

> when the Nationalists launched the Northern Expedition [to defeat the warlords in 1927], the people rejoiced as much as they do today. But the Nationalists did not capitalize on the opportunity; they instead covered up embezzlements, failed to adhere to clear rules of rewards and punishments, and ended up being overthrown today. The people have welcomed the Liberation Army as they do clouds amidst a severe draught. [I] hope that [the Communists] can work hard to the end and do not turn out to be as corrupt as the Nationalists. Science is extremely important to construction, and [I] hope the Communists could pay close attention to it.[10]

Despite their unease, most Chinese scientists seemed genuinely impressed by the strong support the new government provided science. Shortly after the transfer of power, Zhu, Wu, Wang, Qian, and about 200 other prominent scientists in the country were invited by the new government to participate in the making of a new science policy in Beijing.[11] While in Beijing, Zhu took part in the making of a provisional Constitution for China and was apparently responsible for the inclusion of Article No. 43: "[The government] will strive to develop natural sciences, [make them] serve industrial, agricultural, and defense construction, reward scientific discoveries and inventions, and popularize scientific knowledge."[12] On October 1, 1949, Zhu climbed on top the Tiananmen gate to witness Mao Zedong's declaration of the founding of the People's Republic of China. Two weeks later,

---

[7] Xu, Enshi Wang Ganchang. Xu himself later became a translator of Albert Einstein's works and an outspoken human rights advocate in China. See Overbye, Einstein's Man.

[8] Ge, Qian Sanqiang he zaoqi.

[9] Ren, Bochunzhe, pp. 44–45.

[10] Zhu Kezhen diary entries for May 25–27, 1949, in: Zhu, Zhu Kezhen riji, vol. 2, pp. 1255–1256.

[11] Zhu Kezhen diary entry for June 24, 1949, in: Zhu, Zhu Kezhen riji, vol. 2, p. 1265.

[12] Xie, Zhu, p. 537. Zhu Kezhen diary entry for September 1, 1949, in: Zhu, Zhu Kezhen riji, vol. 2, p. 1285.

he was appointed vice-president of the new Chinese Academy of Sciences, as was Wu a year later.[13] The Academy established a new Institute of Modern Physics on the basis of two older physics institutes, one under the Academia Sinica in Shanghai and the other under the Peking Academy in Beijing. Qian was appointed the institute's director and Wang a division leader (later associate director).

Ultimately, most Chinese scientists' enthusiastically received the Chinese Communists because they believed that the Communists shared their nationalist drive, and that a stable government would support the development of science and reconstruction of the country. After all, Mao had declared poetically on June 15, 1949, that:

> China must be independent, China must be liberated, China's affairs must be decided and run by the Chinese people themselves, and no further interference, not even the slightest, will be tolerated from any imperialist country ... The Chinese people will see that, once China's destiny is in the hands of the people, China, like the sun rising in the east, will illuminate every corner of the land with a brilliant flame, swiftly clean up the mire left by the reactionary government, heal the wounds of war and build a new, powerful and prosperous people's republic worthy of the name.[14]

Identifying with these announced nationalist goals of the Communists, Chinese scientists enthusiastically supported the government's decision to establish the Chinese Academy of Sciences (CAS) as the new center of national scientific research. Like the Academia Sinica it replaced, the CAS was a state-sponsored scientific institution, which marked a continuity between the Nationalist and the Communist eras and helped explain the ease with which Chinese scientists embraced the new academy. To many of them, a strong central government willing to both use and support science formed a key part of their nationalist program for a modern China. Their initial, positive interactions with the Communists affirmed their decision to give the benefit of the doubt to the new regime, with hope for a better and more efficient party state to set China on the road to modernization through the application of science and technology. Thus, like their counterparts in two earlier revolutions – the French Revolution and the Russian Bolshevik revolution – they not only made accommodations to the new regime, but actually welcomed it with substantial, if not complete, enthusiasm.

MAO ON SCIENCE AND SCIENTISTS

For their part, Mao and other Chinese leaders both shared and differed from the scientists' view on the role of scientists and science in the Communist party state. On the one hand, Mao agreed with the scientists that natural science was crucial for economic development in general, and for industrialization in particular. On the other hand, Mao, a firm believer in class struggle as a driving force in his-

[13] Xie, Zhu, pp. 539–541.
[14] Mao, Address to the Preparatory Meeting.

tory, regarded scientists, especially those trained in the West, as belonging to the reactionary bourgeois or, not much better, petty bourgeois class. Therefore, in his view, bourgeois scientists were to be used, but not completely trusted politically until they were thoroughly reformed ideologically. Reinforcing his ambivalent attitude toward the scientists was an inherent tension in his attempt to combine and reconcile two potentially conflicting paradigms of Chinese nationalism: revolution and modernization. Was revolution a means toward the end of modernization or vice versa? In many ways, Mao's attitudes toward scientists, which in turn helped shape national science policy, often depended on the manner he tried to solve the tension of the revolution-modernization matrix.

Mao's essentially negative political assessment of Chinese scientists also paralleled his decision to seek and rely on Soviet political, strategic, and technical support during the Cold War. Mao and Josef Stalin had not enjoyed a close working relationship during the Chinese revolution. Even after the Chinese Communist victory, Stalin suspected that Mao and his colleagues might be tempted to align with the United States. "This problem was only solved when we began to fight the Americans in the Korean War," according to Zhou Enlai, a leader of the Chinese Communist Party and premier of China from 1949 to 1976.[15] Yet, despite such clashes, there was little doubt in Mao's mind that he needed Soviet nuclear protection in order to secure the revolution at home and to deter a possible American military intervention. In June 1949 he announced a foreign policy of "leaning toward one side," signaling that China would form a political alliance with the Soviet Union. In late 1949 and early 1950, Mao made a humble pilgrimage to Moscow, during which he asked Stalin for a formal friendship treaty between the two countries. When Stalin initially demurred, citing his Yalta agreement with the Nationalists and the US, Mao stubbornly persisted in his request. Finally, new developments, especially a softening of Western attitude toward China, apparently led Stalin to relent. The new treaty was signed, along with agreements on financial and technical assistance to China.[16]

The outbreak of the Korean War in the summer of 1950 and Mao's decision for China to enter the war later that year dramatically increased China's strategic importance to Stalin. They also led to a large increase in Soviet military-technical aid to China. Indeed Mao's decision to enter the war in the first place may have been influenced by the promise of this assistance.[17] In 1950–1951, the two countries signed agreements to set up four joint companies in the areas of civil aviation, petroleum, metallurgy, and shipbuilding. By the time Stalin had died in 1953, the Soviets had promised to help China build 141 major technological projects in military and civilian industries.[18] The massive scale of Soviet technological transfer was unprecedented in history and greatly accelerated the pace of

---

[15] Chinese Communist Party Central Documentation Research Office, Zhou Enlai nianpu, vol. 1, pp. 620–621.

[16] Jiang/Jin, Mao Zedong zhuan, vol. 1, pp. 28–58; Zhang et al., Sulian jishu, p. 14.

[17] See, e.g. Holloway, Stalin, p. 280.

[18] Mao, Tong Sulian, p. 387; Xie, Dangdai, vol. 1, pp. 14–15.

Chinese industrialization in the 1950s. Like the American Marshall Plan, it was calculated to solidify a strategic alliance. Yet, the Soviet technical aid was not without its hidden cost to the Chinese, as Mao himself soon realized: reliance on Soviet technical assistance encouraged the Chinese government to follow Stalin's model of economic development through emphasis on heavy industries at the neglect of agriculture and consumer industries. The reliance on the Soviets also lessened the importance of indigenous technical resources, including the Chinese scientists themselves, in the eyes of the Chinese Communist party state. Thus it was not surprising that during this period, the government saw only a minimal role for the Chinese Academy of Sciences in national defense, directing it instead to focus on thought reform of the scientists and solving practical problems in industrial production.[19]

The Korean War also dramatically changed the domestic political dynamics with a profound impact on the scientists. Just as the war helped fan the McCarthyist red-scare in the US, it intensified the Chinese campaign to purge domestic enemies. Partly to help pay for the Korean War, Mao started an anti-embezzlement, anti-waste, and anti-bureaucratism campaign. Soon thereafter, targeting the intellectuals and scientists more specifically, Mao and the party launched a Thought Reform movement to remold their thinking along the lines of Marxist ideology and to build up their loyalty to the Communist Party. At the local level, these national movements were widely used by party branch organizations to humiliate and persecute scientists. Without the protection of due process, scientists were often falsely accused of embezzlement or forced to confess crimes that they never committed. Those scientists who had returned from the US faced especially harsh harassment in this period. The movement was the first, but certainly not the last, of the national ideological campaigns in the Mao years from which no scientist or intellectual could escape.[20]

THE BOMB AND TECHNO-NATIONALISM

To many Chinese scientists these ideological campaigns certainly put a damper on their idealistic vision of science under the Communists, but, encouraged by moderates such as Zhou Enlai, they maintained faith in the possibility of using their talent for national reconstruction. In 1956, their moment seemed to have finally arrived. On January 21 of that year, Zhu Kezhen, Wu Youxun, and other leaders of the Chinese Academy of Sciences went to Zhongnanhai, seat of the Chinese Communist Party (CCP) headquarters within the Forbidden City, to give reports on the state of science and technology in China and the world. They were surprised by the attention they received:

[19] Yao et al., Zhongguo kexueyuan, vol. 1, pp. 14, 50–65.
[20] Ibid., pp. 26–30.

Chairman Mao himself was present, as well as President of the People's Congress Liu Shaoqi, Premier Zhou Enlai, and vice premiers Chen Yun, Chen Yi, Li Fuchun, Deng Xiaoping, and others. The audiences consisted of the party chiefs in various provinces and various ministries, totaling between 1,300–1,400, filling the whole auditorium … Today's conference was extremely solemn and grand. [I] did not expect that the people's government would attach so much significance to science.[21]

What accounted for the Chinese government's unexpected surge in interest in Chinese science and what did this interest portend for the relationship between the Chinese scientists and the party state?

The urgent need to develop China's own nuclear weapons made Mao and the party pay attention to Chinese science, especially Chinese physics. Even though Mao had famously called the atomic bomb a "paper tiger," he was enough a realist to realize that the weapon did change the strategic balance in international politics. That thinking was part of the reason he insisted on the signing of the China-Soviet friendship treaty in 1950. In 1953, he also felt the pressure of President Dwight D. Eisenhower's threat of the use of nuclear weapons to force the Chinese and North Koreans to end the war. The next year, Eisenhower again threatened the use of nuclear weapons when Mao ordered the shelling of Nationalist-controlled Jinmen (Quemoy) and Mazu (Matsu) islands off the Chinese coast. Initially Mao sought Soviet help to develop China's own nuclear weapons. However, when he made such a request to the visiting Nikita Khrushchev in Beijing in October 1954, the Soviet leader balked. He tried to convince Mao to give up his nuclear dream because China did not have the industrial infrastructure or financial capability and should instead be content to stay under the Soviet nuclear umbrella. Rebuffed, Mao and Zhou came to recognize that the Soviets would not share all its advanced technologies and that China still had to build its own scientific and technological ability if it was to fend off the American nuclear threat again in the future.[22]

This encounter reinforced a Chinese determination to pursue a techno-nationalist program to enhance China's national strength and international prestige through the building of its own strategic weapons.[23] Two weeks after his meeting with Khrushchev, Mao spoke at a meeting with the National Defense Commission about the need to modernize the Chinese military:

Our industry, agriculture, culture, and military are still not strong. Imperialists dare to bully us because they figure that we don't have much strength. They say: "How many atomic bombs do you have?" But they are mistaken in their evaluation of us in at least one aspect, i. e., the potential power of China at the present will be stunning when it's released.[24]

---

[21] Zhu Kezhen diary entry for January 21, 1956, in: Zhu, Zhu Kezhen riji, vol. 3, p. 641.
[22] Shi, Zai lishi, pp. 571–573. See also Niu, Mao, esp. p. 55.
[23] See Lewis/Xue, China, and Feigenbaum, China's Techno-Warriors.
[24] Mao, Zai guofang.

For the next few weeks and months, the atomic bomb was constantly on Mao's mind even when he met delegations from several Asian countries.[25]

While continuing to call the atomic bomb a paper tiger, Mao and Zhou wasted no time in planning for a Chinese nuclear weapons project, with the expectation that the Chinese scientists would build the bomb with major Soviet technical assistance. On January 14, 1955, Zhou met Qian and Li Siguang, a well-known Chinese geologist and vice president of the Chinese Academy of Sciences, to plan for a major presentation to the top party state leadership. The next day, just about everybody in the Chinese leadership from Mao on down showed up for the scientists' briefing on atomic energy, complete with uranium samples from a uranium mine that had just been discovered in southern China. At the end of the session, Mao spoke of the decision to go forward with the program:

> We now know that our country has uranium mines. After further exploration, we will certainly find even more uranium mines. In addition, we have trained a number of people, laid some foundation in scientific research, and created some favorable conditions [for the nuclear program]. In the past several years, preoccupation with many other things have led to a neglect of this matter. But it has to be taken seriously. Now it's the time to go at it. As soon as we put it on our agenda, focusing on it steadily, we will definitely achieve our goal … Now, with the Soviet assistance, we should make it work. [Even if] we have to do it on our own, we can also definitely get it to work. As long as we have the people and the resources, we can create miracles at will.[26]

A decision was made right at the meeting to pool the country's resources to solve the various technical problems of an atomic energy program. A Department of Technical Physics was established at Beijing University and a Department of Engineering Physics at Qinghua University to work on nuclear physics and engineering. In the Chinese Academy of Sciences, a Bureau of New Technologies was established to handle nuclear, rocket, and satellite projects. Qian and Wang's institute became the center of nuclear research and development.[27] Two days after the big meeting in Zhongnanhai, the Chinese government received a formal notice from the Soviet Union that the Soviets, in an effort to compete with the American Atoms for Peace program, had launched its own initiative to help countries in the Eastern bloc in the development of peaceful nuclear energy.[28] The Chinese Academy of Sciences utilized the Soviet Atoms for Peace offer to build a heavy water reactor (7 megawatt) and a 1.2 meter (2.5 MeV) cyclotron in 1958.[29]

The return in the early and mid-1950s of a large number of Chinese scientists from abroad, mainly the US, but also from Europe, bringing with them cutting-edge science and technology from the West, also contributed to the increasing

[25] See, e.g. Mao, Tong Yindu; Mao, Tong Riben guohui.

[26] Qian, Shenmi.

[27] Chinese Communist Party Central Documentation Research Office, Zhou Enlai nianpu, vol. 1, pp. 440–441; Yao et al., Zhongguo kexueyuan, vol. 1, pp. 366–370; Zhang, Qing lishi.

[28] Chinese Communist Party Central Documentation Research Office, Zhou Enlai nianpu, vol. 1, p. 445.

[29] Yao et al., Zhongguo kexueyuan, vol. 1, pp. 368–370.

confidence of the Chinese party state to launch the atomic bomb project. These included some of the future chief designers of the Chinese atomic bombs, such as Deng Jiaxian who received his Ph.D. in physics from Purdue University in 1950. Shortly after the outbreak of the Korean War, the US government implemented a policy of forbidding the roughly 5,000 Chinese students, especially in science and engineering, from returning to China. But many of these scientists nevertheless managed to find their way home, often by traveling to a third country first.[30]

Once the US and China started direct bilateral negotiations as a result of the Geneva Conference of 1954, more Chinese students and scientists in the US were able to return to China. Among the most famous of these was the aerodynamicist Qian Xuesen (H. S. Tsien) who had been detained by the US as a suspected Communist. Altogether, one estimate puts the number of scientists who returned to China from abroad from 1949 to 1956 at 2,000. Once they arrived in China, the Chinese Academy of Sciences, at Zhou Enlai's direction, had priority in recruiting them, especially for the nuclear weapons program.[31] Thus, from 1949 to 1956, 129 of these returned students went to work in the CAS, and 109 of them achieved the senior status of associate research fellows (equivalent to associate professor), accounting for one quarter of all such positions in the academy.[32] Both individually and as a group, it is difficult to overestimate the importance of these returned students to the Chinese strategic weapons programs, including the atomic and hydrogen bombs, the missiles, submarines, and the satellites.[33] In fact, the Chinese government decided to launch its missile program in large part due to the return of Qian Xuesen, who quickly became its major architect.[34]

Together with the atomic bomb project, the initiation of the first Five Year Plan toward industrialization in 1953 also brought to the attention of the top party state leaders the need for more and better trained technical personnel. In his major Report on the Work of the Government at the First Session of First People's Congress on September 23, 1954, Zhou Enlai called for the building of a strong, socialist, modernized industrial country. "One major problem," he pointed out, was the lack of technical personnel and the imperfection of technical management.[35] Shortly afterward, Mao called attention to the same problem in a speech to the Supreme National Council.[36]

Recognizing that science and technology were the bottleneck for Chinese modernization, Mao, Zhou, and the party state leadership launched several initiatives in 1956. One was the holding of a major national conference on the issue of intellectuals in January. After an extensive survey of the working and living conditions of intellectuals, and discussions within the party, Zhou supervised the

[30] Wang, Technical aliens.

[31] Chinese Communist Party Central Documentation Research Office, Zhou Enlai nianpu, vol. 1, p. 445.

[32] Yao et al., Zhongguo kexueyuan, vol. 1, p. 63.

[33] Ibid., pp. 361, 369.

[34] Zhou, Nie Rongzhen nianpu, vol. 2, p. 765. On Qian, see also Chang, Thread.

[35] Zhou Enlai, Ba woguo.

[36] Mao, Shehui.

drafting of a major policy speech on intellectuals, which he delivered on January 14, 1956, at the CCP Conference on Intellectuals attended by hundreds of national and regional leaders. Zhou detailed the numerous cases where a severe shortage of technical experts had hindered economic and defense programs. He also pointed out the many ways that the party had failed to make optimum use of the existing intellectuals because of political distrust. To scientists, perhaps the most heartening statement in Zhou's speech was his assertion that "most of them [intellectuals] have already become state employees, have been serving socialism, and have been part of the working class."[37]

Even though Mao seemed to agree with Zhou's points in general, in his own shorter speech at the conference, he did not specifically endorse Zhou's assessment of the political standing of the intellectuals. His emphasis was still much more utilitarian: "Now we are carrying out a technological revolution, a cultural revolution. It's a revolution against stupidity and ignorance. For this we can no longer do without the intellectuals and depend only on the uneducated masses [dalaocu]."[38] In another talk he even acknowledged that there need to be more intellectuals and scientists within the party leadership:

> In my view our central committee is still a political central committee, not a scientific central committee. Therefore it is rather reasonable for some to doubt that our party can command scientific work, can command the work in public health, because you simply don't know, don't understand [work in those areas]. Our current central committee is indeed defective in this respect: we don't have many scientists, or specialists.[39]

Although Mao was by no means calling for a technocracy, he was certainly welcoming technocrats who could help speed up Chinese industrialization.

Despite the subtle and still hidden difference between Mao and Zhou in their assessment of the political status of the intellectuals at this point, there appeared to be a remarkable consensus among the party state leadership that the revolutionary paradigm, in the form of socialist reform, and modernization paradigm, in the form of industrialization, reinforced each other in this period, which lasted from 1949 to 1956.[40] As Mao put it, in a speech at the Supreme National Council a few days after the Conference on Intellectuals, "the purpose of socialist revolution is the liberation of the productive force."[41] In this atmosphere, Zhou's speech resulted not only in elevated political status for scientists – in contrast to the recruitment of only one senior scientist into the party, the party conferred such political recognition to ten, including Zhu Kezhen, in 1956 alone – it also led to better living and working conditions for scientists.[42] The conference with

[37] Zhou Enlai, Guanyu.

[38] See Chinese Communist Party Central Documentation Research Office, Zhou Enlai nianpu, vol. 1, pp. 518–525, 538–541.

[39] Mao, Guanyu di ba.

[40] Jiang/Jin, Mao Zedong zhuan, vol. 1, p. 467.

[41] Mao, Shehui.

[42] Yao et al., Zhongguo kexueyuan, vol. 1, p. 69.

the scientists in Zhongnanhai that Zhu Kezhen mentioned above took place the day after the conference on intellectuals had ended.

The next major initiative was the making of a Twelve-Year Plan for Science and Technology. Planning, following the Soviet model, became a key word as China launched its first five-year plan in 1953. The concept of scientific research according to a politically determined master plan had always been part of the new science policy from the beginning of the People's Republic of China, but until then it had remained general and vague. The making of the twelve-year plan under Zhou Enlai involved 787 scientists, engineers, and administrators from the Chinese Academy of Sciences, universities, and other sectors of Chinese science and technology.

Soon a Science Planning Commission was established in the State Council under Marshall Nie Rongzhen to coordinate the making of the science plan. Nie, as vice-premier and head of the CCP science group, was in charge of science and technology in both the party and government; he also headed the newly established nuclear weapons project.[43] Perhaps more than anyone else, Nie not only linked Chinese science with the national security state but also articulated a vision of technonationalism that placed the achievement of sophisticated military technologies at the core of Chinese national independence and modernization drive.[44] Accordingly, the goal of the twelve-year plan was to introduce the state-of-the-art science and technology into China, especially those most urgently needed in defense, and to "approach the level of the Soviet Union and other powers in the world in twelve years." Not surprisingly, nuclear energy, rockets and missiles, wireless electronics, automation, semiconductors, computers, and airplanes were at the top of the list, followed by natural resource survey and utilization, earthquake, ocean, power generation, metallurgy, fuel chemistry, agriculture, and antibiotics. Soviet advisors were also heavily involved in the making of the plan.[45]

Interestingly, in contrast to American scientists' resistance to planning in the US, Chinese scientists seemed to have welcomed the step. At the inaugural meeting in 1955 of the CAS Departmental Committees, whose memberships represented the highest scientific honor in the country, scientists urged the government to make a science plan as soon as possible.[46] Many of these scientists participated in the making of the plan, although the party administrators seemed to dominate the process. The scientists' only complaints, after the formulation of a preliminary list of 55 key tasks, was that it emphasized too much developmental goals at the expense of basic research. Zhou responded by adding a no. 56, on "Key Theoretical Problems," and by asking for the making of plans for the development of each basic science field.[47] As a result of the twelve-year plan, the

---

[43] Yao et al., Zhongguo kexueyuan, vol. 1, pp. 66–79.
[44] Feigenbaum, China's Techno-Warriors, pp. 25–31, 37–39.
[45] Yao et al., Zhongguo kexueyuan, vol. 1, pp. 72–74.
[46] Ibid., pp. 43–44.
[47] Ibid., p. 73.

Chinese and Soviet governments reached 122 cooperative agreements in science and technology in 1957–1958.[48]

The structure of Chinese science and technology was also adjusted as a result of the twelve-year plan. By 1956, as the production ministries and universities established their own research programs, the CAS sought to focus on major problems in basic research, state-of-the-art science and technology, establishing new institutes on electronics, automation, semiconductor, and computers, and tasks that required interdisciplinary and cross-departmental efforts, or "Gong Guan," meaning literally "storming strategic passes." The effectiveness of "Gong Guan," the marshalling of scientists, engineers, equipments, and other resources from all over the country to solve key scientific and technological problems, became a key part of the Chinese science policy, and has been widely claimed to be one of the advantages of socialist planned system. It was not unique to socialism, as the US Manhattan Project, penicillin project, and the postwar missile and Apollo projects also mobilized large numbers of manpower and resources from many different sectors of the country. Yet, it is probably true that in peacetime such coordinated attack was more prevalent in socialist countries than in the West.[49]

A third major initiative in 1956 to promote the development of science and technology was the implementation of the so-called "double-hundred" policy. To guide the party and the nation through the ideological confusion caused by Khrushchev's recent denunciation of Stalinism in the Soviet Union, and to help people vent any discontent that had built up by then, Mao proposed a policy of pluralism in literature ("Let A Hundred Flowers Bloom") and non-interference in science ("Let One Hundred Schools Contend"). To exemplify the new policy, the Propaganda Department of the CCP sponsored a conference on genetics in Qingdao in August 1956 where both Lysenkoists and modern geneticists were allowed to speak and a measure of rehabilitation of the Morgan genetics implicitly took place. [50] Following the example of the Qingdao conference on genetics, scientists in other fields where political ideology had reined also came out to speak for their points of view. For now, politics seemed to retreat from science.[51]

Thus the 1955–1956 period started with a note of high promise for a socialist construction program that could link the revolutionary and modernizational paradigms harmoniously. The launching of the atomic bomb project not only enhanced the practical value of both domestic and returning scientists, but also dramatically improved their political and social capital in the eyes of the party state, resulting in both enhanced political standing and improved living and working conditions. The double-hundred policy even created some space for academic dissent and debate. But this new partnership was built on a shaky, utilitarian foundation. The enhanced status of the scientists did not derive from a recognition of the fundamental principles of academic freedom, freedom of speech, or

[48] Yao et al., Zhongguo kexueyuan, vol. 1, p. 75.
[49] Ibid., pp. 77–78.
[50] See Schneider, Biology, pp. 165–177.
[51] Yao et al., Zhongguo kexueyuan, vol. 1, p. 82.

individual rights. It was bestowed on the scientists because of their new-found utility to the party state; it could be taken away just as easily without institutional guarantees for such rights. As events unfolded in China during the tumultuous post-1956 period, Chinese intellectuals in general and scientists in particular found themselves time and again drawn into the political storms.

PHYSICS AND THE NATIONAL SECURITY STATE

As the Chinese leadership launched the nuclear weapons projects in 1955–1956, national security became the basis for the state support of science, especially physics, in China just as it did in the US and the Soviet Union. At this point, the so-called "advanced" (jianduan) military technologies focused on the atomic bomb and guided missiles, with initial emphasis actually on the latter. The success of both, however, depended on the availability of manpower in the physical sciences, and their initiation fundamentally reshaped physical science research in China, especially nuclear physics. The impact of the atomic bomb project on Chinese physics can be best seen in the evolution of what was initially called the Institute of Modern Physics of the Chinese Academy of Sciences. Its origin can be traced to the Communist Party's first grant in support of nuclear physics – $ 50,000 in cash in US dollars – for Qian Sanqiang to purchase nuclear equipments during a planned trip to Paris in April 1949.[52] Soon thereafter the Chinese Academy of Sciences was founded on November 1, 1949, which established two institutes in physics, the Institutes of Modern Physics (IMP) and the Institute of Applied Physics. During 1950, its first year, the IMP, located in inner Beijing, was headed by Wu Youxiu, with Qian as deputy director, but a year later Wu became a vice president of the CAS, and Qian became director. Qian's wife, He Zehui, who had also received her Ph. D. with the Joliot-Curies, worked on nuclear emulsions at the institute.[53]

As the only Chinese institution in nuclear physics, the IMP soon gathered some of the leading physicists from all over China, including Wang Ganchang and Peng Huanwu who became its deputy directors. A theoretical physicist, Peng had received his Ph. D. in physics with Max Born at Edinburgh in 1940 and returned to China in 1947. In 1950, Zhao Zhongyao, a Caltech-trained experimental nuclear physicist, returned from the US with components for a van de Graff accelerator after having been detained by US forces in Japan for more than a month.[54] He joined the IMP as another deputy director. During the next seven years, about two dozen nuclear physicists returned from the US and Europe to join the IMP, which was renamed the Institute of Physics (IP) in 1952. They not only strengthened the manpower at the institute, but also brought state-of-the art knowledge and skills from the west back with them. Before the party's decision

---

[52] Ge, Qian Sanqiang nianpu, pp. 69–72.
[53] Ibid., pp. 124–125.
[54] Zhao, Wo de huiyi.

to embark on the atomic bomb project in 1955, the institute carried out research in experimental physics, with emphasis on building accelerators and detectors, radiochemistry, cosmic rays, theoretical particle physics, reactor designs, and electronics. It trained a number of young physicists on its own as well as by sending them to the Soviet Union. By the end of 1954, the institute had moved to a larger site in Zhongguancun, a suburban in northwestern Beijing, and its scientific staff had grown from about 37 in 1951 to 90.[55]

Following the party leadership's decision to launch China's atomic bomb project in 1955, the Institute of Physics experienced even more rapid growth. It opened a new division in Tuoli, a suburban area in southwestern Beijing, as the site for the Soviet-provided reactor and cyclotron. In September 1956, the Tuoli division was put under dual control by the CAS and the Third Ministry of Machine Building, which was established in November 1956 to focus on nuclear energy, including the building of the atomic bomb, with Qian Sanqiang as one of the deputy ministers. It was renamed the Second Ministry of Machine Building during a restructuring of the government in February 1958.[56] In March 1956, China joined the consortium of Socialist countries that ran the particle physics center at Dubna in the Soviet Union, contributing about 20% of its operating cost (about 15 million Chinese yuan). In 1956, the Institute of Physics sent Wang Ganchang to Dubna, where he eventually became deputy director and led a group of physicists from China and other countries to discover a new anti-hyperon particle in 1959.[57] By the end of 1957, the institute's scientific staff grew to 560, with about 20 senior researchers and its expenditure for 1957 (2,786,000 yuan), nearly tripling that for 1954 (992,436 yuan). By contrast, the expenditure for the more civilian-oriented Institute for Applied Physics, which stood at about the same level in 1952 (287,876 yuan for the IMP and 284,279 yuan for the IAP) and which also experienced rapid growth, consistently lagged behind the IP's in both 1954 (632,883 yuan) and in 1957 (2,037,000 yuan).[58]

Even though the IP's connections with the bomb project did not spare the institute from the political purges during the Anti-Rightist Movement in 1957 and the Great Leap Forward in 1958, scientists there seemed to have suffered less than elsewhere in the Chinese Academy of Sciences. The "rightists," which eventually numbered around 300,000, were branded enemies of the people and were treated harshly.[59] Under the pretense that without employment they could not survive, the government forced many of the rightists into inhuman labor reform

[55] Yao et al., Zhongguo kexueyuan, vol. 1, pp. 359–365; Ge, Qian Sanqiang nianpu, pp. 81, 114.

[56] Zhou, Nie Rongzhen nianpu, vol. 1, pp. 596, and vol. 2, p. 634.

[57] Li et al., He wulixuejia, pp. 123–156.

[58] On the IP staff, see Ge, Qian Sanqiang nianpu, p. 136. On the expenditures for 1952, see Appendix 3, "Statistical Table of the Expenditures of the Chinese Academy of Sciences in 1952," in: Xue/Ji, Zhongguo kexuyuan shishi huiyao 1952 nian. For 1954, see Appendix 3 in: Wang/Wang, Zhongguo kexuyuan shishi huiyao 1954 nian. For 1957, see Appendix 8 in: Xue/Ji, Zhongguo kexuyuan shishi huiyao 1957 nian.

[59] Spence, Search, p. 572.

camps. For some of them, it was the first time that they recognized that there was no more corners of civil society outside of the harsh control by the state. In the words of Shu Xingbei, one of the best known physicist-rightists and one-time colleague of Wang Ganchang at Zhejiang University, only then did he realize that "the party could do whatever it pleased with me after all."[60] As a result of the campaign, many of the brightest, most outspoken scientists were taken away from science and education and sent to perform physical labor for many years. All suffered grave physical and mental persecution. In the CAS alone, 167 scientists were purged as "rightists," including 11 members of the departmental committees.[61]

Yet, against such a grim background for scientists, the Chinese atomic bomb and missile projects kicked into high gear when the Soviets promised to provide technical assistance in these areas, including prototype bombs and missiles, in an Accord on New Technologies in Defense signed in Moscow between the two countries on October 15, 1957, a week after the Soviet launch of Sputnik. Apparently Khrushchev, like Stalin before him, relented to China's demands in order to maintain China's support in both international and domestic politics.[62] In early 1958, a number of other technical accords ensued to help China implement its twelve-year science plan. Nominally civilian, many of these areas, such as minerals, oceanography, radio electronics, metallurgy, and precision equipments were clearly relevant to military technology.[63] Meanwhile, the Chinese Academy of Sciences, inspired by Sputnik, received approval from the top party-state leadership to launch China's own satellite project under the leadership of the geophysicist Zhao Jiuzhang. It eventually joined the atomic bomb and missiles in the top priority list of advanced technologies.

The mobilization for the bomb took advantage of a centrally planned system that prevailed under the Chinese Communist party state. It made possible, for example, a swift restructuring of Chinese scientific, educational, administrative, and military institutions to facilitate the launching of the bomb and missile projects. Within a year of the signing of the October 1957 agreement, new bureaus were established in the Ministry of Defense for missiles and in the Third (soon Second) Ministry of Machine Building for the atomic bomb. The latter ministry also created a Nuclear Weapons Institute (NWI) to prepare for the reception of the expected bomb and accompanying technical materials. Upon Qian Sanqiang's recommendation, Deng Jiaxian from the Institute of Physics was put in charge of the NWI's technical preparatory work.[64] The Institute of Physics itself was renamed the Institute of Atomic Energy and put under the dual control of the Second Ministry and the Chinese Academy of Sciences. Over the

---

[60] See Liu, Shu Xingbei, esp. pp. 154–181, 240.

[61] Yao et al., Zhongguo kexueyuan, vol. 1, pp. 86–89.

[62] See Zhang et al., Sulian, p. 183.

[63] For a list of 122 areas covered in one major agreement, see Zhang et al., Sulian, pp. 184–204.

[64] Xie, Dangdai, vol. 1, p. 36; Ge, Qian Sanqiang nianpu, p. 139.

next several years, more than a thousand of scientists and technicians from the institute moved over to the NWI. In general, the CAS served as a reservoir of scientists and engineers from which the military R&D system drew its technical manpower in this period. In addition, the CAS created in 1958 a university of its own, the elite University of Science and Technology of China (USTC) in Beijing, to train middle- and lower-level technical manpower for fields relevant to the bomb and missile projects. Zhao Zhongyao added chairmanship of the USTC's Department of Technical (nuclear) Physics to his responsibilities. Continuing to oversee the rapidly expanding nuclear weapons complex on behalf of Mao and Zhou was Marshal Nie Rongzhen, who now directed the vast advanced technologies enterprise from his chairmanship of the powerful National Defense Science and Technology Commission (NDSTC) that was reconstituted out of an earlier, narrower group on aviation and missiles in 1958.[65]

In the highly centralized Chinese military R&D system, state support was central to the success of any project but money was not everything, sometimes not even the primary currency. For one, as the leaders of the Chinese Academy of Sciences recognized, personal relations mattered. In this regard, their ace card was Gu Yu, an administrator in the academy who happened to be married to Hu Qiaomu, one of Mao's political secretaries, and therefore enjoyed easy access to the top leadership. Gu was soon put in charge of the academy's Office for New Technologies to manage its mushrooming satellite project. In 1958, it was she who helped the academy find a suitable site for its new USTC by pulling some strings. That year the academy also received a huge special budget allocation of 200 million yuan for work on the satellite but waited in vain for the money to come in until Gu used her connections to bring in the funds. Of course Gu Yu succeeded not merely because of her personal connections; the leadership recognized the key role that the academy would play in the advanced projects. But the right connections did give the academy an significant edge in the intensified inter-departmental rivalry over not only funds, but also the distribution of manpower, raw materials, and precision equipments. Thus the academy breathed a heavy sigh of relief in 1960 when Gu Yu won it a spot among the top four agencies with the highest national security priorities in the national allocation of materials.[66]

By then the Chinese Academy of Sciences and its scientists had taken on even greater importance to the Chinese advanced technology program. In the summer of 1958, the post-Sputnik Sino-Soviet honeymoon began to end when Mao Zedong refused Khrushchev's request for a joint long-wave-length radio station and a joint nuclear submarine fleet as impinging on Chinese national sovereignty. Ideological cleavage also developed between the two sides over Mao's Great Leap Forward movement and Khrushchev's proposal for peaceful co-existence with the West. A turning point in Sino-Soviet relations came in June 1959 when the Soviets, citing their ongoing negotiations with the US and Britain

---

[65] Feigenbaum, China's Techno-Warriors, pp. 54–57.
[66] Pei/Lu, Liangdan.

toward a nuclear test ban, postponed indefinitely the shipping of the promised atomic bomb prototype to China and began to withdraw their technical advisers. Undaunted, the Chinese leadership decided to go ahead with the project on its own. Another blow to China's advanced technologies program came a year later when the Soviet Union withdrew all its scientific and technical advisers from China.[67]

With these turns of events, the primary role of Chinese scientists changed from that of playing second fiddle to Soviet technical advisers to that of chief designers of the advanced weapons projects themselves. Understandably, this experience of working with the Soviets did not lessen but actually intensified a sense of Chinese nationalism not only on the part of the party state leadership but the scientists as well, putting an interesting twist to the Chinese experience of the Cold War. Increasingly, the Chinese national security policy was re-oriented from a possible conflict with the US to that with the Soviet Union, which quickly affected not only international politics but also Chinese science and technology policy. Among the first Chinese reactions to the Soviet withdrawal was to move a number of senior Chinese physicists, including Wang Ganchang, Peng Huanwu, and Zhu Guangya, whose sense of national and professional pride was once again stimulated by the Soviet actions, fulltime into the bomb project.[68]

In fact, much of the Chinese Academy of Sciences was now devoted to the bomb/missile/satellite projects. In 1960, Gu's Office of New Technologies (ONT) was enlarged into a bureau with 34 institutes under its control, which grew to 47 by 1966, with 11,328 scientists and engineers, comprising nearly half (46%) of all such personnel of the CAS, under their command. Even those institutes not formally turned over to the ONT conducted significant amount of research for the defense programs. Altogether, about half to two thirds of all the scientists and engineers in the academy worked on national security-related projects in this period.[69]

The necessity of relying on Chinese scientists for the making of the bomb and missiles created a protective political effect for them not unlike what happened to their Soviet counterparts under Stalin.[70] In a crucial conference on the political assessment of scientists at the Fifth Academy on missiles, Nie directed the party committee to trust Chinese scientists both politically and technically and also to improve their working and living conditions. "If you ask them to do the work you have to trust them. We have to rely on our own experts."[71] He lost sleep when he heard scientists' complaints that they were not trusted by the power-holding local party committees and that they were treated like technicians and even ordered to do physical labor in the countryside. Further investigations soon

---

[67] See, e.g. Lewis/Xue, China, pp. 60–65.
[68] Ge, Qian Sanqiang nianpu, pp. 146–149, 152–161; Yao et al., Zhongguo kexueyuan, vol. 1, p. 374.
[69] Yao et al., Zhongguo kexueyuan, vol. 1, p. 116; Zhang, Qing lishi.
[70] Holloway, Stalin.
[71] Zhou, Nie Rongzhen nianpu, p. 737.

revealed that such treatment of scientists was not restricted to the Fifth Academy but widespread in the Chinese Academy of Sciences and elsewhere. Nie, who had himself sought to "save China via science" in his youth and who tended to trust the scientists more than most in the party did, then drafted a CCP policy paper to correct the misuse and mistreatment of scientists, with fourteen specific directives – it became known as the Fourteen Points.[72]

Issued by the CCP in July 1961, the Fourteen Points directed that the party should trust scientists politically and do everything to help them fulfill their scientific and technological missions. It specifically demanded that scientists and engineers should be protected from political and other interferences and guaranteed five full days devoted to their work out of a six-day working week. Called the "Constitution of Science," the policy codified scientific professionalism and stabilized the environment of scientific research.[73] In the same vein was the Gunagzhou conference in 1962 where Zhou and Vice Premier Chen Yi announced that intellectuals were part of the working class, a political edict that was of paramount importance to the scientists in a politicized society.[74] Most scientists also received increased pay and promotions in this period.[75] Thus the improvement in scientists' political and living conditions came less from any ideological relaxation than from concerns for the strengthening of the national security state, which culminated in the successful testing of China's first atomic bomb in October 1964.

PARADIGM SHIFT

As China embarked on its long march toward a nuclear deterrent, Mao's thinking about the proper relations between revolution and modernization actually underwent a radical shift that had a profound impact on Chinese science and scientists. The Khrushchev denunciation of Stalin in February and the Hungarian uprising in October 1956 caused Mao to seriously rethink the strategy of revolutionary modernization not only from the point of view of what was best for the nation, for the Communist party state, but also increasingly for his own political standing. As a result of this rethinking, Mao came to increasingly question the possibility of integrating revolution and modernization harmoniously; in weighing the options, he seemed to have come to the conclusion that the earlier, Soviet model of development would inevitably lead to an abandonment of the revolutionary paradigm and to a Chinese version of revisionism that threatened not only the revolutionary course he had shaped for China but also his own position within the Chinese party state.

---

[72] Zhou, Nie Rongzhen nianpu, pp. 739–740, 784.
[73] Suttmeier, Research and Revolution, pp. 93–100.
[74] Yao et al., Zhongguo kexueyuan, vol. 1, pp. 99–114.
[75] Ibid., p. 113.

As a response to these developments, Mao first tried a strategy that he thought could build the revolutionary paradigm so much into the modernization program that when modernization did succeed it would not automatically lead to revisionism. This strategy resulted in the Great Leap Forward (GLF) of 1958, which sought to industrialize China overnight via mass mobilization. Zhou Enlai and Liu Shaoqi initially opposed this policy as dangerous "adventurism" but fell in line when Mao attacked them as aligning with the rightists. As the official biographers of Mao later acknowledged, the debate over "adventurism" on the eve of the GLF "marked the beginning of the breakdown of the party's principle of collective leadership and system of democratic centralization, with perilous negative implications for later historical developments."[76] The GLF failed miserably and as a result China fell into a horrible famine as well as a dire economic situation that threatened the bomb and missile projects. It was Nie who fought for their continuation and his argument, that they would not only provide a nuclear deterrent for China but also offer spin-offs in conventional weaponry, prevailed with Mao's crucial support.[77]

Following the disastrous GLF, Mao retired to the "second front" and pragmatists represented by Liu Shaoqi and Deng Xiaoping came to power. Mao's resentment of their liberal policy grew in the early 1960s. The tension between Mao, on the one hand, and Liu and Deng, on the other, finally broke in 1966 when Mao launched the country into a chaotic, destructive Cultural Revolution to remove the latter from power and to correct what he perceived as a revisionist direction for the country. As with other intellectuals, scientists again were branded as basically bourgeois, and not to be trusted, rather to be cleansed and reformed. Many of them suffered horribly during the decade of unrest. Targeted were not only scientists, but also many of the modern scientific theories, including, as it turned out, Albert Einstein's relativity theories.[78] The seminal circular of May 16, 1966, called for the attack on "the reactionary bourgeois stands of the so-called 'academic authorities'", and the criticism of reactionary viewpoints in the theoretical front in natural sciences.[79]

As with other institutions, the CAS was turned upside down during the first few months of the Cultural Revolution, when Red Guards and other rebels seized power from the leadership and stopped virtually all research. Scientists, especially those in the leadership, were criticized and persecuted. Among the 170 senior CAS scientists in the Beijing area, 131 were attacked, including the vice president Wu Youxiu. Altogether 106 scientists of different levels of seniority in the CAS, including the geophysicist Zhao Jiuzhang, died during the Cultural Revolution due to political repression. In Shanghai, hundreds of scientists were beaten, tortured, and some killed, for dredged-up charge of anti-communist conspiracy during the Nationalist years. By 1969, many scientists who survived the ordeal were

---

[76] Jiang/Jin, Mao Zedong zhuan, p. 789.
[77] See Feigenbaum, China's Techno-Warriors.
[78] See, e. g. Hu, China and Albert Einstein.
[79] Yao et al., Zhongguo kexueyuan, vol. 1, pp. 141–142.

sent to the countryside to do physical labor.[80] The only member of the scientific community spared the worst of the violence were those who worked on the nuclear weapons projects. Zhou and Nie had taken those institutes of the CAS covered by the Bureau of New Technologies out of the academy and put them under the control of the NDSTC for protection.[81] Even there they were not safe as the Cultural Revolution soon spread to the military as well. Following the beating death of Yao Tongbin, a metallurgist with a Ph. D. from the University of Birmingham who returned to China in 1957 to direct its space materials program, in June 1968, Zhou ordered that several hundreds of scientists and engineers in the advanced technology system to be put under the protection by armed guards.[82]

In 1967–1969, Nie himself came under attack by Maoist "rebels" for his role in the formulation of the liberal "Fourteen Points" science policy, his "blind trust" in the experts, and allegedly cultivating his own authority at the expense of Mao's in the NDSTC.[83] In a self-criticism report to Mao, Nie asserted that he "had followed the strategic intent" of the party state leadership, but acknowledged "deficiency" in his dealing with scientists and intellectuals: "I have done more to unite with them and to encourage them than to educate and criticize them."[84] In a second such report, Nie was again forced to reflect on his "serious errors": "In my thinking, I had been single-mindedly pursuing the development of our science and technology, the modernization of industry, agriculture, science and technology, and national defense, so as to accelerate the socialist construction of our country. But I failed to recognize the correct relationship between modernization and revolutionization [geminghua] and failed to let revolutionization take command over modernization."[85]

A turning point of the Cultural Revolution came in 1971 when Lin Biao, Mao's designated successor, fled and died in a plane crash following a failed coupe against Mao. By then, Mao had realized that the revolutionary program he had pursued once again failed to deliver for him the hoped-for results. Instead, the national economy was brought to the edge of collapse. Even though he still worried about the danger of a revisionist cancellation of his revolutionary achievements, his dominance in the party was established and he switched gear yet again: he personally made the decision to open up relations with the United States and supported Zhou Enlai to get the country back on the modernization track. For a brief period, Zhou was able to relax ideological control over science.

Taking advantage of the US-China rapprochement and the recommendation on basic research by some of the visiting Chinese American scientists, Zhou managed to re-legitimate scientific research in the CAS as necessary work on "theory." A conference was held in May 1972 which contradicted the current Maoist or-

    [80] Ibid., pp. 154–157.
    [81] Zhou, Nie Rongzhen nianpu, pp. 1056–1057.
    [82] Chinese Communist Party Central Documentation Research Office, Zhou Enlai nianpu, vol. 3, pp. 313–314. On Yao, see Song, Liangdan.
    [83] Zhou, Nie Rongzhen nianpu, pp. 1047, 1097, 1100–1101.
    [84] Ibid., pp. 1100–1101.
    [85] Ibid., pp. 1102–1103.

thodoxy by affirming pre-Cultural Revolution science policies. Mao's wife Jiang Qing and her influential radical group Gang of Four, however, soon made it impossible for any substantial restoration of the scientific planning, policy, or institutions. When Deng Xiaoping returned to high positions briefly in the early 1970s he also tried to restore order and used the CAS as a starting point. Soon, however, Zhou died and Deng was purged again by Mao in 1976. Only after Mao himself died and after the old guards allied with Mao's designated successor Hua Guofeng crushed the Gang of Four later that year was it possible for a full-scale restoration of pragmatic policy emphasizing the modernization paradigm under the leadership of Deng Xiaoping after 1976.[86] By the 1980s and 1990s, a drive for market-oriented reform began to reshape Chinese science and technology, even though the domination of the state remained. In this regard, the establishment in 1986 of the National Natural Science Foundation, which allowed scientists to make individual proposals outside of the rigid planning system, led the way toward a pluralistic system of science funding that by the early twenty-first century included science funding from private donations and transnational corporations as well.[87]

CONCLUSION

What conclusions can we draw from this examination of the experiences of Chinese scientists under Mao during the Cold War? First, the Chinese case exhibited striking similarities and differences with American and Soviet cases. The willingness of Chinese scientists to join the atomic bomb project recalled similar attitudes of Russian scientists in the Soviet nuclear program and American scientists in both the Manhattan Project and, after the outbreak of the Korean War, in the hydrogen bomb project. Either out of a sense of patriotic nationalism, or a belief that nuclear deterrence was vital to world peace, or a recognition that their professional and personal success depended on their service to the national security state, few Chinese scientists refused the call from the government. As in both the US and the Soviet Union, national security dominated Chinese science and technology policy, in the funding structure and in the allocation of manpower and materials, during much of the Cold War. Likewise, the overwhelming demand for real, "hard" results in the nuclear weapons projects often softened the political and ideological interferences with scientific research in China under Mao as it did in McCarthyist US and Stalinist Soviet Union.

It should be noted that even though the decision to launch the strategic weapons in 1955–1956 was based on widespread consensus within the party-state leadership, even though there was a vigorous debate within a small circle of policy makers over its continuation in the early 1960s in the aftermath of the Soviet withdrawal and the GLF disaster, and even though the bomb test in 1964 and the

---

[86] Yao et al., Zhongguo kexueyuan, vol. 1, pp. 141–171.
[87] On the foundation, see its website at www.nsfc.gov.cn.

satellite launching in 1970 enjoyed overwhelming public support, there was never any open, public, thoroughgoing national debate over these matters. Scientists such as Qian Sanqiang and Qian Xuesen did participate in some high-level discussions over strategic weapons, but their involvement was by and large limited to technical questions, rarely policy issues. Then, of course, China was not alone in the shroud of secrecy around its nuclear weapons policy-making. Such was largely the case with the Soviet Union under both Stalin and Khrushchev. Even in the US, where debates over nuclear policy was much more open than in either China or the Soviet Union and where scientists did participate in policy discussions, strategic weapons policy was often determined more by the policy elite in the Executive Branch than by public sentiment or the Congress.

There were striking differences as well. To a much greater extent than was the case with their counterparts in the US or even the Soviet Union, Chinese scientists suffered political persecution and ideological harassment despite the recognized need for their service to the Chinese national security state. The nuclear weapons projects did provide the scientists, especially physicists, some protection, and were certainly responsible for the immense state support for physics and other relevant disciplines in this period. But even the bomb could not shield the scientists completely from the various political movements that swept the country under Mao. The fact that all Chinese scientists worked for the government meant that there was much less freedom of action when faced with political and ideological pressure than in a more pluralistic science-funding structure that existed, e.g. in the US and Germany during the Cold War. There were no such philanthropic sources of science funding as the Ford and Rockefeller Foundations in the US, or the self-governing DFG in Germany, or even a place like the US National Science Foundation for support of self-generated research.

Of course, a plural system of science funding does not by itself guarantee freedom from politicization of science just as a formally democratic political system did not prevent such abuses of individual rights as the McCarthyist attacks on American dissident scientists. As the historian John Krige has demonstrated, private foundations could often be doing the bidding for the national security state in the US and Europe.[88] But the possibility of getting funding or employment outside of the government did make it easier for scientists and others to take unpopular or politically incorrect stands. Thus, in contrast to the miseries experienced by Chinese scientists purged during the 1957 Anti-Rightist Movement, many of the faculty members of the University of California who refused to sign the anti-Communist state loyalty oaths in the 1950s found employment in other, especially private, universities. The highly centralized funding structure left little room for maneuver by Chinese scientists in terms of their research interests or preference for civilian or military orientations. Happily for at least those Chinese scientists involved in the nuclear weapons projects, their scientific nationalism fit into the national security strategy of the Chinese Communist party state during much of the Mao era, but that changed radically when his political

---

[88] See, e.g. Krige, Ford.

and ideological considerations led to an overturning of the consensus during the destructive Cultural Revolution.

This study also points to both the strengths and weaknesses of techno-nationalism that had been pursued by Chinese scientists, especially physicists, and pragmatists among the Chinese Communist party state leadership such as Wang, Qian, Zhou, and Nie both during and beyond the Cold War. Scientists' sense of nationalism made them to decide to stay and cast their lot with the Communists in the late 1940s and early 1950s. It also drew scientists to return from the US and other countries. Nationalism likewise fueled scientists' devotion to advancing science and technology to achieve either economic benefits, or higher national prestige, as in basic research, or most important of all, a vision of techno-nationalism in the case of the atomic bomb. The new Chinese Academy of Sciences provided crucial institutional and ideological support for much of the scientific research carried out in the Mao years. Thus the success of the Chinese nuclear weapons and other advanced scientific/technological projects during the Cold War shows nationalism could play a powerful mediating role in bringing the Chinese scientists and national security state together and that science and technology could survive and even prosper under an authoritarian regime.[89] Yet, the abuse of the scientists and other intellectuals during the Anti-Rightist campaign and the disastrous consequences of both the Great Leap Forward movement and the Cultural Revolution also demonstrate that techno-nationalism alone could not make for a prosperous and strong China that the scientists and their political allies had dreamed.

BIBLIOGRAPHY

Chang, Iris: Thread of the Silkworm, New York 1995.

Chinese Communist Party Central Documentation Research Office: Zhou Enlai nianpu 1949–1976 (a chronicle of Zhou Enlai, 1949–1976), 3 vols., Beijing 1997.

Feigenbaum, Evan: China's Techno-Warriors: National Security and Strategic Competition from the Nuclear to the Information Age, Stanford 2003.

Forman, Paul: Behind quantum electronics: National security as basis for physical research in the United States, 1940–1960, in: Historical Studies in the Physical and Biological Sciences 18 (1), 1987, pp. 149–229.

Forman, Paul and José M. Sánchez-Ron (eds.): National Military Establishments and the Advancement of Science and Technology, Dordrecht 1996.

Ge, Nengquan (ed.): Qian Sanqiang nianpu (a chronicle of Qian Sanqiang's life), Jinan 2002.

Ge, Nengquan: Qian Sanqiang he zaoqi Zhongguo yuanzineng kexue (Qian Sanqiang and Chinese atomic energy science in the early years), in: Zhongguo keji xiliao (China historical materials of science and technology) 25 (3), September 2004, pp. 189–198.

Holloway, David: Stalin and the Bomb, New Heaven 1994.

Hu, Danian: China and Albert Einstein, Cambridge 2005.

---

[89] For comparison with the case of another field, mariculture, in China in this period, see Neushul/Wang, Between the devil and the deep sea.

Jiang, Xianzhi and Jin Chongji: Mao Zedong zhuan (a biography of Mao Zedong), 2 vols., Beijing 2003.

Kevles, Daniel J.: Cold War and hot physics: science, security, and the American state, 1945–56, in: Historical Studies in the Physical and Biological Sciences 20 (2), 1990, pp. 239–264.

Krige, John: The Ford Foundation, European physics and the Cold War, in: Historical Studies in the Physical and Biological Sciences 29 (2), 1999, pp. 333–361.

Lewis, John W. and Xue Litai: China Builds the Bomb, Stanford 1988.

Li, Rizhi et al.: He wulixuejia Wang Ganchang (nuclear physicist Wang Ganchang), Beijing 1996.

Liu, Haijun: Shu Xingbei dangan (Shu Xingbei's files), Beijing 2005.

Mao, Zedong: Address to the preparatory meeting of the New Political Consultative Conference, June 15, 1949, in: Mao Zedong: Selected Works of Mao Tse-tung, 4 vols., Beijing 1969, vol. 4, pp. 405–409.

Mao, Zedong: Zai guofang weiyuanhui diyici huiyi shang de jianghua (talks at the first meeting of the National Defense Council, October 18, 1954), in: Mao Zedong: Mao Zedong wenji (collected papers of Mao Zedong), 8 vols., Beijing 1993, vol. 6, pp. 354–360.

Mao, Zedong: Tong Yindu zongli Nihelu de sici tanhua (four talks with Indian Prime Minister Nehru, October 1954), in: Mao Zedong: Mao Zedong wenji (collected papers of Mao Zedong), 8 vols., Beijing 1993, vol. 6, pp. 361–373.

Mao, Zedong: Tong Riben guohui yiyuan fanghuatuan de tanhua (a talk with a delegation of Japanese parliamentary members to China, October 15, 1955), in: Mao Zedong: Mao Zedong wenji (collected papers of Mao Zedong), 8 vols., Beijing 1993, vol. 6, pp. 480–487.

Mao, Zedong: Shehui zhuyi geming de mudi shi jiefang shengchanli (the purpose of the socialist revolution is to liberate the force of production), January 25, 1956, in: Mao Zedong: Mao Zedong wenji (collected papers of Mao Zedong), 8 vols., Beijing 1993, vol. 7, pp. 1–3.

Mao, Zedong: Guanyu di ba jie zhongyang weiyuanhui de xuanju wenti (on the question of the election of the 8th Central Committee), September 10, 1956, in: Mao Zedong: Mao Zedong wenji (collected papers of Mao Zedong), 8 vols., Beijing 1993, vol. 7, pp. 100–109.

Mao, Zedong: Tong Sulian zhuhua dashi You Jin de tanhua (a talk with Yudin, Soviet ambassador to China), July 22, 1958, in: Mao Zedong: Mao Zedong wenji (collected papers of Mao Zedong), 8 vols., Beijing 1993, vol. 7, pp. 385–397.

Neushul, Peter and Zuoyue Wang: Between the devil and the deep sea: C. K. Tseng, mariculture, and the politics of science in modern China, in: Isis 91 (1), March 2000, pp. 59–88.

Niu, Jun: Mao Zedong yu zhongsu tongmeng polie de yuanqi (1957–1959) (Mao Zedong and the origins of the breakdown of the Sino-Soviet alliance, 1957–1959), in: Guoji zhengzhi yanjiu (studies in international politics), no. 2, 2001, pp. 53–63.

Overbye, Dennis: Einstein's man in Beijing: A rebel with a cause, in: New York Times, August 22, 2006.

Pei, Lisheng and Lu Shouguan: 'Liangdan yixing xinxue li' – yi Gu Yu tongzhi (you gave your all to the atomic bomb, the missile, and the satellite projects – recollections on Comerade Gu Yu), in: People's Daily, November 23, 1995.

Qian, Sanqiang: Shenmi er youren de lücheng (a mysterious and captivating journey), in: Luo, Rongxing (ed.): Qing lishi jizhu tamen (history, please remember them), Xiamen 1999.

Ren, Xinfa: Bochunzhe: He wulixuejia Qian Sanqiang (the man who sowed the spring: nuclear physicist Qian Sanqiang), Beijing 1989.

Schneider, Laurence: Biology and Revolution in Twentieth-Century China, Lanham, MD., Oxford 2003.

Shi, Zhe: Zai lishi juren de shenbian (at the side of giants in history), Beijing 1991.

Song, Wenmao: Liangdan yixing gongchen Yao Tongbin lieshi liuxia xie shenme (what's the legacy of the martyr Yao Tongbin, a hero of the bomb and satellite projects), in: Yanhuang chunqiu (Chinese chronicles), no. 5, 2003, pp. 16–23.

Spence, Jonathan: The Search for Modern China, New York 1990.

Suttmeier, Richard: Research and Revolution: Science Policy and Societal Change in China, Lexington 1974.

Wang, Shaoding and Wang Zhongjun (eds.): Zhongguo kexuyuan shishi huiyao 1954 nian (major events in the history of the Chinese Academy of Sciences in 1954), Beijing 1996.

Wang, Zuoyue: Saving China through science: The science society of China, scientific nationalism, and civil society in Republican China, in: Osiris 17, 2002, pp. 291–322.

Wang, Zuoyue: Technical aliens: Chinese scientists in the United States and the politics of early Cold War, unpublished paper delivered at the History of Science Society Annual Meeting, Austin, Texas, November 2004.

Xie, Guang (ed.): Dangdai Zhongguo de guofang keji shiye (contemporary China's national defense science and technology enterprise), 2 vols., Beijing 1992.

Xie, Shijun: Zhu Kezhen zhuan (a biography of Zhu Kezhen), Chongqing 1993.

Xu, Liangying: Enshi Wang Ganchang dui wo de qidi he aihu (the inspiration and protection from my mentor Wang Ganchang), in: Hu, Jimin et al.: Wang Ganchang he ta de kexue gongxian (Wang Ganchang and his scientific contributions), Beijing 1987, pp. 208–222.

Xue, Pangao and Ji Chuqing (eds.): Zhongguo kexuyuan shishi huiyao 1952 nian (major events in the history of the Chinese Academy of Sciences in 1952), Beijing 1994.

Xue, Pangao and Ji Chuqing (eds.): Zhongguo kexuyuan shishi huiyao 1957 nian (major events in the history of the Chinese Academy of Sciences in 1957), Beijing 1998.

Yao, Shuping et al.: Zhongguo kexueyuan (the Chinese Academy of Sciences), 3 vols., Beijing 1994.

Zhang, Baichun et al.: Sulian jishu xiang zhongguo de zhuanyi, 1949–1966 (technology transfer from the Soviet Union to China), Jinan 2004.

Zhang, Jinfu: Qing lishi jizhu tamen (history, please remember them), in: People's Daily, May 6, 1999.

Zhao, Zhongyao: Wo de huiyi (my recollections), in: Zhao, Zhongyao: Zhao Zhongyao lunwen xuanji (selected papers of Zhao Zhongyao), Beijing 1992, pp. 198–206.

Zhou, Enlai: Ba woguo jianshe chengwei qiangda de shehuizhuyi de xiandaihua de gongye guojia (build our country into a strong, socialist, modernized industrial power), September 23, 1954, in: Zhou, Enlai: Zhou Enlai xuanji (selected works of Zhou Enlai), 2 vols., Beijing 1984, vol. 2, pp. 132–145.

Zhou, Enlai: Guanyu zhishi fenzi wenti de baogao (a report on the problem of the intellectuals), January 14, 1956, in: Zhou, Enlai: Zhou Enlai xuanji (selected works of Zhou Enlai), 2 vols., Beijing 1984, vol. 2, pp. 158–189.

Zhou, Junlun (ed.): Nie Rongzhen nianpu (a chronicle of Nie Rongzhen), 2 vols., Beijing 1999.

Zhu, Kezhen: Zhu Kezhen riji (diaries of Zhu Kezhen) (1936–1949), 2 vols., Beijing 1984.

ABBREVIATIONS

AEG	German General Electric (Allgemeine Elektricitäts-Gesellschaft)
AMCP	Archives Musée Curie, Paris
BdD	Federation of the Germans (Bund der Deutschen)
BMwF	Federal Ministry for Scientific Research (Bundesministerium für wissenschaftliche Forschung)
BRD	Federal German Republic (Bundesrepublik Deutschland)
CAS	Chinese Academy of Sciences
CCF	Congress for Cultural Freedom
CDU	Christian Democratic Union (Christlich Demokratische Union Deutschlands)
CEA	French Atomic Energy Commission (Commissariat à l'énergie atomique)
CERN	European Center for Nuclear Research (Centre européenne pour la recherche nucléaire)
CMP	World Council for Peace (Conseil mondial de la paix)
CNR	National Resistance Council (Conseil national de la résistance)
CNRS	National Center for Scientific Research (Centre national de la recherche scientifique)
CVIA	Vigilance Committee of Antifascist Intellectuals (Comité de vigilance des intellectuels antifascistes)
DAAD	German Academic Exchange Service (Deutscher Akademischer Austauschdienst)
DDR	German Democratic Republic (Deutsche Demokratische Republik)
DE	Decedent estate
DESY	German Electron-Synchrotron (Deutsches Elektronen-Synchrotron)
DFG	German Research Foundation (Deutsche Forschungsgemeinschaft)
DFR	German Research Council (Deutscher Forschungsrat)
DGaO	German Branch of the European Optical Society (Deutsche Gesellschaft für angewandte Optik)
DHIP	German Historical Institute (Deutsches Historisches Institut), Paris
DM	German Museum (Deutsches Museum), Munich
diamat	dialectical materialism
DPG	German Physical Society (Deutsche Physikalische Gesellschaft)
ECFA	European Committee for Future Accelerators
EURATOM	European Atomic Energy Community
EDA	European Defence Agency
EVG	European Defense Community (Europäische Verteidigungsgemeinschaft)
FAS	Federation of American Scientists
FMTS	World Federation of Scientific Workers (Fédération mondiale des travailleurs scientifiques)
FN	National Front (Front national)
FNU	National Universities Front (Front national universitaire)
Glavnauka	The Main Scientific Administration of the Commissariat of Enlightenment (Soviet Union)
GLF	Great Leap Forward
HG	Helmholtz Foundation (Helmholtz Gesellschaft)
IAP	Institute for Applied Physics (China)
IMP	Institute of Modern Physics (China)

IP	Institute of Physics (China)
IPP	Max Planck Institute for Plasma Physics (Max-Planck-Institut für Plasmaphysik)
ISR	Intersecting Storage Rings
JSC	Science Council of Japan
JSPS	Japan Society for the Promotion of Science (Nihon Gakujutsu Shinkôkai)
KhGU	Kharkiv State University
KPD	German Communist Party (Kommunistische Partei Deutschlands)
KWG	Kaiser Wilhelm Society (Kaiser-Wilhelm-Gesellschaft)
KWI	Kaiser Wilhelm Institute (Kaiser-Wilhelm-Institut)
MPG	Max Planck Society (Max-Planck-Gesellschaft)
MPI	Max Planck Institute (Max-Planck-Institut)
MZWTG	Munich Center for the History of Science and Technology (Münchner Zentrum für Wissenschafts- und Technikgeschichte)
NDSTC	National Defense Science and Technology Commission (China)
NGW	Emergency Foundation for German Science (Notgemeinschaft der deutschen Wissenschaft)
NSDAP	National Socialist German Workers Party (Nationalsozialistische Deutsche Arbeiterpartei)
NTU	The Scientific Technical Administration of the Supreme Economic Council (Soviet Union)
NWI	Nuclear Weapons Institute (China)
ONT	Office of New Technologies (China)
PCF	French Communist Party (Parti communiste français)
PRC	People's Republic of China
PS	Proton Synchrotron
PTB	Federal Physical-Technical Institute (Physikalisch-Technische Bundesanstalt)
PTR	Imperial Physical-Technical Institute (Physikalisch-Technische Reichsanstalt)
RFR	Reich Research Council (Reichsforschungsrat)
RGS	Reiniger, Gebbert & Schall at Erlangen
Riken	Institute of Physical and Chemical Research (Japan)
SDS	Socialist German Student League (Sozialistischer Deutscher Studentenbund)
SFIO	French Socialist Party (Section française de l'internationale ouvrière)
SN	Donors' Union of the NGW (Stifterverband der Notgemeinschaft)
SPD	Social Democratic Party (Sozialdemokratische Partei Deutschlands)
STA	Science and Technology Agency (Japan)
TH	Technical University (Technische Hochschule)
UFF	National Committee of the Union of French Women (Union des femmes françaises)
UFTI	Ukrainian Physical Technical Institute
UNAEC	United Nations Atomic Energy Commission
UNI	National Union of Intellectuals (Union nationale des intellectuels)
USSR	Union of Soviet Socialist Republics
USTC	University of Science and Technology of China
VDLF	Handbook of German Teaching and Research Institutions (Vademekum deutscher Lehr- und Forschungsstätten)
VDW	Federation of German Scientists (Vereinigung deutscher Wissenschaftler)
WFSW	World Federation of Scientific Workers

LIST OF CONTRIBUTORS

Helmuth Albrecht is Professor and Director of the Institute for Industrial Archaeology, History of Science and Technology at the Technical University Freiberg, Germany.

Richard Beyler is Associate Professor of History at Portland State University in Portland, OR (USA).

Cathryn Carson is Associate Professor of History and Director of the Office for History of Science and Technology at the University of California at Berkeley, CA (USA).

Karl Hall is Associate Professor of History at the Central European University in Budapest, Hungary.

Paul Josephson is Professor of History at Colby College in Waterville, ME (USA).

Morris Low is Senior Lecturer in Asian Studies at the University of Queensland in Brisbane, Australia, and Adjunct Associate Professor at Johns Hopkins University, Baltimore, MD (USA).

Gerhard Rammer is a Researcher (Wissenschaftlicher Mitarbeiter) at the Interdisciplinary Center for Science and Technology Studies at the University of Wuppertal, Germany.

Yury Ranyuk is a Senior Scientist at the Ukrainian Physical Technical Institute in Kharkiv, Ukraine.

Martin Strickmann is Postdoc Researcher in History and Humanities, who worked at the Universities of Cologne and Paris-Sorbonne, the Deutsches Museum, the German Historical Institute in Paris and the German Archive of Literature in Marbach, Germany.

Helmuth Trischler is Head of Research at the Deutsches Museum in Munich und Professor for Modern History and History of Technology at the Ludwig-Maximilians-University Munich, Germany.

Ivan Tsekhmistro is Professor of philosophy at the Kharkiv State University in Kharkiv, Ukraine.

Alexander von Schwerin is a Researcher (Wissenschaftlicher Mitarbeiter) at the Department for the History of Science and Pharmacy at the Technical University Braunschweig, Germany.

Mark Walker is the John Bigelow Professor of History at Union College in Schenectady, NY (USA).

Zuoyue Wang is Professor of History at California State Polytechnic University, Pomona, CA (USA).

Index

ZUR GESCHICHTE DER DEUTSCHEN
FORSCHUNGSGEMEINSCHAFT – BEITRÄGE

Herausgegeben von **Rüdiger vom Bruch** und **Ulrich Herbert**

1. Isabel Heinemann / Patrick Wagner, Hg.
   Wissenschaft – Planung – Vertreibung
   Neuordnungskonzepte und Umsiedlungs-
   politik im 20. Jahrhundert
   2006. 222 S., kt.
   ISBN 978-3-515-08733-9
2. Wolfgang U. Eckart, Ed.
   Man, Medicine and the State
   The Human Body as an Object of
   Government Sponsored Medical
   Research in the 20th Century
   2006. 297 S. m. 4 Tab., kt.
   ISBN 978-3-515-08794-0]
3. Michael Zimmermann, Hg.
   Zwischen Erziehung und Vernichtung
   Zigeunerpolitik und Zigeunerforschung
   im Europa des 20. Jahrhunderts
   2007. IV, 591 S. m. 10 Abb., geb.
   ISBN 978-3-515-08917-3

4. Karin Orth / Willi Oberkrome, Hg.
   **Die Deutsche Forschungsgemeinschaft
   1920-1970**
   Forschungsförderung im Spannungsfeld
   von Wissenschaft und Politik
   2010. Ca. 485 S., kt.
   ISBN 978-3-515-09652-2
5. Helmuth Trischler / Mark Walker, Eds.
   Physics and Politics
   Research and Research Support in
   Twentieth Century Germany
   in International Perspective
   2010. 285 S., kt.
   ISBN 978-3-515-09601-0

FRANZ STEINER VERLAG STUTTGART

ISSN 1861-1478